Renewable Energy

The Power to Choose

Norton/Worldwatch Books

Lester R. Brown: *The Twenty-Ninth Day: Accommodating Human Needs and Numbers to the Earth's Resources*

Lester R. Brown: *Building a Sustainable Society*

Lester R. Brown, Christopher Flavin, and Colin Norman: *Running on Empty: The Future of the Automobile in an Oil-Short World*

Lester R. Brown et al.: *State of the World 1984: A Worldwatch Institute Report on Progress Toward a Sustainable Society*

Daniel Deudney and Christopher Flavin: *Renewable Energy: The Power to Choose*

Erik P. Eckholm: *Losing Ground: Environmental Stress and World Food Prospects*

Erik P. Eckholm: *The Picture of Health: Environmental Sources of Disease*

Denis Hayes: *Rays of Hope: The Transition to a Post-Petroleum World*

Kathleen Newland: *The Sisterhood of Man*

Colin Norman: *The God That Limps: Science and Technology in the Eighties*

Bruce Stokes: *Helping Ourselves: Local Solutions to Global Problems*

Renewable Energy

The Power to Choose

Daniel Deudney
and
Christopher Flavin

A WORLDWATCH INSTITUTE BOOK

W. W. NORTON & COMPANY
New York London

Published simultaneously in Canada by Stoddart,
a subsidiary of General Publishing Co. Ltd,
Don Mills, Ontario.
Printed in the United States of America.

The text of this book is composed in Avanta, with display type set in Baker Signet. Composition and manufacturing by The Haddon Craftsman, Inc.

First published as a Norton paperback 1984

Library of Congress Cataloging in Publication Data
Deudney, Daniel.
Renewable energy.

"A Worldwatch Institute book."
Includes index.
1. Renewable energy sources. 2. Energy policy. I. Flavin, Christoper. II. Title.
TJ163.2.D476 1983 333.79 82-14465

ISBN 0-393-30201-6

W. W. Norton & Company, Inc., 500 Fifth Avenue,
New York, N.Y. 10110
W. W. Norton & Company Ltd., 37 Great Russell Street, London
WC1B 3NU
2 3 4 5 6 7 8 9 0

To our parents

Contents

Preface

The decade since the 1973 Arab oil embargo has been a remarkable one for renewable energy. The major shift in the economics of energy that began in the early seventies harnessed the technological capacities of the late twentieth century. Since then the world's scientists, engineers, businessmen, and ordinary citizens have been hard at work in a vigorous search for new sources of energy. New ideas and inventions continue to command attention in technical journals and newspapers almost daily. Increasingly, solar collectors, wind machines, biogas digesters, and many other renewable energy technologies are becoming practical everyday devices used throughout the world.

Renewable Energy: The Power to Choose charts the progress made in renewable energy in recent years and outlines renewable energy's prospects. The focus is on practical here-and-now technologies, and our intention is to be realistic yet hopeful. We suggest a strategy for making the transition to renewable energy and evaluate the impact these changes could have on different parts of the world.

Great progress has been made in thinking about energy in the last decade. Prior to 1973, energy analysis seemed to consist mainly of drawing exponential curves that were intended to forecast future trends by assuming that past trends would continue. Today's world is much more complex and uncertain, and major strides have been taken in understanding the underlying factors at work in energy trends, a fact from which we have benefited greatly. The pioneering work of Amory Lovins has contributed particularly to our thinking.

We are grateful to the U.S. Solar Energy Research Institute and the George Gund Foundation for supporting the research and writing of this book. The Worldwatch Institute provided an ideal setting for the project with its access to a wide array of information sources as well as a bright and capable staff. Lester Brown, the president of Worldwatch, originally suggested the writing of this book, and he provided ideas and enthusiastic support throughout.

Other members of the Worldwatch Institute staff who reviewed the manuscript and made helpful criticisms are Kathleen Newland, Pamela Shaw, and Bruce Stokes. Much of the research for the book was carried out by Worldwatch research assistants Ann Thrupp, Paige Tolbert, and Edward Wolf. And special thanks are owed to the entire Worldwatch support staff for its immense help throughout this project.

Dozens of people outside of Worldwatch provided comments and suggestions as the book progressed. The entire manuscript was reviewed by Todd Bartlem, Erik Eckholm, Jose Goldemberg, Denis Hayes, James Howe, Ron Larsen, and

Vaclav Smil. Individual chapters were reviewed by David Anderson, Carl Aspliden, Thomas Cassel, Bill Chandler, Joe Coates, Jeffrey Cook, Kenneth Darrow, Darian Diachok, Ron DiPippo, Peter Fraenkel, Calvin Fuller, Jon Gudmundsson, Keith Haggard, Michael Holtz, Mark Lyons, Leonard Magid, Paul Maycock, Scott Noll, Carel Otte, Alan Postlethwaite, Mortimer Prince, Vasel Roberts, Robert Schreibeis, Dianne Shanks, Scott Sklar, Jeffrey L. Smith, Barrett Stambler, and Ben Wolff. Their critical insights have been invaluable.

We also benefited from having a great editor. Kathleen Courrier's literary skills were strengthened by her detailed knowledge of renewable energy sources; both helped bring this book to life. David Macgregor pitched in and did an excellent job of editing the footnotes. All remaining omissions and errors are, of course, our responsibility alone.

Daniel Deudney and Christopher Flavin
Worldwatch Institute

Renewable Energy
The Power to Choose

1

Introduction
The Power to Choose

Celebrating a new spirit of global coexistence, the industrial nations in 1972 established an international research center where the world's best scholars could gather to study humanity's most pressing problems. This unique venture in cooperative global forecasting—dubbed the International Institute for Applied Systems Analysis (IIASA)—opened its doors in a sumptuous Viennese palace in 1974. Soon the Institute focused its computer models on the subject of energy.

For four years an international team of distinguished scientists and analysts studied, conferred, and wrote. Their work, *Energy in a Finite World,* appeared in seven languages in 1981.[1] It laid forth more comprehensively than ever before a planetary energy future—the future implicit in the conventional wisdom guiding many of the world's energy officials. In IIASA's view humanity will use three to four times as much energy in the year 2030 as it did in 1975. Coal, oil shale, and nuclear breeder reactors figure most centrally in this supply-side extravaganza. Renewable energy resources do not.

The IIASA researchers may have lost touch with reality in their years of labor, for most of the important developments on the world energy scene in the decade since the 1973 oil embargo contradict their study. The report largely ignores the potential for energy conservation and fails to take into account important resource, environmental, and health limitations that now make a fossil fuel- and nuclear-powered future more threatening than desirable. Moreover, the energy sources that the IIASA researchers expect to make the largest contribution have failed to grow as rapidly as projected, while the energy sources they ignore have soared.[2]

The safer, more modest energy future charted in this book reflects a very different perspective. Its starting point is the end-use approach to energy pioneered by Amory Lovins in the mid-seventies, focusing first on the myriad needs for energy and then on meeting those needs economically.[3] The role of individual factories, communities, and individuals is emphasized, and the perspective is clearer since the actual motivation and constraints that determine energy trends are apparent. Another difference is that the focus here is on the major energy developments of the last ten years, particularly those that could affect future directions the most. One is energy conservation. The other is renewable energy.

Energy conservation has been the lifeboat in a decade-long storm of energy problems. Conservation's short "lead times" and modest costs make it the ideal response to sudden oil price increases. With few exceptions, industrial countries have increased energy efficiency by 10 percent or more since the early seventies. Developing countries too have begun to realize energy conservation's immense potential. In the halls of government and in industry boardrooms around the world, the central role of energy conservation is now accepted.

The past decade has also witnessed a quieter energy revolution. More than a dozen renewable energy sources have been explored, and many harnessed. Wood fuel and hydropower have been used for centuries and today provide nearly one-fifth of the world's energy. Passive solar design, wind power, alcohol fuels, and geothermal energy also have been used in the past, but not on the large scale they soon will be. Such new technologies as solar photovoltaic cells and solar ponds now appear to have a huge, untapped potential. All of these energy sources will last indefinitely, and all except geothermal power are based on sunlight—which annually delivers to the earth more than 10,000 times as much energy as humanity uses.[4]

The progress of the last several years marks a coming of age for renewable energy. Technical advances have brought wind machines and solar cells to the edge of the commercial market for electricity in some countries. Over 3 million solar water heaters have been sold in Japan and 5 million wood stoves in the United States. Government commitments have been demonstrated by an ambitious alcohol fuels program in Brazil, wind and solar programs in California, and geothermal and wood energy programs in the Philippines. Dozens of communities around the world have developed their own renewable energy and conservation plans, no longer relying exclusively on the programs of distant bureaucrats.

Equally important are the false starts and wrong turns now on record. Some attempts to introduce solar cookers in developing countries and solar concentrating systems for electricity generation in industrial countries, for instance, were oversold and did not meet initial hopes. But many of the social and technical problems encountered have been instructive, and few of these mistakes will have to be repeated. In some cases the technology simply needs to be introduced more carefully. In others a new approach or a new technology is needed.

The aim of this book is to draw on the decade's experience with renewable energy and critically assess its potential. Ten years of trial and error have weeded out the less promising technologies, so the emphasis here is on the major sources of renewable energy with the most potential. Passive solar design, active solar collectors, solar photovoltaic cells, wood fuel, energy from other plants and wastes, hydropower, wind power, and geothermal energy are covered at length, while such limited—or limiting—options as wave power and solar satellites are discussed briefly. Although obstacles still surround the use of these eight major sources, their collective potential is enormous.

Of course, no energy transition can unfold overnight. Switching from wood fuel to coal during the industrial revolution took most countries a century or more, while several decades were needed to introduce oil and natural gas. The key to a viable renewable energy-based future is that the world find means to make the transition gradually—phasing in new fuels before the old ones run out and simultaneously reshaping economies and societies. The most encouraging aspect of the progress made in the last decade is that it has cleared the way for gradual change. Energy conservation has provided breathing room while new technologies are developed that will allow a meshing of renewable and conventional energy sources during the decades of transition. Change will be continuous and the challenges enormous, but this process of historic change will also provide opportunities for creativity and growth for generations to come.

One misconception that seems to spring up again and again is that energy sources must come in large packages. Early on some energy analysts did take solar energy to mean large arrays of collectors strung across the world's deserts and connected by long-distance power lines to cities and factories—solar power based on the nuclear model. While the opposite view—that renewable energy meant an exclusively "soft," decentralized energy path—also found adherents in the seventies, a middle ground is emerging today. Large and small, centralized and decentralized energy technologies all appear to have their place. Wind power can be harnessed by the megawatt at "wind farms" and also by small turbines that supply individual homes. Solar power can be captured at large solar ponds on vacant land and by photovoltaic cells on rooftops. Renewable energy has appeal for growth-oriented economists and safe-energy advocates alike.[5]

Fifty years from now historians may well look back at the world's heavy reliance on one fuel as an unhealthy anomaly born of decades of low oil prices. In the future differences in climate, natural resources, economic systems, and social outlook will determine which energy sources will be used in which regions. Already Brazil is making alcohol fuels from sugar cane, and China is converting agricultural wastes to fuel in community biogas digesters. In Iceland geothermal energy is now the most popular means of heating homes, whereas in Canada fuelwood and passive solar design are providing a large share of residential heat. The United States and Japan are meanwhile applying their technical muscle to a promising space-age technology—photovoltaics. Even within nations energy supplies will vary by region. Some countries will make use of five or six major sources of energy—true energy security.

Of course, as energy supply patterns change so will economies and societies. Industries will tend to locate near large rivers, geothermal deposits, and other "lodes" of renewable energy since the new fuels are less portable than oil. New patterns of employment, new designs for cities, and a revitalized rural sector could all emerge with renewable energy development. Less welcome changes might include increased land-use pressures and shifts in the balance of economic power among regions.

For individuals and the environment the changes would be

rejuvenating. Because "renewables" are less polluting than coal, people will breathe easier as energy systems change, as will crops and forests. And renewable energy offers people who are interested the chance to take more direct control over their energy supply. For others, relying on renewable energy will simply involve flipping a switch. As for housing, people will be able to choose between free-standing homes that harness their own energy or energy-efficient district-heated apartments. For many people in the Third World, renewable energy development will bring electric lights, running water, and space heating for the first time.

Renewable energy is the power of choice. It works in a rural or urban setting, in centralized or decentralized systems. Renewable energy development is a gradual process that unfolds with many small investments. A mistake today does not foreclose another option tomorrow. Banking on renewable energy and energy-efficiency is fundamentally the most conservative energy course we can take. Risks are minimized, options preserved.

The risky course is sticking mainly with coal and nuclear power. The investments needed to buy the new technologies and the environmental controls they require are too big to allow investments in alternatives too. Mines, ports, railroads, and synthetic fuel plants will have to be abandoned when coal runs out or becomes too environmentally damaging to use so heavily. To ensure that nuclear technologies do not fall into the wrong hands, governments will have to police their use, circumscribing civil liberties to do so. Environmental damage from coal and nuclear plants will eventually make some areas off limits and pose serious threats to human health. Equally disturbing, the imposing institutions needed to guide these megasystems could become too entrenched to respond to the public's needs and desires.

Of course renewable energy will not flower on its own no matter how powerful the logic behind its use. Important changes will have to be implemented by national governments, communities, utilities, and businesses. And those whose income and profit is tied to existing energy sources will fight the changes that the majority so needs. Yet the new policies required will not turn our world upside down. They need be only improved versions of the research programs, financial incentives, and community projects already afoot in many parts of the world. "Renewables" already enjoy broad-based grassroots support in many countries. And as more people seize the political power to choose, that popular base is growing.

In the long run, humanity has no choice but to rely on renewable energy. No matter how abundant they may seem today, eventually coal and uranium will run out. The choice before us is practical: We simply cannot afford to make more than one energy transition within the next generation. We have not money enough or time.

2

Energy at the Crossroads

For most of the last decade, the world has been stranded at an energy crossroads. The shocks to the world economy caused by the oil price increases of the seventies have set in motion complex reactions and adjustments that are still unfolding. In 1980 alone, scores of national and corporate energy forecasts were torn up and discarded, their ten-year predictions rendered irrelevant by a year of real-world developments. Since then analysts have again been caught flatfooted by the sudden slack that developed in the world oil market. Today confusion and hardship seem to typify the new energy era. The two global recessions triggered in part by the rise in oil prices have been a blow to nearly all countries, but especially to the poorest nations that now find it difficult to meet their most basic needs.

Rising oil prices, engineered in part by the Organization of Petroleum Exporting Countries (OPEC), are a blessing as well as a curse, however. Petroleum cannot support civilization indefinitely. The oil-price rises prepared people for the inevitable and set in motion the wheels of change.

Energy progress achieved so far has come largely from energy conservation. During the last several years, the energy saved via millions of small efficiency improvements by businesses and individual citizens has outstripped the impact of all new sources of energy supply combined. Between 1979 and 1982 energy use fell 10 percent or more and oil consumption was down 20 percent in many industrial nations, only partly owing to the recession.[1] In fact, without energy conservation the oil "glut" of the early eighties would not exist. A sign of hope, this trend toward efficiency opens up the possibility that the energy transition can be smooth and gradual. In contrast, if exponential growth in energy demand were to resume, that transition would perforce be disruptive, even brutal.

More vexing questions about which energy sources the world will rely on remain clouded in uncertainty. Oil and natural gas will play an important but diminishing role for some time, but how long is less than clear. Coal will likely grow in importance, but how much we should burn considering the serious side effects of its use is a tough question. In the ongoing debate over nuclear power, economic, health, and safety uncertainties continue to come to light. Answering these thorny questions, however difficult, has become an obligation for ourselves and our children.

The Oil Rollercoaster

Oil is a remarkably versatile and valuable fuel. It contains more energy per volume than any other major fuel, and it is easy to extract and transport. What's more, petroleum refining is so highly evolved that the same barrel of oil can power jet airplanes, light a peasant household, or serve as a feedstock in plastics production. Technological progress and inexpensive oil went hand in hand in shaping industry, agriculture, and lifestyles during the twentieth century.

As recently as 1950 oil supplied less than 30 percent of the world's "commercial" energy. At that time industrial economies relied heavily on coal, which was the major fuel everywhere except North America. Oil's rise came rapidly. Petroleum extraction expanded by over 400 percent between 1950 and 1973. (See Figure 2. 1.) Soon nations that had never used oil before and possessed no domestic reserves were using it to run their industries and vehicles. From the United States the petroleum economy spread rapidly to Europe, Japan, and the Soviet bloc countries, and later to the developing world. In Japan oil imports increased eightfold between 1960 and 1973, making the country briefly the largest oil importer in the world, dependent on the Middle East for half its energy.[2]

Altogether oil now supplies 44 percent of the world's commercial energy and 38 percent of total energy (including biomass), but even these numbers understate its impact on societies.[3] Industries built thousands of new plants that relied on oil and natural gas, and consumers began using oil and gas to heat and cook. The "car culture" took longer to spread outside the United States, but since 1970 the world automobile fleet (now consisting of over 300 million vehicles) has been the most rapidly increasing oil consumer in many regions. Electricity use also rose dramatically during this period. In the past, electrification had been based mainly on hydroelectric dams, but the 5 to 10 percent annual growth rates of the sixties to a large extent reflected the contribution of new oil- and gas-fired plants. Huge amounts of capital were sunk into equipment that could be powered only by petroleum.

As reliance on oil continued to rise in industrial countries, petroleum use in the Third World increased too. Yet even

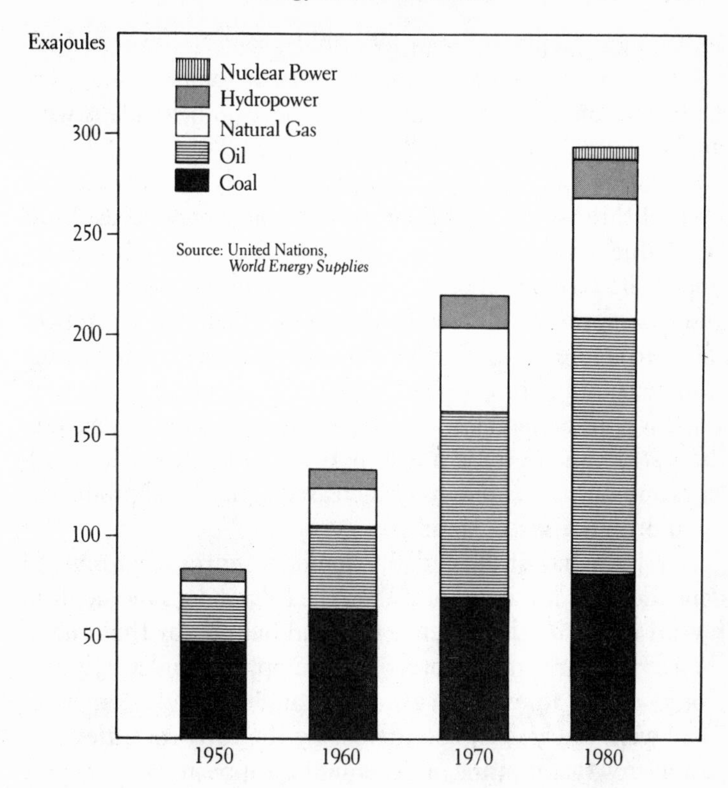

Figure 2.1. World Commercial Energy Use by Source 1950–1980.

today the developing countries, which contain three-quarters of the world's population, consume just one-quarter of the oil used each year. Many developing countries use less than one barrel of oil per person annually, compared to over twenty barrels per person each year in some rich nations.[4] Of course, what makes this comparison striking are the more than 2 billion people who still rely mainly on such traditional fuels as crop wastes and wood.

Oil long seemed the ideal fuel for development. Using it requires relatively modest investments in transportation and combustion facilities. Then too, unlike some traditional fuels oil can be used with equal ease in cities or rural communities. Until the late seventies, virtually all development plans were predicated on the availability of cheap oil.

Today, sixty-seven developing nations rely on imported oil to meet three-quarters of their commercial energy needs. Most face a fuelwood shortage as well.[5] Modern housing, industry (especially cement, chemical, and pulp and paper producers), and transportation all rely heavily on petroleum. Even in poor rural areas, the oil era has left its mark. Kerosene is becoming an important lighting and cooking fuel, particularly where fuelwood is scarce. Diesel-powered generators and pumps have in the last decade become a common sight in Third World villages and farms—emblems of increased agricultural productivity and higher living standards.

For both industrial and developing countries, current oil dependence is less important than the tremendous momentum toward *increased* dependence that had built up by the time of the 1973 oil embargo. World oil consumption consistently rose 6 or 7 percent annually, in good years and bad, and alternatives to oil were rarely even considered. By the early seventies economic growth and rising oil consumption appeared inextricably linked.

The Arab oil embargo of 1973 and the Iranian revolution of 1979 will enter the history books as watershed events that brought about some of the most important changes in the twentieth century. As oil prices rose from $2 per barrel in the early seventies to $12 per barrel in the mid-seventies to $35 per barrel by the end of the decade, the initial impacts were economic. Inflation became a global epidemic, reaching an average rate of 11 percent in the Western industrial countries by 1981. Inflation subsided in 1982, but slow economic growth and soaring unemployment were other legacies of the new era. In Western Europe alone, over 16 million people or 10 percent of the labor force were without work in 1982, a particularly grim new record. Few economists expect a return of the vibrant economic growth that provided adequate jobs until the early seventies.[6]

Though they were not the only difficulty facing the world's economies in the seventies, oil prices were nonetheless a critical variable. They turned good economic performances into mediocre ones and put marginal economies on the intensive care list. Even at current prices and with continuing slack in the market, the cost of oil will cause economic problems for years to come. As a 1980 report by the International Energy Agency concluded, the oil upheavals of the seventies "signalled a fundamental change in the ability of the industrialized nations to chart their own economic destinies."[7]

For developing countries that need economic growth to alleviate poverty, the situation is particularly bleak. Although the oil requirements of Third World nations are small by industrial world standards, oil vulnerability is even greater. Net oil imports in oil-importing developing countries doubled during the seventies, and the cost of those imports rose nearly fifty times, reaching an estimated $47 billion in 1980. Today oil imports eat up more than a third of export earnings in most developing countries. As Costa Rica's economic minister observed, "In 1970, one bag of coffee [Costa Rica's chief export] bought 100 barrels of oil, but today, one bag of coffee buys just three barrels of oil." The oil-import bill in Turkey in 1980 exceeded the country's *total* export earnings, and in Bangladesh, India, Sudan, and Tanzania, the figure was over 50 percent.[8]

In much of the Third World industrialization has slowed and agricultural productivity is stagnating—problems that the high price of oil greatly exacerbates. In many rural areas reliant primarily on traditional biomass fuels, the end of cheap oil means that fuelwood will continue to be used up faster than it can be replenished and crop wastes will not be returned to the undernourished soil. The tropical forests of developing countries shrink by 1.2 percent annually (some 10 to 15 million hectares or an area the size of Cuba each year), and fuelwood shortages have become a major Third World energy problem.[9] Without doubt the need for a rapid energy transition is more critical and the issues raised more fundamental in developing countries than in the richer nations.

Predicting the adequacy of oil supplies is well nigh impossible these days. Geological uncertainties, OPEC connivings, political instability, and the shifting responses of consumers have left oil analysts in disarray. One five-year forecast made in 1982 concluded that oil prices would be between $15 and $150 a barrel and "the probability that the price could be anywhere in that range is about equal."[10] With so many forces at work on the world oil market, instability is bound to continue, and this in itself presents a tremendous threat.

Global proven reserves of oil now stand at approximately 650 billion barrels, and perhaps another 600 billion remain to be discovered. Although together these supplies equal 2.5 times the amount of oil the world has used so far, they could be used up rapidly if demand grows. Assessments of future oil-production levels made in the seventies that were based on reserve figures and assumed escalating demand led to the conclusion that production would peak in the early nineties at 50 percent above current levels and then fall precipitously.[11]

Geological estimates of oil reserves have changed little since the early seventies, but most other aspects of the oil prospect have. Energy conservation combined with a global recession has caused world oil use to fall dramatically between 1979 and 1982. Not since oil became the world's largest energy source has there been a continuous three-year decline. In the major industrial countries oil demand appears unlikely to regain the 1979 peak level in the foreseeable future. While this slack in demand will help relieve pressure on the world oil market, those developing countries that can afford to claim a more equal share of the world's petroleum will provide a counterforce. Led by rapidly industrializing countries such as Brazil and South Korea and by oil exporters that still keep domestic oil prices low, such as Mexico and Nigeria, the Third World is likely to more than double its petroleum needs in the next two decades, accounting for most of the additional pressure on the oil market.[12]

The outlook for oil supply is meanwhile dominated by geological considerations in countries that have limited reserves and by political uncertainties in the few oil exporting countries that have ample resources. In the United States, the world's first major oil producer, oil production in all areas but Alaska has fallen 25 percent since 1970. Oil-price decontrol has briefly slowed the decline, but the petroleum yield per foot of exploratory well continues to fall. The United States, much of Europe, and parts of the Soviet Union are dependent mainly on over-the-hill oil fields.[13]

During the eighties oil production declines in the United States and a few other nations should be offset by small increases in China, Mexico, and one or two Middle Eastern countries. Significant global increases could stem only from improbable decisions by the major oil exporters, improbable political stability in the Middle East, and improbable turnabouts in the findings of petroleum geologists. On the other hand, one or two minor wars or national revolutions could reduce world oil production considerably. On balance, world oil production will probably never rise more than 10 percent above the 1980 level of nearly 60 million barrels per day.[14]

Today the Western industrial countries and Japan consume more than 60 percent of the world's oil, but produce less than one-quarter of the total. In fact, the resource base has shifted to the developing world even more rapidly than these figures indicate. Approximately 80 percent of proven oil reserves lie in the Third World, three-quarters of that in the Middle East and North Africa. In contrast, the Soviet Union has 10 percent and North America and Western Europe combined have just 9

Table 2. 1. World Oil Production, Consumption, and Reserves, 1980

Region	*Production*	*Consumption*	*Proven Reserves**
	(million barrels per day)		*(billion barrels)*
Middle East	18.2	1.6	362
Africa	6.0	1.5	55
Asia-Pacific	4.9	10.8	40
West Europe	2.5	13.9	23
Latin America	5.6	4.6	70
North America	10.1	18.3	33
USSR & E. Europe	12.4	10.9	66
Total**	59.7	61.6	649

*Figure for year end.
**Production and consumption totals differ due to different accounting methods.

Source: *Basic Petroleum Data Book, Oil and Gas Journal,* and *BP Statistical Review.*

percent of global oil reserves. (See Table 2. 1.)[15] It is these figures—not the absolute size of oil reserves—that will largely determine the adequacy of world petroleum supplies. Whether there is oil enough to continue production at current levels for fifty years matters little if just two or three countries control it. Reliance on so extremely concentrated a resource is an invitation to crisis. Although the current slack in the oil market is well entrenched, it is far from permanent. Unless the transition away from oil dependence continues to gather momentum, another oil disruption by the end of the decade is worth betting on.

Natural Gas: A Temporary Buffer

One possible cushion against oil shortages is natural gas, a relatively new and underexploited resource. As recently as 1972, the United States used half of all the world's natural gas and only a few nations used it in significant amounts. Since then natural gas has been one of the fastest growing energy sources. It now supplies 20 percent of the world's commercial energy and 18 percent of its total energy—about half as much as oil does.[16]

Most natural gas is found together with oil deposits. Until recently, it was often simply flared—burned for no purpose. Indeed, without pipelines and related facilities, this precious fuel is of little value. In many places where natural gas abounds, only a few industries or private consumers are in a position to use it.

Yet flaring will go by the way as more people recognize natural gas's value as a clean and efficient fuel and as a feedstock for petrochemicals. Already some countries limit oil production to reduce the amount of gas being flared, and many companies have recently begun exploring for natural gas. Another sign of the growing value of natural gas is its rising price. Once far cheaper than an equivalent amount of oil, gas now costs almost as much wherever a competitive energy market exists.[17]

It is easier to be optimistic about gas than about oil supplies. Many as yet untapped areas hold great promise. Deep reservoirs as well as such unconventional sources as geopressured aquifers, coal seams, and Devonian shale may all yield gas. Huge, easy-to-tap reserves in the Middle East and other oil-producing regions will be exploited as soon as the necessary facilities are built. In contrast to oil production, natural gas extraction is likely to rise 20 to 30 percent during the next two decades.[18]

Unfortunately, the world's natural gas reserves are as unequally distributed as its oil reserves. Most of the increase in output will occur in just four regions—Mexico, the Soviet Union, the Middle East, and North Africa. A few other developing nations have ample reserves, but most poor countries do not. Among industrial countries, natural gas is a severely limited resource. Most U.S. reserves have been tapped, and the United States will be lucky to maintain current production levels for the next decade. Western Europe will obtain large amounts of natural gas from the North Sea during the eighties. But Europe's chief gas resource, which is in the Netherlands, will diminish steadily. On balance, natural gas will be a major energy resource for just a few nations.[19]

For the world as a whole, however, even expanded natural gas supplies do not spell energy salvation. The costs and safety problems of transporting large quantities of liquefied natural gas overseas cannot be dismissed lightly, and geography will limit pipeline exports of gas to such natural connections as that between the Soviet Union and Western Europe and between Mexico and the United States.[20] Put bluntly, natural gas is not oil's equal. It can never be widely traded on the world market. Nor can it be put to all the tasks oil performs. Although it is ideal for heating homes and for use in the manufacture of nitrogen fertilizer, it cannot replace oil in the world's automobile fleets or in remote Third World villages. At best, natural gas can help cushion us from oil shocks and help us buy time to develop indigenous, sustainable energy sources.

King Coal

Eclipsed by oil since mid-century, dirty old coal is well on its way to being king again, according to some energy analysts. World coal use is expanding by roughly 3 percent yearly in the early eighties, after more sluggish growth in the sixties and seventies.[21] In Australia, India, the United States, and other coal-producing nations, huge investments are going into coal mines, transport facilities, and coal-fired power plants. Even virtually coal-less nations such as Japan and Sweden are gearing up to use large quantities of this resource.

Part of coal's appeal is its abundance. No other fossil fuel is so plentiful: Recoverable reserves are estimated at 660 billion metric tons, 270 times the amount extracted each year. Today coal supplies 27 percent of all commercial energy used and 24 percent of total energy. Almost certainly it will overtake oil as the world's largest source of energy by the nineties.[22]

The most thoroughgoing evaluation of the coal prospect is the World Coal Study, a decidedly bullish assessment completed by a team of coal experts from sixteen countries in 1979. Assessing likely demand for coal in various regions and then projecting supply availability, the international team forecasts that coal use will double or triple in the next two decades. (Over the last twenty years, coal use has increased only 40 percent.) "In the industrialized countries coal can become the principal fuel for economic growth and the major replacement for oil in many uses," the study concludes.[23]

The World Coal Study is far from the last word on the coal outlook, however. It underestimates the potentially enormous economic constraints on the use of such large amounts of coal. Nor does it take proper account of the environmental and health consequences of using coal to replace oil, much less the widespread public opposition to further increases in coal use they could ignite. And it does not acknowledge fully that coal is at best a second-rate substitute for oil in many applications. Indeed, even if production triples, many nations will be hard-pressed to make coal serve their most essential energy needs.

Transportation figures centrally in the economics of coal since ten countries possess 92 percent of the world's reserves and three nations—China, the Soviet Union, and the United States—own 57 percent. (See Table 2. 2.) Today only 8 percent of the world's coal is exported. To triple world coal use, world trade in steam coal (which is used for everything but steel production) would have to rise approximately twelvefold.[24] That amount of shipping could raise coal's price significantly, since transporting it requires large investments in port facilities, barges, railroads, and slurry pipelines.

Transportation is by no means the only big expense in the coal business. Power plants and industrial boilers require huge investments. And if synthetic fuels facilities are eventually

Table 2. 2. Coal Reserves and Annual Production for Major Coal-producing Countries, 1977

Country	*Economically recoverable reserves*		*Production*	
	(billion metric tons)	*(Percent)*	*(million metric tons per year)*	*(Percent)*
United States	167	25	560	23
Soviet Union	110	17	510	21
People's Rep. of China	99	15	373	15
Poland	60	9	167	7
United Kingdom	45	7	108	4
South Africa	43	7	73	3
West Germany	34	5	120	5
Australia	33	5	76	3
India	12	2	72	3
Canada	4	<1	23	1
Other Countries	56	8	368	15
World	663	100	2,450	100

Source: World Coal Study.

built to transform coal into liquid and gaseous fuels, they will boost the cost of using this energy source dramatically. In isolation no single investment seems unmanageable. But added together they make a doubling or tripling of coal production staggeringly expensive to producers and consumers alike.

The largest costs of expanded coal use are health and environmental. Increased coal use likely means more deaths among miners, more air pollution, more land degradation, and more carbon dioxide build-up in the atmosphere. New technology and additional money can alleviate some of these problems, but such expenses hurt coal's economic viability. Other problems —such as carbon dioxide—may elude control altogether.

Mining coal is a deadly occupation. While major coal producers such as China and the Soviet Union do not publish statistics, an estimated 15,000 to 20,000 coal miners are killed on the job each year. The majority of these deaths are in China,

India, and the Soviet Union, where most of the coal is extracted manually rather than by large machines. Indeed, though China and the United States produce roughly the same amount of coal, between 3,500 and 5,000 Chinese miners are killed yearly compared to 150 in the U.S. Clearly mechanizing the coal industry and adopting safe operating procedures makes a difference, but given the long governmental neglect of these problems and the high cost of mechanizing Third World mines, a major increase in coal extraction is likely to take a heavy toll in miners' lives.[25]

The localized health effects of coal burning are like mining casualties—preventable in theory but not always in practice. The pollution controls now used in Western industrial countries have made coal burning much cleaner than it was in the early industrial period. In particular, pollution-control technology has removed the sooty particulate matter that once covered many cities. Yet large amounts of sulfur and nitrogen oxides and other pollutants are still emitted. In developing countries, where most coal is burned in small boilers, pollution-control technologies often cost too much to use at all.

Exactly how many people coal burning kills is difficult to tell, but a convincing 1980 study found that doubling coal use in the Ohio Valley (as the U.S. Government proposed to do to reduce oil use) would shorten lives of 45,000 people over a five-year period even if the $3.2 billion needed to meet pollution-control standards is spent. Given that 50,000 people already die prematurely from coal pollution each year in the United States alone, the worldwide count is probably around half a million a year. Unless stringent and expensive controls are widely imposed, increasing coal use would probably shorten the lives of several million people in the next two decades.[26]

The dimensions of another form of coal pollution—acid rain—are just coming to light. Caused when sulfur and nitrogen oxides released from fossil fuel combustion combine with atmospheric water, acid rain is of growing concern in such industrial regions as northern Europe and eastern North America. In these areas acid rain is destroying aquatic life and damaging historic buildings, monuments, and other manmade structures. Unlike mining deaths and local air pollution, the effects of acid rain are often experienced hundreds of miles from the pollution source, making regulation difficult and tension between bordering states and nations likely. Exacerbated by the use of tall stacks to disperse local pollutants, the acid rain problem reinforces the need for expensive pollution-removal systems.[27]

Carbon dioxide emissions from coal burning may prove a more far-reaching and intractable pollution problem. Since the Industrial Revolution, the level of carbon dioxide in the atmosphere has increased by approximately 20 percent—partly as a result of coal burning, which releases substantially more carbon per unit of available energy than oil and gas do. Scientists estimate that tripling coal production by the century's end could double carbon dioxide concentrations in the atmosphere by the year 2025. If, as many scientists suspect, carbon dioxide accumulation causes the atmosphere to warm up, weather patterns could be altered, probably reducing rainfall in some agricultural areas. Ocean levels would rise as Antarctic ice melted. While technically possible, removing carbon dioxide from stack emissions is prohibitively expensive.[28]

Carbon dioxide poses unique dilemmas. Conclusive evidence about its effects could well come only after the problem is beyond repair, and few politicians make careers of attacking the next generation's problems. Moreover, given the global scale of carbon dioxide pollution, unilateral efforts to halt its release would have little effect. So great are the complexities surrounding the carbon dioxide issue that some argue that effective action is impossible and that we should begin planning for the "warm up." They would be right if carbon dioxide buildup were coal's only drawback. But it is not. The need for "a breather" so that scientists can continue to assess the magnitude of the carbon dioxide problem is only one element of a powerful case for slowing the growth of coal use.

Coal's final drawback is its limited utility. Today fully 60 percent of the world's coal is used to generate electricity, and another 23 percent (high-grade metallurgical coal) is used for steel production.[29] Most of the remaining 17 percent is consumed for other industrial purposes. The contrast with oil could scarcely be greater. The most common (and valuable) uses of oil are as a fuel in transportation and buildings and as a feedstock in petrochemical production.

Most likely, the uses to which coal is put will not broaden significantly in the near future. The economic constraints are simply too large, a point even the most bullish coal forecasters recognize. The World Coal Study concludes that most of the huge increase it forecasts will be used in power plants. Coal's role in industry could increase substantially where coal is accessible, but the many small industries far from coal mines or in areas that already have heavy air pollution will have to find other alternatives. In residential and commercial buildings, coal has little place: It is simply too expensive to transport and too dirty to use.

Coal would hold more promise if it could be converted into a liquid or gaseous fuel cheaply and effectively. But while coal chemistry has become sophisticated after more than a century of research, coal-conversion processes remain complex and inherently energy inefficient. Cost estimates for synthetic fuels plants have escalated as quickly as oil prices since the mid-seventies, extinguishing early optimism about "synfuels." Accordingly, ambitious synthetic fuels programs in the United States and West Germany have been scaled back greatly. By most reckonings, synthetic fuels will play only a minor role by the end of the century. The most economical synthetic fuels are likely to be methane and methanol rather than the more complex hydrocarbons.[30]

For the foreseeable future, coal will be used mainly for

electricity generation. Already coal-fired power plants have a significant economic edge over oil-fired and nuclear plants in areas where coal is abundant. But coal addresses only a small slice of the energy problem in most areas. Less than a third of the world's electricity is currently generated using petroleum; in such countries as France and the United States, substituting coal for oil completely in electricity generation would reduce oil imports by a mere 10 percent. Meanwhile, the growth rate in electricity demand has fallen off precipitously in much of the world, making a mockery of extravagant forecasts for coal's use in power generation.

Nuclear Power: Too Bleak to Meter

Nuclear power has had a short and meteoric history. No other new energy source has received as much government support or stirred such controversy. Originally conceived as safe and "too cheap to meter," nuclear power enthralled scientists and the general public alike during the postwar period. Several governments, led by the United States and the Soviet Union, supported large nuclear research programs, and the technical breakthroughs of the fifties soon became the "commercial success" of the mid-sixties as governments persuaded utilities to begin investing in nuclear power.[31]

During the sixties and seventies, utilities in Canada, France, Great Britain, Japan, the Soviet Union, and the United States committed billions of dollars to this new technology. These nations were soon followed by many other industrial countries and a few developing countries. Beginning in 1970 the number of operating nuclear plants increased rapidly. By 1981 some 256 nuclear reactors in twenty-two countries were supplying approximately 8 percent of the world's electricity (2 percent of total energy supplies).[32] Energy planners foresaw a rosy future in which nuclear plants not only supplied most electricity but also began to displace residential and industrial fuels. Literally thousands of plants would be required to meet those goals, but nuclear experts from the Soviet Academy of Sciences and the U.S. Atomic Energy Commission alike were genuinely confident that their goals could be met. Construction, they argued, would become easier and costs would fall as the industry acquired experience.

The prospects for atomic power began to dim almost as soon as the first large nuclear plants were completed. As a theoretical prospect materialized into a concrete reality, important unanswered questions related to public safety, long-term waste disposal, and weapons proliferation emerged. Political opposition to nuclear power began to grow. By the mid-seventies individual plants in Europe, Japan, and North America had become targets of local public protest, and by the early eighties many government officials and nuclear scientists had joined the growing anti-nuclear movement.

Since the first days of civilian nuclear power, disposing of spent nuclear waste has been a major concern. Twenty-five years after the first commercial power plant began operation, it still is. Early hopes that nuclear wastes could be stored in extremely stable geological formations for millennia have been dashed by the realization that extensive tunneling and drilling destabilize rock structures. And our ability to predict the paths of subterranean water flows seems more questionable as we learn more about the earth's inner complexities. Nevertheless, some pronuclear countries—notably France—have moved ahead with retrievable storage systems that rely on the capacity of future generations to monitor the materials effectively, repair the containment vessels, and prevent their theft. Such measures are obviously expensive, and their long-term effectiveness can never be guaranteed.[33]

Born of warfare and then transferred to civilian power production, the two uses of nuclear energy have never been securely separate. The early belief—central to commercial nuclear power's acceptability—that civilian reactors and bomb-making capabilities could be kept apart grows less plausible each year. New technologies for making nuclear power more efficient are further eroding this thin line, and the International Atomic Energy Agency's proliferation safeguards system is generally recognized as too weak to prevent the diversion of nuclear materials from power plants to warheads. In the wake of India's surprise detonation of a bomb made from materials from a civilian reactor, several countries now appear to be developing nuclear bombs behind the façade of a "peaceful" nuclear power program. While a few additional nuclear weapons in a world with over 50,000 warheads might seem a small additional risk, the possibility that irresponsible governments may acquire nuclear materials makes nuclear power an extraordinarily dangerous way to generate electricity.[34]

Meanwhile, some fundamental economic problems have also begun to plague nuclear power. Irvin Bupp of the Harvard Business School observes that "the nuclear plants that were being sold in the mid-sixties on the *promise* of cheap power would not actually begin to operate until the early seventies. But there was little or no effort by reactor manufacturers, by the purchasers or by the government itself to distinguish fact from fiction on a systematic basis."[35] It turned out that these original cost estimates were low by a large margin, a fact that became painfully apparent as cost overruns accelerated throughout the seventies.

The most thorough economic study done so far is by Charles Komanoff, a U.S. energy analyst. He found that in the United States between 1971 and 1978 real capital costs for nuclear plants (after accounting for inflation) rose 142 percent—13.5 percent per year or nearly twice as fast as costs for coal plants. Komanoff's analysis indicates that these increases were not, as the industry alleges, caused by licensing delays. Rather, cost increases reflect design changes needed to resolve important safety problems discovered as earlier commercial plants began to operate. Komanoff concludes that similar increases can be expected in the eighties as we face still unresolved safety issues, including those raised by the accident at Three Mile Island. As a result the simple economic viability of nuclear power is now uncertain at best.[36]

A related obstacle confronting nuclear power is one that it shares with coal—slowing growth in demand for electricity. Nuclear power is used virtually entirely for electricity generation, and electricity demand slumps have been one of the major reasons for power plant cancellations in recent years. In France it now appears that the country will have expensive excess nuclear capacity by the late nineties, a problem the government could solve only by dramatically lowering electricity prices.[37] Yet in France, as elsewhere, cost overruns on current nuclear plants are partly to blame for electricity price increases. For the remainder of the century, coal and nuclear power will be competing mainly against each other in a severely limited electricity market, and coal has a decided edge in most countries.

The combined effects of cost overruns, slowing growth in electricity demand, high interest rates, and widespread public opposition are showing up in utility construction programs, particularly in the United States. Although most U.S. utilities still outwardly express enthusiasm for nuclear power, many are simultaneously pulling the plug on the industry. From a peak of twenty to forty new plants per year in the early seventies, new orders fell to an average of three per year between 1975 and 1978 and then ceased entirely. Meanwhile, nuclear plant cancellations mounted steadily, reaching a total of fifty-eight for the years 1977 through 1982, a figure that represents more than the total installed nuclear capacity in the country in 1982. Once it was assumed that the United States would have 300,000 to 500,000 megawatts of nuclear capacity by 1990 with even faster growth in later years. More likely now, U.S.

nuclear capacity will be less than 120,000 megawatts in 1990, with little further growth in the nineties.[38]

Canada, Great Britain, Japan, and West Germany have scaled back their nuclear programs, too. Rising costs are part of the reason, but even more important is mounting public opposition to nuclear power. In Germany a de facto moratorium on new plant orders has been in place since the early seventies. And in Sweden, which gets fully 15 percent of its electricity from nuclear plants today, a 1980 referendum banned further orders for new plants and decreed that nuclear power will be phased out by 2010. France is perhaps the only Western country likely to rely heavily on nuclear power in the coming decades. France now gets over 40 percent of its electricity from nuclear power, but French nuclear critics charge that the country's program survives largely through taxpayer subsidies.[39]

Nuclear programs in Eastern Europe have followed a similar path—surprising, considering the differences in the political systems of those countries. There, too, nuclear plant construction has been more costly and slower than expected. Although Soviet leaders continue to support the nuclear program, actual capacity today is less than half the level forecast in the early seventies. During the early eighties, projections for 1990 were trimmed by more than 40 percent.[40]

In the Third World nuclear power has had a mixed welcome. Today a handful of developing countries are operating nuclear plants, and about a dozen more have nascent nuclear power programs.[41] For many developing countries, technologically sophisticated nuclear power has important prestige value. In the Third World nuclear power's financial problems, however, appear intractable since huge capital investments are needed.

Another problem that impedes nuclear plants in the Third World is their size. Even the smallest reactors marketed in industrial countries are too large to be used in the electricity grids of most developing nations. If a single power plant provides too high a proportion of generating capacity, shutting it down knocks out the entire system. Canada and France have attempted to get around this difficulty by marketing "mini-reactors" in the Third World, but electricity from these "small fry" costs much more than that from larger power plants.

Compelling evidence suggests that nuclear power will supply just 3 to 4 percent of the world's energy during the closing years of this century. By 1990 the world will likely have about 300,000 megawatts of nuclear capacity. (See Table 2. 3.)[42] The outlook for the year 2000 is more uncertain, but growth rates are likely to slow further, since many of the recent cancellations have been for plants once scheduled for completion in the nineties. Given continuing cost overruns and the long lead times for nuclear plant construction, nuclear power cannot possibly soon provide the massive contributions to the world energy supply that were envisioned a few years ago.

Table 2. 3. Estimated World Nuclear Power Capacity, 1981 and Projections to Year 2000

Region	*1981*	*1990*	*2000*
		(1000 megawatts)	
Western Europe & Japan	57	115	150
North America	60	120	130
Soviet Union & Eastern Europe	16	50	75
Developing Countries	3	18	25
Total	136	303	380

Source: U.S. Atomic Industrial Forum and the *Financial Times Energy Economist*. The projections are the authors'.

Some optimists still cling to the hope that new nuclear technologies will one day resurrect this problem-plagued energy source. In particular, many hopes have been pinned on the breeder reactor. The attraction of the breeder reactor—which is being developed in France, the Soviet Union, and the United States—is that it would produce more nuclear fuel than it consumed, unlike conventional reactors that consume precious enriched uranium. Yet today uranium is plentiful and its cost is falling. More to the point, breeder technology is likely to come up against most of the economic problems that confront light water reactors. On a commercial scale, breeder plants would likely be extremely complex and expensive and would raise safety and proliferation hazards. As commercial operations increased, traffic in plutonium, a raw material used to manufacture nuclear bombs, would inevitably rise, greatly increasing the likelihood of nuclear war or terrorism.[43]

Even ignoring these formidable problems, the earliest substantial energy contribution breeder reactors could make would come in 2010. Meanwhile, breeder technology absorbs well over a billion dollars of government research funds each year —funds that could be far more productively spent on other energy sources.

Nuclear fusion is another technology under extensive research. To explore the attractive possibility of producing inexpensive power by fusing isotopes of superabundant hydrogen, hundreds of millions of dollars are being spent. Some fusion enthusiasts speak of the technology with a messianic zeal, having transferred to it the old hope of unlimited and environmentally benign energy. But research efforts have yet to demonstrate even the technical feasibility of commercial fusion power, and energy technology's history is strewn with theoretically brilliant devices that never made the jump to economic viability. Fusion technology, in contrast to breeder reactors, does deserve continued government-backed research. But for all its promise, this technology remains speculative, and the bets won't be called in until the year 2025 at the earliest.[44] Fusion therefore offers no answers to the most pressing energy problems of the near future.

The Conservation Revolution

The prospects for oil, natural gas, coal, and nuclear power afford little optimism about our energy future. The inexpensive and convenient energy sources are running out, while the abundant sources are dangerous or poorly matched with the world's energy needs. The conventional "supply-side" approach to energy planning appears increasingly uneconomic and antisocial.

But amid these disappointments, conservation is a shining light. From tiny bungalows to steel mills, improved energy efficiency has been the most successful response to rising oil prices. Today saved energy costs less than energy produced from new sources almost everywhere, a development that has brought many economists up short. Truly a "conservation revolution," this radical departure from established trends provides hope for resolving the world's energy dilemmas.

The most common way of gauging energy conservation or energy efficiency is to compare the rate of growth of energy use with that of national economies. After World War II the two tended to grow in parallel, and conventional wisdom held that they were inalterably linked. But since the early seventies economic growth has been three times as rapid as energy growth in the United States, and in Europe and Japan it has been twice as high. By 1981 the economies of the Western industrial countries were already 19 percent more energy efficient than they were in 1973. Conservation's contribution to meeting additional energy needs during this period was several times the size of all new sources of supply combined. Between 1979 and 1981 alone, oil use fell by 14 percent in the United States, 15 percent in Japan, and 20 percent in West Germany, almost twice the declines that occurred during the 1974–75 recession.[45]

Energy conservation is taking hold in various forms virtually everywhere. In Nairobi, Kenya, a major hotel cut its electricity use for air conditioning in half during a four-year period. In

Japan most household appliances purchased today are 40 to 60 percent more efficient than Japanese appliances were in the mid-seventies. The fuel used per passenger-mile in the U.S. airline industry has been cut by 30 percent. Many of these improvements stem from modest technology improvements and simple "housekeeping" measures. Yet a vast range of slightly more complex and expensive innovations are now economical. As they are introduced, conservation's momentum will build.[46]

In major energy studies in Denmark, Sweden, the United States, and other nations, energy analysts have recently surveyed the potential for further energy conservation. By far the most comprehensive of these analyses was that completed by the U.S. Solar Energy Research Institute in 1981. According to the SERI report, even with rapid economic growth, energy use could be cut by 25 percent by the year 2000. In fact, so many inviting opportunities for investing in energy efficiency were identified that SERI concluded that lowering energy use will actually improve economic prospects. The advent of less energy-intensive "service economies" will accentuate these trends.[47]

Already conservation has become a $10-billion a year business in the United States.[48] Similar though less dramatic results have been obtained in other countries where energy waste was lower at the outset. In the industrial nations, a general consensus holds that growth in energy use will not exceed 1 to 2 percent per year and that it could be even less if conservation is embraced wholeheartedly.

The conservation revolution has more than upset the projections of economists, however. It has fundamentally changed the context in which energy systems operate. No longer can energy be seen as a single commodity needed in predetermined amounts. Today, with few inexpensive energy alternatives available, the emphasis is on conserving energy wherever possible and using whichever energy resources are most economical in particular applications. As a result, energy growth will tend to be much more varied and "use-specific," a development with important implications for renewable energy's future.

When supply availability alone guided energy assessments, new energy sources were compared primarily with oil or coal for large-scale conversion to electricity. Since it was imagined that world energy demand would inevitably multiply and that the main choice was between thousands of nuclear reactors or hundreds of millions of solar collector systems, renewable energy advocates were soon branded as unrealistic. But with the current pressing need to conserve and to pay attention to end-uses, the competitiveness of renewable energy sources with conventional energy sources on a case-by-case basis has become all important.[49]

Many renewable energy technologies appear to fit current energy needs quite well. Most industrial countries, for instance, need small amounts of additional electricity generating capacity, most of it centered in a few rapidly growing regions. By the late eighties and early nineties (when the new capacity is needed), small-scale hydropower plants, wind turbines, and wood-fueled cogenerators will be among the cheapest power sources available to meet that additional need. All can be built quickly, and an additional unit or two can easily be added as demand dictates. Similarly, households and industries that use energy efficiently and carefully calculate their future needs are finding that renewable energy technologies are on the verge of competitiveness.

Just as energy conservation has revolutionized energy economics, so has it encouraged far-reaching changes in the geographical energy balance. In the past most energy growth occurred in a relatively small number of industrial countries. In the future a much larger share will occur in the Third World. Although developing countries can make substantial cost-effective investments in energy efficiency, their need for new energy sources is certain to grow more quickly than that same need will grow in the industrial world. According to World Bank estimates, energy needs in the Third World will grow at 5 percent per year in the eighties (compared to 7 percent growth in the seventies). Developing countries simply cannot afford to meet most of their growing needs with imported oil.[50]

The case for renewable energy clearly rests on more than oil price forecasts and the economic prognosis for coal. More important than either is a clearheaded assessment of the evolving world energy situation and its underlying subtleties—exactly what's missing from supply-oriented energy studies such as the one conducted by the International Institute for Applied Systems Analysis. Supply-side studies take an exajoule for an exajoule no matter whether the energy in question is for use in automobiles or air conditioning, which now makes about as much sense as saying that human beings can live exclusively on carrots or anchovies as long as their need for calories is met.

Of course, supply-side studies cannot be dismissed lightly as long as they continue to dominate energy policy making. But their shortcomings underscore the need to widen the energy debate to take account of the diverse uses of energy and the wider social and environmental implications of the course we choose.

3

Building with the Sun

To the surprise of many technologists, the oldest and simplest use of solar energy is proving to be among the most successful in the 1980s. Just a decade ago, it was commonly thought that residential solar heating had to mean the use of pump-driven "active" systems employing solar collectors. Yet today, passive solar or climate-sensitive design is one of the most rapidly growing uses of solar energy despite a minimum of government support. The reason is simple: Passive solar buildings use relatively simple, inexpensive changes in design and construction techniques to maintain comfortable temperatures. By combining design concepts that have been known for centuries with modern building materials and technologies, builders are constructing houses that use 75 to 90 percent less fuel than conventional ones at only small additional cost.[1]

The principles being applied in climate-sensitive design are quite simple, since they are based on the idea of using natural conditions to the best advantage. The designs are intended to admit sunlight during the winter but keep it out in the summer. Insulation and thermal mass are used to prevent rapid temperature changes. Of course, the emphasis is on maximum "solar gain" in cold, sunny areas and on keeping the building cool in tropical regions. Because passive solar design incorporates both energy conservation and the use of renewable resources, it exemplifies the twin energy strategies with the most potential in the decades ahead.

Knowledgeable observers predict that the next decade will see some of the most rapid and far-reaching architectural changes in history. As a pioneering solar architect noted in 1980, "traditionally, architecture has been a response to the times, and energy conservation is the issue of our time." Passive solar buildings are already catching on in many industrial countries, particularly the United States where over 60,000 have been built since the early seventies. So far there is little activity in the Third World, but the long-run potential there is equally large. Buildings are a growing part of the energy problem in most countries, and improving designs today would greatly enhance the comfort and economic appeal of the world's buildings well into the next century.[2]

Energy and Architecture

In this age of standardized buildings and mechanical heating and cooling systems, it is easy to forget that passive solar design was once the norm. In some parts of the world, it still is. Built without the aid of architects or engineers, many traditional buildings make clever use of sunlight and natural convection for heating and cooling. Over 2,000 years ago Socrates observed that "in houses that look toward the south, the sun penetrates the portico in winter, while in summer the path of the sun is right over our heads and above the roof so there is shade."[3] This basic idea—that the sun describes a lower and more southerly arc in winter than in summer (a more northerly one in the southern hemisphere)—is applicable everywhere but near the equator. Two to three times more sunlight strikes a south-facing wall in winter than in summer, making it the logical side for windows.

As Ken Butti and John Perlin point out in their history of solar architecture, *A Golden Thread,* the Greeks were among the earliest passive solar designers. Many of their buildings were oriented to the south and had thick adobe or stone walls that kept out the summer heat. Passive solar heating was also employed by the Romans. By the fourth century A.D., the pressure of firewood scarcity had become a strong incentive for solar heating, and Roman architects slowly adapted solar design to the various conditions found throughout the Roman Empire. Access to the sun was actually made a legal right under the Justinian Code of Law adopted in the sixth century A.D.[4]

In other cultures other climate-sensitive building styles prevailed. Most homes in ancient China were built on the north side of courtyards, facing south, and sunlight was admitted through wood lattice windows and rice paper. Even today, millions of passive solar houses are found throughout northern China. The Anasazi people of the American Southwest lived in mud or stone buildings constructed against overhanging cliffs that faced south. Solar-heated in the winter and shaded in the summer, these earth-sheltered dwellings were built without benefit of modern building materials or theories. In northern Spain many apartment buildings built in the nineteenth century have glass-enclosed south-facing balconies called galerías that provide effective solar heating.[5]

The world over, traditional architecture also incorporates simple passive cooling techniques. Throughout tropical Asia and South America, open-sided pole and thatch buildings allow ample ventilation and protection from the heat. Thatch, which rivals fiberglass as an insulator, is also found atop mud and straw buildings in sub-Saharan Africa. For thousands of years in Moslem Asia, cooling towers have been used to draw air into buildings, providing ventilation and relief from the hot summer climate.[6]

Since the onset of the Industrial Revolution and the urban migration that accompanied it, many traditional architectural forms have been abandoned. Climate-sensitive building designs were not easily adapted to cities, and standardized architectural styles took over as the need for low-cost housing grew. Architect Richard Stein writes that "during the 1920s many of the most prophetic and influential architects projected the form of the future as being freed from the rigorous demands of climate and orientation."[7]

This revolt against nature combined with growing populations more than tripled the fuel requirements of buildings worldwide between 1950 and 1980. New buildings use much more energy per square foot than those of the past since they have energy-intensive central heating and air-conditioning systems. Furthermore, only half the residential buildings in Europe, for instance, have any insulation at all, and storm windows are a rarity. In the United States close to one-third of the residential housing stock is uninsulated, and another 50 percent is underinsulated. The buildings in many countries, particularly the homes of the poor, are loosely constructed and "leaky": Cracks around windows and in walls and attics let too much heat out and in.[8]

Turning our backs on climate-sensitive design and construction techniques has proved costly. Consider the typical modern office building. With glass façades and mechanical "climate-control" systems in use every day of the year, its energy appetite is enormous. Commonly, a quarter of an acre of lights must be turned on to illuminate a few square feet surrounding a desk.

In private houses and apartments the rapid spread of air conditioning has upped energy use more than any other factor in recent years. Together, residential and commercial buildings account for between 20 and 40 percent of national energy use in most industrial countries. (See Table 3. 1.) Of this energy, approximately four-fifths is used to heat, cool, and light buildings and the rest runs water heaters and other appliances.[9]

Table 3. 1. Energy Use in Residential and Commercial Sectors in Selected Industrial Countries, 1978

Country	*Residential and commercial energy use*	*Share of total national energy use*	*Residential and commercial energy use per person*
	(million barrels of oil equivalent)	*(percent)*	*(barrels of oil equivalent)*
United States	3256	33	14.8
Canada	338	33	14.3
Sweden	96	38	11.5
Netherlands	154	39	11.0
West Germany	581	39	9.5
France	375	35	7.0
United Kingdom	331	31	6.0
Italy	235	30	4.1
Japan	419	21	3.6

Source: Organisation for Economic Co-operation and Development, *Energy Balances of OECD Countries.*

Fuel use per person in homes and commercial buildings is nearly twice as high in the U.S. and Canada as in most of Europe. European cities are laid out more compactly, and Europeans prefer to keep their buildings relatively warmer in summer and cooler in winter. The industrial country with the best record is Japan. There, per capita fuel use in buildings is only one-quarter of the U.S. level, because most Japanese buildings are compact and few have central heating. Even in northerly Sweden, the fuel requirements of buildings are 25 percent lower than in North America. With traditionally higher fuel prices and lower per capita incomes, Europeans and Japanese treated energy use in buildings less nonchalantly than did North Americans.

Few such generalizations hold with regard to the Third World. The developing nations located in the humid tropics have traditionally relied entirely on the sun for heating and on natural ventilation for cooling. In more temperate developing countries in Central Asia and Latin America, firewood and charcoal have been the heating fuels of choice. However, in the last decade Western-style office and residential buildings have sprung up in the developing world's cities. Flagrantly climate-insensitive, most of the new buildings require electricity-hungry mechanical cooling systems designed in the West. Since many developing countries lack both engineers and the spare parts needed to keep the systems running, the air conditioning systems are often broken down and the buildings stifling hot. So far, the heating, cooling, and lighting of buildings account for less than 10 percent of the energy used in most developing nations, but a major future challenge will be to improve the miserable housing conditions without compounding an already severe energy problem.[10]

Awareness of the energy problems of buildings has, of course, blossomed since 1973. Surveys indicate that energy has become a primary concern to most homebuyers, and residential energy-conservation measures are becoming popular the world over. Newly energy-conscious Americans brought the rate of growth in energy use in the U.S. residential and commercial sectors down from 5 percent annually in the sixties to less than 2 percent in the late seventies and early eighties. Energy use in buildings is now increasing at only 1 percent annually in West Germany, while it has leveled off in Great Britain and fallen slightly in Sweden.[11]

The fuel savings so far achieved in buildings must be kept in perspective, however. They have been quite modest, deriving mainly from simple conservation improvements such as the addition of insulation, and they come at a time when building owners and renters throughout the world are being hit hard by high fuel prices. Worse, these energy savings have not yet helped the poor much. In 1979 the Tennessee Valley Authority reported that one of its customers paid her electric bill with a Social Security check "and walked out to face the month of February with less than $30," a situation that has become all too common in many parts of the world.[12] Even commercial building owners are hard-pressed to make ends meet. Electricity bills now constitute the biggest operating expense in most large structures, and they have helped boost rents at a record pace.

Climate-Sensitive Design

Fortunately, the options now available for lowering the fuel requirements of buildings go well beyond simple conservation measures. The field of architecture has been turned inside out in the last several years as everything from office towers to mobile homes has been redesigned for a new era. According to R. Randall Vosbeck, president of the American Institute of Architects, "Energy will rank with the elevator and the masonry arch as having a major influence on architecture. . . ."[13]

Behind modern passive solar heating are glass and plastics. These substances readily transmit sunlight but impede thermal radiation—in effect, trapping heat in the building. Known as the "greenhouse effect," this phenomenon is familiar to anyone who has left a car in the sun on a cool day and returned to find it overheated. In its simplest form passive solar heating consists of placing most of a building's windows on its sunny side because windows on the east and west tend to lose more heat than they gain in winter and because they can cause overheating problems in the summer. Taking passive solar architecture one step farther, many architects now design buildings that are elongated on an east-west axis so that the area available for "solar gain" on the sunny side is maximized. Properly siting a solar building is almost as important as the design itself since correct positioning helps assure access to the winter sun and protection from cold winds.[14]

The first modern solar house was built in Chicago in the thirties. From the outside, it looked conventional enough. But it was carefully sited to take full advantage of the sun, and it had a large expanse of window on the south side. Similar experimental buildings were constructed over the next two decades, attracting considerable attention and convincing some onlookers that a new physical principle had been harnessed. *Business Week* suggested in 1940 that the Chicago house rivaled the newly discovered Middle East oil reserves as the "newest threat to domestic fuels."[15]

As solar architectural research proceeded it became clear that retarding heat loss was as essential as admitting sunlight. The walls, roofs, and windows of conventional houses lose heat rapidly during cold weather through radiation and convection. When heated only by the sun, such houses cool rapidly after dark. In comparison, solar houses developed more recently in Europe and North America have included more than twice as much wall and attic insulation as conventional dwellings have. Most windows are double- or triple-glazed, and the use of vestibules prevents the loss of warm air when someone opens a door. These buildings are also tightly constructed—important since in conventional buildings up to half of all heat loss occurs through direct infiltration of cold air.

Also integral to the success of a passive solar building is heat storage. Built of materials that hold heat well, a building can remain warm even after a day or two of cold, cloudy weather. Such traditional building materials as brick, concrete, adobe, and stone all serve as "thermal mass," greatly reducing temperature fluctuations. Thermal storage materials are typically incorporated in fireplaces, walls, or floors. Though somewhat difficult to use in a building, water is one of the best materials for storing warmth. Sometimes used in "water walls," it can also be used in fish-pond heat storage. Researchers at the New Alchemy Institute in Massachusetts maintain that aquaculture tanks located inside a greenhouse can pay for themselves in heat-storing capacity alone.[16]

Besides providing heat during the winter, successful climate-sensitive buildings are also cool in summer. Fortunately, the same high-grade insulation and thermal storage that retain heat in winter help keep a building cool in warm weather. Cooling-only passive features include shades that protect south-facing windows from the high summer sun and ventilation systems that keep air moving continuously through a building. Deciduous vegetation is ideal for protecting a house from the summer sun only and can keep the "microclimate" several degrees cooler than surrounding areas.

One of the more ingenious solar designs—the Trombe wall—involves using a thermal-storage wall placed several centimeters inside a large expanse of glass on a building's south side. The wall, usually constructed of masonry, is painted a dark color to absorb heat from the sun during daylight hours. The wall then radiates the collected heat to the rest of the house for many hours after sundown. Extremely effective and versatile, the Trombe wall has in recent years been used in everything from office buildings in the United States to peasant huts in Ladakh, India. The Trombe wall and its variations have just one main drawback: Considerable heat is lost through radiation via nearby glass. Special thermal shades that are closed at night are needed in cold climates.[17]

A related but distinct method of passive solar heating is the use of a greenhouse or "sunspace" on a building's south side. An attached greenhouse serves as a natural solar collector that can easily be closed off from the rest of the building at night, and it can extend the gardening season as well as provide heat. As with other passive solar systems, the importance of double or triple glass, tight construction, thermal mass, summer shading, and ventilation is clear. Well-designed and properly sited, a greenhouse can supply more than half a building's heat in sunny climates.

Some solar designers, particularly those in Israel and the United States, are catching the sun by moving under ground —which only seems like a contradiction. In earth-sheltered buildings, earth serves as a natural insulator. If a building is exposed to inclement weather only on the sunny side, it can effectively collect and store the sun's heat. Earth-topped roofs also provide natural evaporative cooling in the summer, an important advantage. Still unclear, though, is whether earth-sheltered buildings can be built cheaply and whether they can overcome their undeserved reputation for gloominess. Right now building under ground costs 25 to 50 percent more than it does above ground, but some builders are convinced that the cost can be reduced substantially.[18]

Other types of passive solar buildings are also springing up, the fond labors of enterprising architects. A house developed by Harold Hay in California uses an enclosed pool of water on the roof for heat collection and radiative cooling. Another interesting concept, developed independently in California and in Norway, is the double-envelope house. It incorporates a greenhouse on the south side and a continuous air space running through the roof, north wall, and basement to supply heated air throughout the building. Both the roof pond design and the double-envelope house have fared well in the custom-built market, but their broad economic appeal remains to be determined.[19]

Some Canadians and northern Europeans are taking quite a different tack in designing climate-sensitive buildings. Since the sun in these climates makes only brief appearances at midwinter, a solar house designed for sunny conditions would be a cold house in Canada or northern Europe. Architects there are thus designing superinsulated, very tightly constructed houses with relatively few windows. Typically, a "low-energy house" features an air-to-air heat exchanger—a small unit resembling an air conditioner that ventilates the building but prevents heat loss. Its use keeps the air from getting stale or even unhealthy as pollutants like cigarette smoke or the radon found in concrete slowly accumulate. Pioneered primarily in Austria, Canada, Denmark, and Sweden, these prototypical homes have performed impressively so far. The Saskatchewan Conservation House in Canada, for example, uses 90 percent less energy than does a well-constructed conventional home.[20]

Even more challenging is the development of passive solar designs for climates where cooling is needed. Passive cooling research has been relatively neglected, though Australia, Israel, and the United States have made promising gains. Evaporative coolers have proved effective in hot, dry climates, and designs that enhance air flow help greatly in most areas. Also essential to comfort in warm weather is insulation and a means of shading building surfaces from the sun. Jeffrey Cook, professor of architecture at Arizona State University, notes that "of all the cooling strategies, heat avoidance provides the most for the least." In most climates such measures can reduce fuel requirements for cooling greatly.[21]

Further research in passive cooling will have to meet the difficult challenge of designing buildings for hot, humid climates where evaporative coolers do not function well and dehumidification is essential for comfort. Japanese and American researchers are working on passive dehumidifiers using desiccants, but such efforts are preliminary at best.[22] Active solar air conditioners may turn out to be one answer to this sticky problem. Another is to lower air conditioning needs as much as possible via careful design and use smaller, less expensive electric or gas-powered air conditioners on the muggiest days.

The cooling needs of the poor majority in the Third World

have received even less attention. Hundreds of millions now live in warm, humid climates without benefit or hope of getting air conditioning. In many developing countries past efforts to upgrade traditional housing actually made the structures less livable. The tin roof that has spread throughout much of Africa, for instance, is inexpensive and long-lasting, but it is less effective than a thatch roof in combating heat buildup. Minor design changes to encourage ventilation and the use of locally available insulating materials could greatly improve comfort. Furthermore, such changes could be implemented by the buildings' owners, who in developing countries tend to do much of their own construction. Additional work on this problem is badly needed, preferably at the village level so that the techniques developed make use of local resources and meet local needs.[23]

One of the beauties of passive solar design is diversity. Although the basic principles are simple, they can be applied in a great number of ways. In solar architecture constant innovation is the rule. Darian Diachok, who in 1980 conducted an international solar architectural survey, notes: "Passive research is taking on a distinctly regional flavor. Individual countries are now making major strides in developing buildings that are economical in their climates."[24]

While some architecture critics describe solar buildings as dull or gimmicky, the inherent limitations of solar design are less in question here than the creativity of architects and the preferences of homebuyers. Whether a solar house is conventional or breathtaking depends on the designer. Some architects are already looking forward to a time when buildings include solar features as matter-of-factly as buildings today have plumbing and electric wiring. One day solar buildings may be as diverse as architecture itself.[25]

Flexibility has become the watchword for designers interested in cost-effective solar buildings. From a financial viewpoint, relying exclusively on just one design principle is unlikely consistently to yield the "right" answer. Douglas Balcomb of the Los Alamos Scientific Laboratory, a leading expert in the performance of passive solar systems, has found that a mix of passive solar and conservation methods usually represents the best economic bet. Based on data from a house in Kansas, Balcomb's analysis suggests that a nearly equal investment in conservation and passive solar measures yields the lowest total cost over a building's life.[26]

An important aspect of this flexible approach to design is that passive systems need not be 100 percent passive. In many cases some form of auxiliary heating system makes sense. Just how big the system needs to be depends on the climate and the local fuel costs. And in many cases adding such "active" features as a fan that moves heated or cooled air to other parts of a building can make climate-sensitive buildings more effective at only a small additional cost.[27]

Off the Drawing Board

The combined work of architects, builders, and engineers over the last decade has laid the foundation for a transition to climate-sensitive, fuel-conserving buildings. The principles are simple, the necessary materials readily available, and the buildings cost-effective at today's prices. But the transition will be gradual and complex all the same. A whole generation of design and construction professionals needs to be educated. Solar and conservation designs must be integrated into mass-produced and low-cost buildings. And the commercial building industry needs to shed its laggard's reputation.

Solar design is just beginning to enter the architectural mainsteam. Until recently, heating, cooling, and lighting were the concerns of engineers, not architects. No more. In Europe and America today architectural plans for a custom-designed solar or low-energy building are not much harder to come by than those for a conventional one. A recent U.S. government listing of solar designers included over 1,000 firms and individuals, and the American Institute of Architects has enthusiastically embraced climate-sensitive design. Architecture schools are also taking to solar architecture. For the first time many are teaching passive solar design.[28]

If solar buildings are ever to become widespread, they must be accepted by builders as well as architects. In the United States several hundred thousand builders, subcontractors, and suppliers erect more than 1 million single-family homes, apartments, and commercial buildings each year. Most are "tract" homes built as part of large suburban developments, and only 10 percent are custom-designed by architects. These builders have large investments at stake, and they are very sensitive to the fears of the sizable number of people who until recently saw solar buildings as unconventional and costly.[29]

In truth, most passive solar buildings are an economic bargain. Consistently, financial analyses show that well-thought-out passive solar features quickly pay for themselves in reduced fuel costs. After that, they in effect produce wealth for the occupants, yielding a lower "life-cycle" cost than a conventional building would. The owners of climate-sensitive buildings are their most fervent boosters, making frequent references to the fact that only a half a cord of wood or a couple of nights of electric heat was needed to weather a particularly frigid winter. The fuel bills of these buildings are usually ridiculously low—witness the figures compiled for the Saskatchewan Conservation House and Village House I, a passive solar home built in New Mexico. (See Table 3. 2.)[30]

A useful rule of thumb is that for a 10 percent higher initial cost, climate-sensitive designs can reduce fuel bills by a full 80 percent.[31] A south-facing window costs no more than one that faces north, and a concrete floor that can store heat costs about as much as a wooden one. Options such as using two-by-six inch wall studs rather than two-by-fours to allow space for extra

Table 3. 2. Annual Heating Costs According to Different Building Standards*

Structure or standard	*Annual cost (1980 dollars)*
U.S. average house, 1978	680
U.S. building standards, 1978	360
Swedish building code, 1977	230
California building code, 1979	220
Saskatchewan Conservation House	20
Village House I, passive solar	15

*Assumes similarly sized houses using oil heat in a similar climate.

Source: A. H. Rosenfeld, Building Energy Use Compilation and Analysis.

insulation or employing triple-glazed windows or night shades add only marginally to building costs. Other design possibilities —extensive glazing, a Trombe wall, or a large amount of thermal storage material—can be quite expensive. But most can be sound investments nonetheless. In many cases the additional cost of solar design features is offset by immediate savings because large air conditioners or central heating systems are not needed.

The day when only "chics or freaks" lived in passive solar houses is now ending as builders warm to the new designs and further lower costs. In the United States 40 percent of builders are now building at least some passive solar houses, a clear indication that the designs are entering the mainstream of the housing market. There were an estimated 60,000 to 80,000 full-fledged passive solar houses up already in the U.S. in 1982, and 11 percent of new housing starts incorporate some passive solar features. All of this has occurred amidst a record-breaking slump in the construction industry, and a passive solar boom may occur as the recession ends. No other country has moved so quickly to change its building styles, although it appears that several European countries may be following a similar learning curve. Passive solar homes are becoming popular in West Germany, while in Scandinavia low-energy houses are finding a place in the coop-dominated housing market. France is a few years behind, but since 1980 there has been an explosion of interest among architects there.[32]

An emerging frontier in passive solar architecture is incorporating climate-sensitive features into apartments, offices, and other high-density urban developments. These present unique design problems: Their occupants and appliances often have a larger impact on the building's temperature than do outside weather conditions, and lighting and cooling usually use more energy than does heating. Typically, both heating and cooling systems in such buildings are operating even while people outdoors are strolling in shirt-sleeves in ideal weather.[33]

Many architects are now developing appropriate solar designs for large buildings. "Passive daylighting," as engineer Douglas Bulleit notes, "is becoming the champion of passive design techniques." Another design challenge is integrating the new passive solar features with buildings' mechanical systems, which in most cases cannot be eliminated entirely. According to architect George E. Way, a leader in this field, "we design to provide comfort and lighting in a passive way for at least 50 percent [of the energy load] and then use the mechanical systems to handle only the extremes." Recently developed microelectronics-based control systems are a big help since they automatically adjust artificial lighting according to the availability of natural light and enable heating and cooling systems to take maximum advantage of both indoor and outdoor weather conditions. The U.S. corporate giant IBM has taken climate-sensitive design to heart and is building skyscrapers in several parts of the country that use half as much energy as conventional buildings do.[34]

Along with mammoth buildings, old buildings also present a trying energy challenge. Only 1 percent of most nations' buildings are torn down each year, and annual construction accounts on average for 2 to 3 percent of the building stock.[35] So even if all the homes and commercial structures built between now and the year 2000 were solar buildings, not quite one-third of the total stock at the turn of the century would be solar. Obviously something must be done with the buildings we have.

Most of the impressive energy savings achieved in existing buildings so far have come from simple conservation measures rather than from "solarizing." Adding passive solar features to an existing house is more complicated and expensive than working with a new structure, but passive solar "retrofits" do make sense in many situations. In the United States such retrofits have become one of the most popular forms of home improvement. The most common passive solar retrofit is a solar greenhouse. Such greenhouses can be attached to the south side of a building without replacing existing walls, thought it often makes sense to vent the walls and add a fan to circulate the captured heat. Since a number of firms now market prefabricated solar greenhouses, it is possible to "solarize" a house for a few thousand dollars.[36]

Other types of passive retrofits are also wise buys in many cases. A Trombe wall can be created by glazing the outside of a south-facing masonry wall. Adding clerestory windows to the roof to admit more sunlight is easy and effective under some conditions. Many older schools, factories, and warehouses in the northeastern United States have uninsulated south-facing brick walls that would make ideal Trombe walls. Another popular low-cost strategy for existing buildings is the use of fan-driven, air-filled solar collectors mounted on the ground on a building's sunny side. Although not technically "passive," these are very simple devices that usefully complement a climate-sensitive design. In the cold, impoverished San Luis Valley in Colorado, hundreds have been built, bringing solar heating to people with incomes below the poverty line.[37]

New Policies for New Buildings

Climate-sensitive architecture has a strong foothold after just one decade's progress. Most impractical designs have been weeded out, the economic promise of the better designs has been proven, and homebuyers' and developers' interest is rising. But economic, political, and institutional hurdles stand in the way of a true architectural revolution. The world's building industries, ever conservative, have been in recent years under considerable financial pressure too. More important, builders do not pay the fuel bills of the houses they construct, so unless governments and potential buyers encourage them to build solar homes, the transition could be slow.

Until recently, governments have done little to help climate-sensitive architecture, and they have tended to favor active solar technologies when allocating research funds or providing tax incentives. In the United States some consumers choose more expensive active solar systems rather than passive systems simply to take advantage of the tax breaks for solar collectors. While active systems clearly deserve market support, even greater fuel savings would result if similar amounts of money were invested in promoting the use of passive solar design.

Some governments have taken the passive cause to heart, however. Canada, China, Denmark, France, the United States, and West Germany have started small but growing passive solar research programs since the mid-seventies. They include a variety of research and demonstration projects. But more is needed. If climate-sensitive design is to take hold, governments will have to work with the building industry—clearly the main vehicle of the solar transition. In some nations passive solar design competitions have been used to spur the private sector's interest. In France a small village of solar homes—Nandy—built in 1981 as part of a design competition triggered interest among French architects and builders. In the United States the Solar Energy Research Institute gave funds to Colorado developers to hire architects to design passive solar homes to add to their list of models. Some excellent designs came of this program. So did a regional solar building boom.[38]

Educational programs for consumers, builders, real estate agents, and others are proving very successful at erasing some of the myths surrounding passive solar buildings and so speeding their acceptance. This is an area where trade associations, community groups, and local governments probably have the largest role to play. In the United States groups such as the National Association of Home Builders and the Home Improvement Council have quickly gone from being skeptics to enthusiastically sponsoring the workshops and newsletters that have helped launch solar buildings.[39]

A complementary approach is to label the fuel requirements of buildings for sale. Expected fuel use and price could be noted along with the likely life-cycle cost of the building. Buyers could thus compare the efficiency of different buildings. Already the fuel bills of solar homes are displayed during real estate transactions in some parts of the U.S., a practice that local governments may want to require.[40]

Financial incentives are probably most likely to send passive solar building on its way. Many climate-sensitive design innovations require a slightly higher initial investment than that for a conventional building. No matter how cost-effective these changes ultimately are, builders who are under immense pressure these days to cut initial costs to bare bones levels tend to shy away. Both builders and owners can have trouble getting loans to pay the extra costs, due to high interest rates and the fact that most bankers are still unfamiliar with climate-sensitive design.

Educating the financial community about the common sense and cost-effectiveness of energy-saving buildings is one key to solar architecture's future. To assess a homeowner's mortgage-paying ability, loan officers need to know that passive solar buildings have negligible fuel bills so their owners have more income available to repay a loan. The San Diego Savings and Loan Association in California is one of several U.S. banks that offer slightly reduced interest rates on passive solar houses. This program brings monthly payments below what they would be for a conventional home, adding to the homeowner's savings from reduced fuel costs. Similarly, the Hanover Insurance Company in the United States has offered a 10 percent discount on homeowner insurance rates for passive solar homes—in recognition of the fact that they are less prone to destruction by fire.[41]

Tax incentives also encourage energy-saving homes. In much of Europe, Japan, and the United States, there are now tax credits for solar collectors. Conservation improvements are also eligible for tax credits in many nations. Unfortunately, passive solar design seldom qualifies taxpayers for these benefits. Because passive features also serve nonenergy functions, most governments do not allow individuals to write them off as energy investments.

To get around this serious shortcoming—which works against some of the most cost-effective means of reducing buildings' fuel needs—many U.S. states added to the tax code detailed standards for determining what constitutes a fuel-saving measure. Another approach that is being considered by the U.S. Congress is simply to give builders of climate-sensitive buildings a tax break of up to $2,000 for each energy-efficient building constructed, depending on the building's performance.[42]

Luckily, government programs to encourage climate-sensitive building need play only a limited, transitional role. Tax incentives and information packages that persuade builders to take climate-sensitive architecture seriously will become unnecessary as passive buildings soon start selling themselves. It may be that the entire package of government programs—including financial incentives and demonstration projects—can be phased out after only a decade, the job completed.

Building for the Future

Worldwide, there are now over 100,000 passive solar buildings, over half of them in the U.S., and rapid growth is continuing. The U.S. Department of Energy's goal is to have a half million climate-sensitive buildings standing by 1986, and the National Association of Home Builders expects to see passive solar systems in 30 percent of all new houses by the year 2000.[43] And even these are arguably conservative figures. Based on current growth rates, a reasonable worldwide target is to have 10 million passive solar buildings in place by 1990 and between 50 and 100 million by 2000. By the end of the century most countries should aim to use energy saving designs in all new buildings.

Unfortunately, measuring the precise energy contribution of climate-sensitive design is difficult. Since a solar building does not produce a fuel that can be measured by a meter, it makes more sense to calculate the amount of additional heating and cooling fuel that would have been used by a comparable conventional building. Yet conservation and solar technologies are fused so tightly in a good climate-sensitive building that solar collection gains and conservation gains are hard to distinguish.

Even without the benefit of precise measures, it can be estimated that constructing new passive solar buildings will save at least half of the fuel currently used to heat and cool similar structures. If a Baltimore house's energy use is taken as the average, that means that 10 million solar buildings in 1990 would in effect yield 0.7 exajoules of energy or enough to run all of the cars in Canada for over six months. Fifty to 100 million passive solar buildings by the century's end would yield 3.7 to 7.3 exajoules or 6 to 12 percent of the energy currently used to heat, cool, and light the world's buildings. Together, climate-sensitive design for new buildings and conservation measures for existing structures should reduce the fuel needs of the world's buildings by 25 percent by the turn of the century, despite substantial growth in the housing stock.[44]

The potential of passive solar architecture is no longer in doubt. Nor are the benefits of more rational design and construction for people at all income levels and in all climates. We can learn something from the architecture of the ancients: As one solar designer recently observed, "Our buildings would be more beautiful if they responded to energy concerns and had a more natural configuration."[45]

4

Solar Collection

The idea of harnessing the sun's heat and light has for centuries inflamed the human imagination. Besides employing various passive solar architectural techniques, the ancient Greeks, Romans, and Chinese from the second century B.C. on experimented with "burning mirrors" that could concentrate the sun's rays onto an object and make it burst into flames. The Greeks used their knowledge of geometry to build sophisticated parabolic dish concentrators. To conserve scarce and expensive firewood, the Romans heated their public baths by running water over sun-exposed black tiles. Yet burning mirrors and solar water heating largely remained objects of scientific curiosity rather than of widespread practical use.[1]

The technology used today to harness the sun's heat owes much to the work of the eighteenth-century Swiss scientist Nicholas de Saussure. Working with the ingeniously simple notion that sunlight penetrating glass can be absorbed by a black surface and trapped as heat, Saussure designed a variety of heat-trapping boxes—the prototypes for solar collectors that today heat water, warm buildings, and power machines.[2]

Active solar technology leapt forward again in the nineteenth century when a French scientist, Augustin Mouchot, modified these simple collectors to create solar cookers, stills, pumps, and steam engines. By applying his knowledge of glass heat-trap principles to burning mirror technologies, Mouchot achieved temperatures high enough to roast food, distill liquids, and boil water. Mouchot's solar steam engine included a clock mechanism that moved the collectors to follow the sun's course.[3]

Despite the technical success of these early solar technologies, the availability of cheaper and more reliable coal-fired equipment blocked their widespread use. As fossil fuels became cheaper and more abundant, furnaces and industrial boilers grew more advanced. As a result solar-thermal devices remained experimental curiosities during the late nineteenth and early twentieth centuries.

Despite this general eclipse, the simple solar water heater—a collector box and a metal water storage tank painted black—found a large following early in this century in parts of the United States, Australia, South Africa, and Argentina where conventional fuels were scarce and expensive and sunlight abundant. In California several thousand such contraptions were in use until the advent of cheap natural gas in the 1920s. In the 1930s a solar industry bloomed in Florida. By 1941 approximately 60,000 solar hot water heaters were used in Miami, supplying more than half the city's population with hot water. But the wartime freeze on civilian copper use crippled the industry, which vanished completely when cheap electricity generated using oil became available.[4]

In the meantime, another solar industry flourished in Israel. In 1940 Ruth Yissar—wife of scientist Levi Yissar—painted an old tank black and put it out in the sun to warm bath water. Struck by his wife's common sense, Yissar began developing solar-heating technology. His company began manufacturing collectors in 1953 and sold 1,600 units the first year. Between 1953 and 1967 Israeli solar companies built and installed over 60,000 solar water heaters. Cut off early from cheap oil supplies, Israel built a solar industry that is today a leading exporter of advanced solar heating equipment.[5]

Heating Water and Buildings

The global increase in oil prices in 1973 set off a worldwide boom in solar heating. Overnight the economics of solar energy use were revolutionized. Today momentum is still gathering. In Israel, Japan, and parts of the United States, high fossil fuel and electricity prices, abundant sunlight, and strong government support aimed at reducing petroleum imports have heated up the solar market. While most government attention has focused on research and development (R&D) programs for innovative solar technologies, simple systems based on proven technology account for most of the growth in solar energy use. Indeed, for all the talk about solar energy's role in the future, solar's present role rests on simple technology from the past.

Almost all the solar collectors in use today are solar panels that heat water or buildings. The typical flat-plate collector consists of a rectangular box with wooden or metal sides, a blackened insulated bottom, a copper absorber plate, and a cover made of transparent glass or plastic. When operating, water, some other fluid, or air circulates from the panel to a tank, carrying the sun's heat to where it is needed. Designs differ widely in terms of cost, efficiency, and durability. For instance, most systems built for use in extremely cold climates feature extra insulation and some means of draining water from the pipes at night.[6]

Solar radiation's intermittent nature presents difficulties for all solar energy systems. Since energy may be required when the sun is not shining, it is necessary either to store heat from sunny days or to use a conventional heater as a back-up. While storing high-temperature heat is costly and relatively ineffective, this approach does make sense for water heating. A well-insulated, somewhat larger-than-average water heater tank is usually all that is needed. To store larger quantities of heat, such technologies as underground tanks filled with heat-absorbing rocks can be used, though not always economically.[7]

The consumer costs of using today's flat-plate collectors are determined principally by the costs of materials, labor, transportation, and installation. The materials—glass or plastic for the cover; aluminum, wood or steel for the frame; and copper for the tubes and backing—are widely available, and their costs are set in markets much larger than that for solar equipment. Labor typically accounts for more than half the cost of fabricating collectors. Installation and transportation can easily double the cost to the consumer. Proper installation—a key to solar equipment's efficient operation—requires skills akin to those in the plumbing and heating business. Transporting bulky collectors costs so much that local manufacturers have an edge over distant competitors, and do-it-yourself collector kits have found a market. (Unfortunately, though, collectors built from kits are less efficient and less durable than factory-built collectors, factors that offset their initial economic appeal.)[8]

Although solar energy is free, using it requires investing relatively large sums. Unlike conventional energy systems, most of whose costs are spread out in fuel bills paid over a period of years, solar systems have high initial costs and minimal operating expenses. Thus, meaningful economic comparisons of solar and conventional systems must take into account the total costs of both systems over time. Although making such "life-cycle" cost comparisons involves estimating future fuel costs and interest rates, it is the least biased way to judge solar systems' economics. Still, when interest rates are high and future fuel prices uncertain, consumers and industry pay less attention to life-cycle costs and more to the payback period—the time it takes fuel savings from a solar collector to pay for the cost of the collector. Most consumers insist on a payback time of less than five years with today's high interest rates.[9]

Another determinant of solar economics is the cost of alternatives, principally electricity and natural gas. Solar water heaters can compete with gas, electric, or oil water heaters nearly everywhere natural gas and electricity price controls are not in force. Even where price controls are extensive, as they are in the United States, solar water heaters can still compete in many areas.[10]

Overall, about 90 percent of all flat-plate collectors used today heat water. For perspective, home hot water use in the United States takes one-fifth as much energy as the entire automobile fleet—some 4 percent of the nation's end-use energy. In most developing countries less than 5 percent of residential energy is used to heat water, but hot water use is growing rapidly. Depending on how hot and sunny the region, today's solar panels can heat between 30 and 100 percent of the water a typical home, business, school, or hospital uses.[11]

Although solar water heaters have a much firmer foothold in the market, the public tends to equate solar energy with active space heating. Yet solar space heating systems are still plagued by storage problems because air is typically used instead of water to transfer heat from the collector to the room and because larger quantities of heat are involved. Many are also too large to install in any but new buildings. A third drawback is that their use entails maintenance, weathering, and freezing problems commensurate with their size.[12]

It is not only for these reasons that the market for active space heaters is much more limited than the market for hot water heaters. Quite simply, the demand for space heating is less widespread than that for hot water. In areas such as the southern United States, Brazil, and southern Europe, space heating is required only a few weeks a year so solar systems do not replace enough fuel to become economical. Even in northern areas with cold but cloudless winters, active solar heating may prove less economical than investments in conservation, passive solar design, or heat pumps since active systems may be too expensive to meet postconservation demand.[13]

The use of solar panels to heat water and buildings has grown most rapidly in Israel, Japan, and California. Common to all three areas are a highly educated populace, high energy prices, and government support. On a per capita basis, Israel leads the world in active solar heating—33 percent of all Israeli households have solar water heaters, and active solar systems now meet 1 percent of all energy needs. By the mid-eighties, some 60 percent of Israel's households are expected to have solar-heated water—enough to reduce national electricity consumption by 6 percent.[14]

Israel's success derives partly from the simplicity and inexpensiveness of the technology being used. Typically, the systems cost $500 and require only $25 worth of supplemental electric heating per year. In comparison, a gas or electric water heater costs about $175 initially and at least $120 a year to run. The combination of mass production and simple design has kept costs low.

Second to Israel in per capita use of active solar equipment is Japan, truly the land of the rising sun. As of late 1982 some 3.6 million houses, or 11 percent of the total in Japan, were using solar systems, most for heating water. Japanese companies are now manufacturing over half a million solar hot water heaters a year, more than any other country. The Japanese government expects 4.2 million buildings to be solar equipped by 1985 and 8 million by 1990.[15]

In absolute terms, the United States leads the world in using active solar energy systems. Between 1974 and 1980, annual collector production has increased twenty times. (See Figure 4. 1.) Yet a significant share of all U.S. collector sales have been

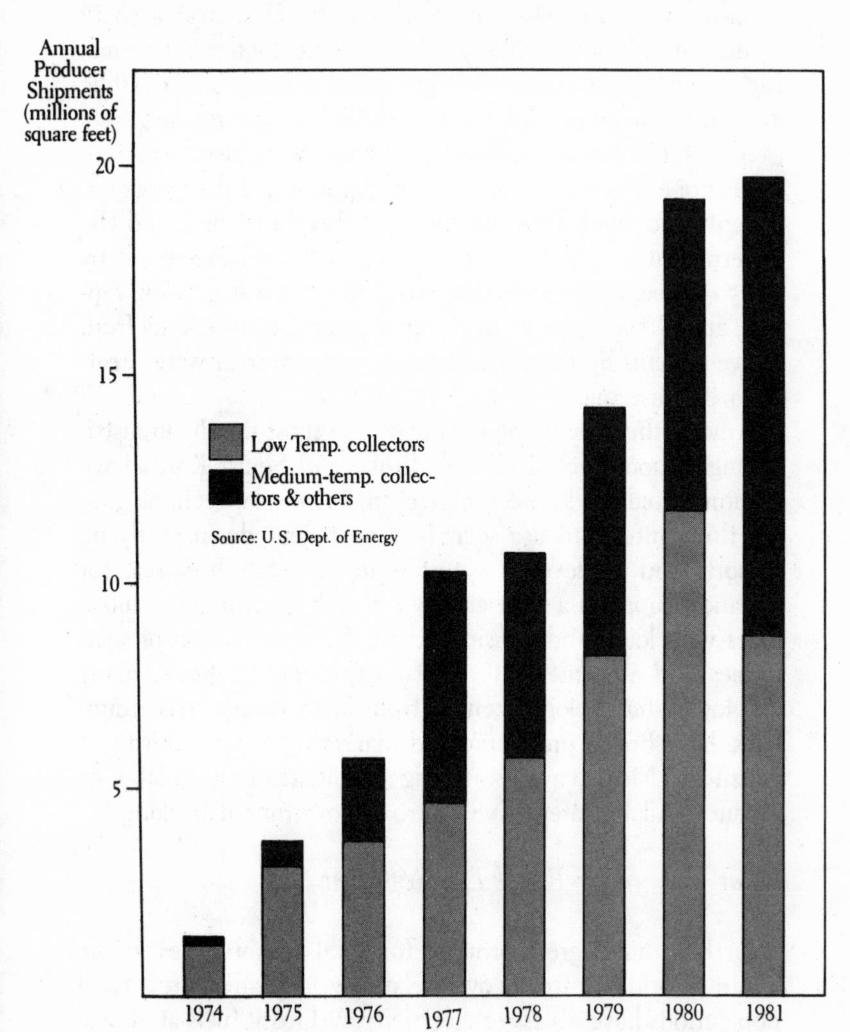

Figure 4.1. U.S. Solar Collector Manufacturing, 1974–1981.

low-temperature systems used to heat California's swimming pools, so a more meaningful measure of solar activity may be the increase in the number of solar collector manufacturers from a few in 1973 to over 500 in 1980. U.S. solar activity centers in California, where a sunny climate, high conventional fuel prices, vigorous government support, and broad public awareness have given rise to a $1 billion a year market.[16]

Solar technologies are developing a following elsewhere, too. In Europe France has the most aggressive solar program. Twenty thousand water heaters have been installed, and the government's ambitious goal is to see half a million in use by 1985. Throughout sunny southern Europe use is growing rapidly, and Greece expects to obtain 2 percent of its energy from active systems by 1985. In Australia 100,000 solar water heaters are in use mainly in western Australia.[17]

Among the developing countries, the most rapidly industrializing nations such as Brazil, India, and South Korea have demonstrated the keenest interest in active solar technologies. All three intend to use solar heating to cut down costly oil imports and to develop export industries. South Korea, for instance, supports a solar energy research institute, and subsidizes with loans and housing bonds the construction of solar homes and apartments. Over twenty domestic firms, many employing technology licensed from firms in industrial countries, have begun producing and marketing exports. Firms in Brazil and Mexico are also taking advantage of cheap labor to pursue similar strategies with strong government backing.[18]

Solar Energy for Rural Development

Solar heat holds great promise for rural communities in the Third World. Scattered over relatively large areas, few rural populations have access to electricity and fossil fuels at affordable prices. Then too, most rural energy needs are for low-temperature applications such as drying crops, cooking food, and pumping water—processes well matched with solar energy.

Some simple, easy-to-use technologies could substitute for firewood in much of the Third World. A wide variety of small focusing collectors has been used successfully over the last century to cook food, while the solar "hot-box" or oven—an insulated box oven with a transparent window on the side exposed to the sun—has been developed more recently. In bright sunlight solar cookers rival an open fire for heat, and solar ovens can keep food warm for hours. Although ultraefficient collapsible reflector units and elaborate high-temperature ovens have been developed, simple and effective collectors made of polished metal can be produced for between $10 and $30 each.[19]

Despite these advantages, solar cookers are not the cook's choice. They do not work when the sun is not shining, and the cook must stand in the heat when it is. A small solar cooker industry in India in the fifties and a four-year project to introduce cookers in three Mexican villages during the early seventies both failed because villagers did not take to the unfamiliar technology. One overriding cultural factor—mealtime—severely limits the prospects for solar cookers in many areas, and no effort can succeed fully without the involvement of the women largely responsible for food preparation. Still, a recent effort by a Danish church group to introduce cookers into Upper Voltan villages did work because the villagers helped adapt the cookers to local needs and conditions. And China has not given up on the cooker: More than 10,000 units are reportedly used there.[20]

Of all the direct uses of the sun's heat, crop drying is probably the most ancient and widespread. Throughout the developing world, farmers still spread crops on the ground or hang them on open-air racks to dry. According to the U.N. Food and Agriculture Organization, 225 million tons of food is dried in this traditional way. But open-air drying does expose food to dirt, animals, insects, molds, and bad weather—all of which result in significant crop losses. Thus, reducing postharvest food loss through the use of closed-cover dryers has become a critical part of the efforts of many Third World countries to feed their growing populations.[21]

Partly because the gas and electrical dryers widely used in developed-country agriculture are becoming less economic as fuel costs rise, probably no active solar application for rural development is receiving more attention than solar drying. Many types of solar crop dryers are being tried, most with success. Simple rice dryers work in Thailand, while more complex grain dryers have been used effectively in Saskatchewan. But problems remain. Particularly in closed systems designed for use in colder climates, dust buildup is one. Volume poses another. Because space for collectors is limited, solar dryers are seldom as cost-effective for drying large volumes of grain in centralized facilities.[22]

Yet solar dryers are appropriate for on-farm drying. Fine tuning the technology for this purpose, the Brace Research Institute of Canada has built corn dryers in Barbados, fish dryers in Senegal, and lumber dryers in Guatemala. The key here is the full cooperation of agricultural extension services in disseminating information about solar dryers.[23]

Solar technology could also help meet critical needs for fresh water. Indeed, an inexpensive method of removing the salt from saline water would find almost unlimited application in agriculture and industry in arid regions, mainly because the cost of heat plays such a decisive role in shaping the economic viability of distillation. Among the simplest and easiest to construct solar technologies, solar stills have black bottoms to evaporate saline water and glass tops to admit the sun and collect condensing fresh water. With slight modifications, the glass or plastic covers of this simple basin-type still can double as a rain-collection system. As early as 1872, a 4,000-square-meter solar still was built in Chile's Los Salinas desert to provide fresh water for mules. Unfortunately, solar stills would have to cover vast areas to provide the quantities of fresh water industry and agriculture need. In Algeria's dry clime it would take one square kilometer of solar stills to produce enough fresh water to irrigate three square kilometers of cropland.[24]

Solar stills do, however, hold considerable potential in isolated rural communities. Where less than 50,000 gallons of water per day is required, they are the cheapest source of distilled fresh drinking water. On islands, where fuel costs are high and fresh water supplies are limited, they are ideal. Several Caribbean, Pacific, and Aegean islands currently employ solar stills to provide drinking water. The most extensive solar still usage is in the dry central Asian regions of the Soviet Union and the interior of Australia where livestock are watered from solar stills.[25]

Since solar stills are easily fabricated by low-skilled labor using locally available materials, their use is particularly appropriate in Third World villages. Yet, efforts to adapt solar still technology for use in such villages has met with mixed results to date. In Source Phillipe, a small deforested island off Haiti's coast, community support and voluntary labor made a project work, but in several Indian villages projects failed because the villagers had grown accustomed to drinking the brackish unhealthy water. Even with community support, outside financing is typically needed—one reason that several aid groups are exploring the use of cheap plastic substitutes for glass.[26]

Solar water heaters also have a place in rural development. Few poor villages now have the hot water needed to make rural health clinics and schools sanitary, much less to put to use in communal showers and lavatories. Still, simple systems made of inexpensive local materials have proven economically and technologically appropriate in many developing countries. In Peru some simple solar water heaters sell for about $12.50 each, while thirty Chinese factories turned out 50,000 square meters of solar collectors during 1980, mostly for use in hospitals, apartment complexes, and schools.[27]

Active solar systems can also pump water for irrigation and household use. In many rural areas irrigation pumps are principal users of electricity. In California the state water-supply agency consumes more electricity than any other user. And in rural India 87 percent of the electricity consumed is used in water pumps. Increasingly, hopes for raising food production in poor countries hinge on the greater use of pumped irrigation.[28]

Since the need for water pumping is typically greatest where sunlight is abundant, solar water pumps seem a logical choice. Indeed, successful solar thermal pumping systems have been built using concentrators for large pumps (25 to 150 kw), while flat-plate collectors work for smaller units (1 to 25 kw). In solar pumps collector-heated water is used to turn an easy-to-boil liquid such as freon into a gas, whose expansion drives pumps.[29]

For twenty years the leader in developing solar-powered irrigation systems has been SOFRETES. This French company has installed more than thirty-six water-pumping irrigation systems in Africa and Mexico, and it has also begun to develop solar electric pumping machines. In the United States the world's largest solar-powered irrigation system—a 50-horsepower pump capable of delivering up to 10,000 gallons of irrigation water per minute—was built in Arizona in 1977. Several other large systems using trough concentrators are being built in Israel and the Soviet Union.[30]

Although solar pumps hold promise based on operating experience, the overall outlook is not encouraging. They are less efficient than diesel engines, and few are economical. (Capital costs range greatly, from $6,000 to $78,000, depending on size.) Even where fossil fuels are unavailable, solar thermal pumps compete economically only with photovoltaic systems, whose price is falling steadily.[31]

An Evolving Technology

Even as current solar technologies catch on, researchers the world over are making solar collectors more versatile by improving performance and lowering costs. The use of new glasses and plastics, in particular, looks to improve the economics of using conventional solar designs. So too, the development of such new solar design concepts as concentrating collectors, evacuated tubes, Fresnel lenses, and solar air conditioners is making it possible to use solar energy to meet the rapidly growing demand for industrial process heat and cooling buildings. Bewildering in its multiplicity, all solar research does have at least one common aim—lowering the delivered cost of solar energy by improving performance, using cheaper materials, or merging storage and collection systems.[32]

The most visible engineering trend is replacing solar collectors' corrosion-prone, expensive metal parts with plastics. Lightweight plastics cost less to transport, install, and support. They do not conduct heat well, but they can be configured to compensate for that drawback; and although plastics are made from fossil fuels, plastics production requires less energy than do mining and refining metals. The major challenge in plastics work is extending longevity since plastics degrade faster in sunlight than metals do.[33]

Another materials innovation, the use of plastic thin-film on collector surfaces, may revolutionize solar heating technology. The new "solar sandwich" collector being developed at Brookhaven National Laboratories features layers of highly heat-absorbent plastic films suspended by lightweight steel. With installed costs of $5 per square foot and manufacturing costs as low as $1 per square foot, these films offer strength, durability, good performance, and short paybacks. Experiments show they may also be able to endure the high temperatures industry uses.[34]

The use of "super-glazing"—a type of glass—could also

speed solar evolution. A U.S. company, Corning Glass Works, developed a process that minimizes the internal heat loss of regular glass, and the Solar Energy Research Institute is testing this glass in solar collectors. SERI's hunch is that this glass may work better in solar collectors and be easier to handle than glass originally designed for windows.[35]

Entirely new collector designs are also emerging. Of the lot, evacuated tube collectors come closest to widespread commercial application. Resembling fluorescent lightbulbs, the tubes consist of a blackened air-filled glass cylinder enclosed within an outer protective cylinder from which the air has been removed. A vacuum insulates perfectly, so the high winds and cold weather that reduce flat-plate collector performance do not affect the tubes. Because air is used as the heat-transfer medium, freezing is not a problem either. Evacuated tubes can also deliver heat at higher temperatures than flat-plate collectors can. Indeed, attaining temperatures of 82°C or above (116°C with reflectors), they can be used in a broad range of residential and industrial applications. They are also light weight, versatile (of use in space heating, water heating, and cooling), and easy to mass produce in highly automated factories.[36]

Evacuated tubes are, however, fragile and easily broken. Expert opinion on the prospects for this technology is sharply divided between promoters and those who question whether the tubes' high price and breakage problems undercut their advantages. Private companies like Philips Electronics of the Netherlands, General Electric in the United States, Sanyo Electric in Japan, have done the most work to develop evacuated tube collectors.

In industry, which uses roughly one-third of the energy consumed in industrial countries, new solar technology will be used increasingly. However, that use will be constrained by industry's need for high temperatures, since the cost-effectiveness of solar heating decreases as the temperature increases.

While very high temperatures can be obtained when large expanses of mirrors reflect sunlight to a small receiver, the smaller and simpler low- and medium-temperature systems are for now more cost effective. In the United States 27 percent of industrial heat use is below 287°C, a temperature that can be met with commercially available solar systems.[37]

Solar technologies for achieving the spectrum of temperatures needed by industry are here or on the way. For heating water and for low-temperature drying, flat-plate collectors using air or water are appropriate. Linear concentrator collectors can best provide low- and medium-temperature industrial process heat. Parabolic troughs that track sunlight and focus it on a black liquid flowing through a long, narrow pipe or tube are being marketed by several dozen firms.[38]

The technology of concentrating collectors is evolving rapidly. Films that preserve reflectiveness, thinner more durable reflectors, more efficient heat-transfer systems, and cheaper tracking mechanisms—all are at the forefront of concentrating-collector design. Research is also focused on substituting plastics and reflective foils for costly metals and on lowering the weight of concentrating collectors so smaller, cheaper motors can be used to track the sun. Now handmade, concentrating collectors will also grow cheaper when mass produced. Recognizing that the market will belong to the company that first culls enough sales to justify the investment in automation, several governments (most notably France, Japan, and Israel) are heavily subsidizing the concentrating-collector industry.[39]

An entirely new type of concentrating collector made of cheap plastic—the Fresnel lens—offers high efficiency at a modest price. Transparent grooved sheets of plastic that bend light rays much as prisms do, Fresnel lenses can concentrate sunlight by as much as fifty times. An experimental Fresnel lens has achieved temperatures of 550°C, and plastic sheets costing only $3 to $4 a square meter have attained temperatures in excess of 300°C. With farm uses in mind, U.S. Department of Agriculture scientists project a two- to ten-year payback period for these lenses.[40]

Reaching the still higher temperatures needed to smelt metal and produce superheated steam requires using "solar central receivers" that can concentrate onto a small spot the sunlight falling on several acres. Temperatures to 750°C can be obtained with such receiving towers, which concentrate the sunlight reflected off hundreds of flat mirrors. Although this technology has not been pursued with industrial users in mind, a recent study found it to be economical in the smelting industry at present energy prices. Looking at the giant Hidalgo copper smelter in New Mexico, a New York engineering and architectural firm found that a multi-million dollar solar system covering a square mile of desert could displace almost a half million barrels of oil annually and pay for itself in less than two years.[41]

To date, the research on solar power towers has emphasized electricity production. The largest power tower, with a 10-megawatt capacity, stands in the Mojave Desert in southern California near the town of Barstow. This plant, known as "Solar One," relies on 1,818 flat sun-tracking mirrors, each 430 square feet in size, to concentrate sunlight on a central receiver atop a 300-foot tower. The Barstow power tower has been repeatedly criticized by U.S. solar energy advocates who question the economic feasibility of the technology and who object to the project's dominance of the federal government's solar research budget. While the budgetary priority granted Solar One makes little sense, the technology will have application both for utilities and industries in desert regions. A southern California utility is seeking bids for the construction of a 2,000-acre, 100-megawatt power tower.[42]

In industry, solar energy probably has the best prospects for early and rapid growth in food processing. Two-thirds of the heating needs of this industry are for heat under 100°C, and food processing now takes 10 to 15 percent of all industrial energy. Under a U.S. Department of Energy experimental program, solar technology is being employed for such diverse tasks as frying potatoes in Oregon, washing soup cans in California, processing sugar in Hawaii, and drying soybeans in Alabama. So far, such experiments have been expensive but technically sound.[43]

Solar energy also has a place in the oil industry. As now practiced, enhanced oil recovery involves injecting steam into wells to loosen highly viscous oil. In California, the world's leading producer of heavy oil, it takes one barrel of oil to heat enough steam to extract three additional barrels. Although oil-fired systems are currently cheaper than solar concentrators, rising oil prices and pollution from burning heavy unrefined oil in highly polluted areas are making solar energy increasingly competitive. When oil-producing countries are forced to turn to enhanced oil recovery to extract petroleum from their old or low-quality fields, solar collectors could be extensively employed.[44]

For all its merits, putting solar heating technology to work in industry has turned up problems. Government-funded industrial process heat projects, for instance, never achieved expected efficiencies. Among other things, dust builds up on concentrators used near polluting factories and pipes freeze and burst in cold weather. The uneven output of solar systems can also pose problems in factories that depend on a steady source of heat. A final difficulty is that of retrofitting some factories. While none of these problems is insurmountable, businesspeople are not likely to invest heavily in solar technologies until reliable cost and performance data accumulate. How long solar heating systems will last in real-world operating conditions is another unknown.[45]

As for solar air conditioning, it holds particular promise for displacing the fast-growing use of electricity. Use of the sun's heat for cooling is particularly appealing because demand for air conditioning is highest where sunlight is most abundant.

Need matches supply, and storage and backup systems are less critical because periods without sunshine require much less air conditioning. Already, several distinctly different types of active solar air conditioners are commercially available. One design marketed by a U.S. firm, Zeopower, makes use of a water-absorbing material called zeolite to provide cooling during the day and warmth at night. A factory capable of manufacturing 100,000 units a year is being built in Texas, and the firm hopes units selling for $12,000 to $20,000 apiece will capture 1 percent of the U.S. market by 1985. An entirely different design is already being marketed by Yazaki of Japan and Arcla in the United States.[46]

Solar air conditioners are large, technically complicated, and expensive, but so too are the conventional systems solar units must compete against. If these systems can gain commercial acceptance, an enormous market awaits them. Worldwide, air conditioning accounts for a large and rapidly growing share of electricity use. In the United States air conditioning uses 20 percent of all energy expended to heat and cool buildings. In some tropical developing countries, air conditioning uses more than half the electricity produced.[47]

Sun on the Waters: Solar Ponds and Ocean Thermal Energy Conversion

Two extremely simple technologies, salt gradient ponds and ocean thermal energy conversion (OTEC), rely upon abundant and cheap salt water to economically collect and store heat from the sun. Little more than elaborate plumbing systems, these technologies convert relatively small differences in water temperature into useable energy. Although they convert only tiny percentages of the water's heat, the low cost of the collectors and storage media make these systems economically comparable, if not superior, to metal or plastic-based solar collectors. Because solar ponds have low conversion efficiencies, sunlight must be collected over large areas to obtain appreciable quantities of energy. However, this is not a significant constraint to the use of either technology since the preferred locales—tropical oceans and desert salt flats—are abundant.

Salt gradient ponds, or solar ponds as they are known for short, work by trapping solar heat in very salty waters at the lower levels of shallow ponds. Since salty water is heavier than fresh water, the heated water fails to rise and evaporate. Insulated from heat loss into the air by the water above it, solar pond water can reach the boiling point, and its energy is available throughout the coldest winters. Because the basic materials of such salt-gradient ponds—water, salt, earthen walls, and plastic lining—are so cheap and widely available, solar ponds could be used almost anywhere.[48]

Solar ponds are being successfully employed in several countries to generate electricity, desalinate water, and provide heat. Israel has one solar pond that produces 150 kilowatts of electricity, and a pond several times that size is being built on the Dead Sea's shores. If this larger model proves as cost-effective as expected, Israel plans to build 2,000 megawatts of pond capacity, enough to meet 20 percent of national energy demand by the year 2000. To convert hot water into electricity, the Israelis employ Rankine engines containing freon, which boils at 50°C. Another experimental project, on the Salton Sea in arid Southern California, will produce 5 megawatts. If this pilot plant works as planned, a 600-megawatt plant large enough to provide power for a city of 350,000 may be built. A 2,000-square-meter pond in Alice Springs, Australia, is successfully supplying heat and electricity to a restaurant, vineyard, and winery complex.[49]

Most solar pond development is being pursued in very sunny regions with natural salt lakes. But an experimental salt-gradient pond is heating a municipal swimming pool and a recreational building in Miamisburg, Ohio, for about as much as it would cost to buy the necessary heating oil. At Hampshire

College in the United States, researchers have laid out a detailed plan to show how Northampton, Massachusetts, a town of 30,000, could economically meet all of its space- and water-heating needs from community solar ponds connected to buildings by underground pipes. Distributing heat from solar ponds to whole neighborhoods in this way appears to cost no more than using dispersed, household-sized solar water heaters and considerably less than using active solar space heaters simply because the storage and the collection systems are one in the same.[50]

Few insurmountable barriers stand in the way of the large-scale use of solar ponds. Desert salt lakes have virtually no other development value, and land requirements are reasonable: Salt ponds can be used everywhere except densely populated center cities. (Northampton, for example, could meet all its needs by turning just 1.8 percent of its land area into solar ponds.) As long as liners are used to prevent salt water intrusion into land or water tables, solar ponds are also environmentally benign. Surprisingly, the solar ponds being built at the Salton Sea will actually reduce the salt build-up now threatening fish life.[51]

The sun's energy can also be tapped from natural bodies of salt water by a technology known as ocean thermal energy conversion (OTEC). The earth's oceans absorb vast amounts of sunlight, most of which is radiated back into the atmosphere or dissipated as currents. Yet a small fraction of this heat—in absolute terms, several times total human energy use—can be harnessed in areas of the ocean where the temperature difference between warm water and cooler water 1,000 meters below is at least 18°C.[52]

OTEC plants operate like a common household refrigerator, only in reverse. Heat from the warm surface water first evaporates a working fluid, usually ammonia. The ammonia vapor drives a turbine attached to an electric generator and is then condensed by cold water brought up from the deep sea. Virtually all the ocean area within the tropics has a sufficient temperature gradient to tap with OTEC. Altogether some sixty-two countries, most of them Third World nations, have national or territorial waters capable of supporting an OTEC plant. Obtaining significant quantities of energy from OTEC plants will, however, be a herculean undertaking. A 250-megawatt plant would use a pipe 30 meters in diameter through which would flow a volume of water comparable to the Mississippi River.[53]

Although OTEC is simple in principle, several basic problems cast doubt on the practicality of the technology. Corrosion of pipes from salt water, growth of algae and barnacles on heat exchangers, and tropical storms all pose major, as yet unsolved, engineering hurdles. Aluminum pipes that last no more than fifteen years in salt water could be replaced with titanium, but at prohibitive cost. Colonizing sea organisms must be scraped off heat exchangers of experimental OTEC plants once a week, imposing potentially significant maintenance costs. The tropical seas with the highest thermal gradients are periodically swept by devastating hurricanes and typhoons generating hundred-mile-per-hour winds and thirty-foot waves. The first OTEC system was sunk by a hurricane off Cuba in 1922, setting the technology back a half century. So great are the engineering challenges to stabilizing a thousand-meter pipe in rough seas that several experts believe that OTEC will be feasible only where the pipes can be securely fastened to sloping ocean floors.[54]

Despite these obstacles, OTEC has strong supporters. A panel of OTEC experts assembled in 1981 for the UN Conference on New and Renewable Sources of Energy estimated that 10,000 megawatts of OTEC capacity would be built by the year 2000, a projection not likely to be realized. The principal OTEC researchers, Japan and the United States, have both spent more than $100 million on OTEC research. The first U.S. OTEC unit, built on a barge off the island of Hawaii, was ruined in 1981 when its piping was torn away by strong ocean currents. Japan has assisted the Pacific island nation of Nauru in building a 100-kilowatt facility firmly anchored on the seabed. One result of this project is to dramatically alter the surrounding aquatic environment by bringing the nutrient-rich subsurface water to the clear nutrient-starved surface waters. The resulting luxuriant plant and fish life is seen by environmentalists as a serious disruption of coral reef ecosystems but by OTEC advocates as a major side-benefit to energy production. Indeed, elaborate designs for giant open-ocean OTEC plants envision using the energy to process and refrigerate fish that are caught in the area.[55]

Because solar ponds and OTEC plants are such inefficient energy converters and require such large areas, extensive reliance on them could alter weather and perhaps climate patterns. Extensive networks of solar ponds would probably raise the ambient temperatures of desert regions, with difficult to envision effects on precipitation patterns and wildlife. By altering ocean currents and surface temperatures, large-scale OTEC use could affect tropical storms and fisheries in ways that are not easy to project. However, given the major engineering challenges still ahead, it will be many years before those large-scale environmental constraints come into play. In the meantime they should be carefully assessed.[56]

Barriers and Incentives

To realize solar energy's promise fully, many governments have begun providing incentives and reducing the barriers to solar energy use. Most visibly, R&D funding has multiplied over the last decade. In the United States spending passed the $400 million mark in 1980 but has since declined to less than $200 million. French spending increased from $12 million in 1975 to $63 million in 1978—a 400 percent increase in just three years. Several large R&D centers funded by the International Energy Agency have been set up in Spain, which is rapidly emerging as the hub of solar development in Europe. Japan and the Soviet Union also have extensive R&D programs under way. Other countries have specialized in particular technologies: Israel in solar ponds, France in high-temperature concentrators, and Australia and Mexico in solar distillation.[57]

Yet spending is still too meager to compensate for past neglect, to match government research on conventional fuels, or to exploit the most promising technological leads. Many important applications—industrial process heat, solar ponds, and advanced materials research among them—deserve vastly expanded financial support. In the countries with mixed economies, where most R&D is occurring, government programs must be carefully tailored to augment rather than duplicate or displace corporate activity. Although governments have more resources and more incentive to fund long-term projects with distant payoffs than private corporations do, they are relatively less attuned to what will be commercially viable. Where governments hold the patents to all inventions growing out of publicly supported research, inventions reach the marketplace slowly at best. The interruption of government-sponsored R&D projects in midstream for political reasons also causes problems.[58]

Another pitfall of government R&D programs is the tendency for agency officials to award research grants to large established firms instead of new, potentially innovative small firms. Such untoward caution clearly retards technical innovation. The pattern is particularly visible in the United States, where 80 percent of government solar R&D funding has been channeled to large firms. Yet smaller firms tend to be much more innovative and to create more new employment.[59]

Balancing near- and long-term applications is another problem with no simple solution. Too many government scientists and corporate researchers have tended to pursue technological perfection as an end in itself, focusing on long-term high-technology applications of more intellectual than practical interest. This approach might be advisable if the private sector were perfecting and refining current technologies at the same time. But the energy crisis made the overemphasis on long-term R&D only too plain.[60]

Apart from research, development, and demonstration programs, several national governments have also sponsored such consumer-financing initiatives as tax breaks and direct grants. In the United States a tax credit enacted in 1978 and expanded in 1980 offsets as much as 40 percent of the cost of buying and installing a solar system. Almost every U.S. state offers some sort of solar tax incentive, ranging from sales and property tax exemptions to a 55 percent credit deducted from the state income tax in California. France and Spain took the U.S. approach in 1981, while Japan provides direct cash grants covering one-half the cost of purchasing and installing solar heaters.[61]

The second approach to accelerating the use of active solar systems is direct regulation, which works remarkably well when implemented by a local government mindful of local conditions and needs. Since two years ago when Israel began requiring all new residential structures of less than ten stories to install solar hot water heaters, 250,000 solar water heaters have been installed. San Diego, California, has also required all new buildings to install solar water heaters if they would otherwise make use of natural gas or electricity. Still, the simplicity, economics, and popular support that underpin the market success of solar hot water heaters cannot be exaggerated, and federal government attempts to require the use of other solar technologies could well backfire.[62]

Greater use of solar energy in industry means overcoming a different set of barriers. Even where solar equipment can compete economically with conventional energy sources, industry is likely to consider other claims on its investment capital as more important and less risky. Researchers at the Harvard Business School contend that manufacturing firms require a much higher threshold of profitability for investments that do not directly relate to their product than for those that do. Where annual rates of return of 10 to 20 percent are enough to trigger investments in the company's product line, rates of return approaching 30 percent are needed to get firms to invest in money-saving energy conservation and solar collectors. Unlike oil, gas, or electricity (which do not entail an initial capital investment by an industrial firm), solar equipment must be purchased directly by the company—an added risk.[63]

The key to overcoming this barrier may be in a new type of solar marketing strategy based on leasing solar systems or selling their output. An Israeli firm, LUZ International, Ltd., has set up subsidiaries that have negotiated several twenty-year, multimillion dollar contracts with textile manufacturers in Georgia and North Carolina for steam produced from highly efficient solar collectors. LUZ must make sure the collectors are operated and maintained properly, and the textile companies do not have to tie up their capital in unfamiliar technologies. Solar leasing is being pioneered by a small Southern California firm, PEI, Inc. Under the conditions of the first signed contract, fifty-two PEI-installed, owned, and maintained collectors will enable a laundromat to save $165,000 in energy over seven years. Widely used in the information-processing and office machine industries, leasing offers customers the advantages of solar heating without a large capital commitment or the risk of obsolescence.[64]

A second critical but artificial constraint to the greater industrial use of solar energy is tax policies that continue to favor conventional fuels. Since the costs of heating fuel are tax deductible for commercial businesses and industries while the "fuel" for solar systems is not because it is free, much of the economic incentive to use solar energy is negated by the tax system. Either a deduction for the amount of oil being saved by using solar energy or the abolition of the business deduction for fossil fuels would eliminate this bias.[65]

The widespread use of solar energy systems will also profoundly affect the electric utility industry. As providers of backup power for solar-equipped buildings on sunless days and as disseminators of solar equipment, utilities will play new roles in the energy economy—ones that will affect both the economic viability of solar systems and the rates all electricity users pay. Although solar systems will reduce both total demand and peak demand on a typical day, on a rainy day in a peak-demand season every backup system may have to draw on the grid at once. Since law requires utility companies to maintain power-generating capability to meet any reasonably expected demand, the widespread use of solar equipment with electric backup systems could leave utilities with expensive excess idle capacity.

In Western Australia, where 15 percent of all households have solar water heaters, 4 percent of the winter peak can be attributed to solar hot water boosters. One study of U.S. utility customers with solar heaters found that the typical user of solar heat had a load factor 40 to 50 percent lower than that of a conventional customer. Since servicing a solar-heated home costs the utilities as much as servicing a conventional home, this means that current electric rates do not cover the costs of serving solar homes. In Colorado one utility has unsuccessfully attempted to impose a $40 a month surcharge on customers who have solar hot water heaters.[66]

Rather than charging solar equipment owners special rates, utilities should charge all users of peak electricity equally high rates that reflect the added costs the system incurs as a result of their demand. As experience with "time of day" pricing in West Germany shows, demand peaks can be shaved if users have an incentive to curb power use at certain times. In cases where backup power for solar water heaters increases peak demand, simply installing extra storage capacity usually makes more economic sense than foregoing the use of solar equipment.[67]

Utilities may also find it smart to finance, install, and maintain solar heating systems for customers. Because utilities have a highly developed service network and longstanding relations with all energy users, they are in an ideal position to bring about a rapid growth in solar collector use. Several innovative utility programs to finance solar water heaters are currently under way in the United States. In the Solar Memphis Project, the Tennessee Valley Authority is loaning consumers $2,000 at 3 percent interest rates for twenty years. The consumer pays a set monthly fee to the utility and the utility arranges the installation, certification, and maintenance of the system. Some 10,000 water heaters will be installed under this scheme.[68]

An even more ambitious utility solar-financing scheme was launched in California in 1979 when the state's Public Utilities Commission ordered the state's four largest private utilities to provide cash rebates and low-interest loans to customers who purchase solar equipment. Under this plan, utilities will make financial incentives worth $182 million available for the purchase of an estimated 375,000 solar water heaters. According to PUC calculations, this expenditure will save the utilities $615 million in power plant construction costs, for a net savings of $433 million. California consumers will be spared the high initial expense of buying a solar water heater.[69]

Many solar energy advocates oppose utility involvement in solar energy. The fear is that utilities will reduce competition in the solar industry, drive up costs to the consumer, or attempt to give solar energy a bad name. In truth, the attitude of the U.S. electric utility industry toward solar energy has been unenthusiastic. While more than 100 U.S. utilities are experimenting with solar energy, it has fallen to publicly owned utilities such as TVA or heavily regulated ones such as those in California to actually promote its use. Still the profit motive has led some utilities to embrace solar energy and they may one day become good sources of financing and promotion for its use.[70]

Solar advocates have also protested the entry of some of the world's largest conventional energy corporations into the solar industry. During the mid-seventies, several major oil companies purchased major shares in solar collector firms. Other large firms, ranging from General Motors' radiator division to the glass conglomerate Libby-Owens-Ford, have also moved rapidly into the emerging industry. The objection voiced here is that giant energy conglomerates would slow the pace of solar development to protect huge investments in conventional fuels. In fact, some conspiracy theorists suggested that Exxon's acquisition of Kennecott Copper was a move to monopolize copper, a key raw material for making solar collectors.[71]

Such fears appear exaggerated. After the initial flurry of acquisitions, oil and aerospace firms began selling the solar subsidiaries, few of which have made profits. Stung by several years of disappointing returns, Exxon, the world's largest oil company, in 1981 sold its solar hot water heater subsidiary (Daystar) to an independent solar company (American Solar King). The new owner sees profit to be made, chiefly by reducing the highly paid administrative staff. In general, most large, high-technology corporations are recognizing that marketing solar water heaters requires a semiskilled work force, attention to small separated markets, and settling for profit levels typical in small business. An industry more akin to plumbing than oil drilling simply doesn't need a large corporation's technological and managerial force.[72]

The Solar Prospect

The solar technologies already for sale will contribute ever more to meeting the world's energy needs in the years ahead. The well-established solar hot water heating industry will grow rapidly. Government programs, the momentum of the growing industry, and economic forces will bring solar water heaters to one-fourth of the homes in Japan, two-thirds of the homes in

Israel, and one-sixth of all U.S. homes by the year 2000. (See Table 4. 1.)[73]

Table 4. 1. Use of Solar Water and Space Heaters, 1982–2000

	1982		*2000 Midrange Projections*	
Country	*Number of units*	*Share of homes*	*Number of units*	*Share of homes*
United States	1.5 million	1 in 75	15 million	1 in 9
Japan	3.6 million	1 in 10	10 million	1 in 4
Israel	300,000	1 in 3	1 million	3 in 4
Western Europe	60,000	1 in 2000	7 million	1 in 15

Source: Worldwatch Institute.

The prospects for industrial and agricultural process heating and solar air conditioning are harder to gauge. But these technologies could displace use of oil, gas, and electricity even more dramatically than solar water and space heaters do, even if no major technical breakthroughs occur or users do not congregate in sunny areas just to use solar technologies. (See Table 4. 2 for midrange estimates.)

Table 4. 2. Worldwide Active Solar Energy Potential

	1980	*2000*	*Long-range potential*
		(exajoules)	
Residential/commercial water & space heat	<0.1	1.7	33%–50% of total
Industrial/agricultural process heat	<0.1	2.9	25%–50% of total
Solar ponds	<0.1	2.1–4.2	10–30^{+}

Source: Worldwatch Institute.

Despite a slow start, applying solar technology to industry's needs could spawn a new industry. The InterTechnology Corporation asserts that tracking parabolic concentrators could command a third of the process heat market by the year 2000, assuming a 15 percent rate of return, and the 1979 U.S. Domestic Policy Review on Solar Energy predicted that 2.8 exajoules of solar industrial energy use is technically and economically feasible for the year 2000. Supplying this much energy will require between 700 and 900 million square meters of collectors at a cost of about $400 billion and will probably not occur until well after the turn of the century. It will also require installing solar equipment in most new industrial facilities.[74]

No detailed surveys of the worldwide potential of solar ponds have been carried out, but scattered national and regional assessments indicate these ponds are a world-class energy resource. One survey of fourteen sunny countries puts energy capacity from naturally saline lakes alone at between 30,000 and 160,000 megawatts by the year 2000. Analysts at the University of Sydney estimate that Lake Torrens, one of many saline lakes in southern Australia, could yield over thirty times as much electricity as the state now consumes. And in the most detailed large-area survey yet performed, Jet Propulsion Laboratory researchers found that 8.9 quads of heat and electricity (more than 10 percent of total U.S. energy use in 1980) could be economically produced by solar ponds in the U.S. Sunbelt by the year 2000. Large areas of Soviet and Chinese Central Asia, the Middle East, and northern Africa also appear well suited for salt ponds.[75]

5

Sunlight to Electricity: The New Alchemy

If some renewable energy technologies are workaday devices, photovoltaic cells excite the imagination. Developed during the semiconductor revolution of the fifties, these ingenious devices convert sunlight into electricity in one simple and nonpolluting step. By changing one of the world's most abundant and widespread energy sources into one of the most versatile and valuable forms of energy, photovoltaic solar cells work a feat of near alchemy. Steam turbines and other conventional technologies powered by fuel combustion appear clumsy and inefficient by comparison.

Without moving parts, photovoltaic systems are reliable and need little maintenance—claims that can be made for few new energy technologies. And solar cells could well be the ultimate decentralized energy technology. Unlike most energy systems, the cost of harnessing photovoltaic electricity falls only modestly as system size increases, so solar cells can be used in small quantities on rooftops, on farms, in rural communities, and even in cities. Photovoltaics offers individuals an unprecedented opportunity to generate their own electricity. In Third World villages, solar cells could provide small but vital amounts of electricity for the poor majority.

But such changes are still around the corner. The main problem is not technological: Solar cells have worked well in various applications for over two decades. Rather, the problem is cost. At current prices, a photovoltaic system can easily increase the cost of an electricity-guzzling modern house by 50 percent. Indeed, most solar electricity systems installed so far are tiny and are located on microwave repeaters, fire lookouts, and similar remote facilities. The approximately 10,000 houses equipped with small photovoltaic systems worldwide are virtually all in regions without conventionally generated electricity.[1]

Still, photovoltaics development has been so rapid that economic constraints could rapidly fall away. Between 1977 and 1982, the worldwide production of solar cells expanded more than tenfold and their cost fell approximately 50 percent.[2] During that period approximately fifty companies worldwide entered the photovoltaics business. Such spectacular advances cannot continue indefinitely, but significant progress is expected throughout the coming decade. In fact, there are now several technologies in the world's photovoltaic laboratories with the potential to revolutionize the solar cell industry if they prove feasible for commercial production.

A Space-Age Technology

Solar cells are of modern science born. They have no rich history, no traditional uses. Photovoltaic technology rests on solid state physics, a science barely understood until the mid-twentieth century. Like microelectronics, photovoltaics is based on the use of semiconductors—materials that have properties in between those of a metal and nonmetal and so conduct electricity only slighty. Also like microelectronics, photovoltaics could become one of the twentieth century's great technological success stories.

While French scientist Edmund Becquerel discovered in 1839 that when light strikes some materials it causes an electric spark, it took scientists many years to understand the cause of this "photoelectric effect"—that "photons" of light can dislodge the electrons that orbit all atoms. In silicon and a few other semiconducting materials these dislodged electrons can be turned into a tiny electric current. For decades, the utility of this phenomenon went unrecognized.[3]

In 1954 scientists at Bell Laboratories in the United States discovered that single crystals of silicon could be made into practical photovoltaic cells. Within a year experimental silicon cells made in Bell Labs were converting 8 to 11 percent of incoming sunlight into electricity. Briefly, Bell considered using the newly developed solar cells to power telephone systems in remote areas. *Business Week* let its imagination run wild, envisioning an automatically controlled solar car in which "all the riders could sit comfortably in the back seat and perhaps watch solar-powered TV."[4] Such dreams were soon dashed by economic reality, however. Costs for the newly developed solar cells were sky high (perhaps sixty times current prices).

Were it not for the U.S. space program, photovoltaic energy might have faded from the scene. But when satellite scientists in the mid-fifties began searching for a very light and longlasting power source that could be boosted easily into orbit, the newly developed solar cell emerged as the best candidate. When the space race began in earnest a few years later, the U.S. government devoted considerable funds to solar cell development, bringing into being a photovoltaics industry that supplied power panels for hundreds of American satellites. Today solar cells power virtually all satellites, including those for defense as well as scientific research. Solar electricity is particularly important to the growing world information economy since solar cells are used on satellites that relay long distance telephone calls, computer hookups, and television transmissions.

Yet space program research did not lead directly to the development of photovoltaics of practical terrestrial use. The space program's needs were for light, efficient, and reliable cells operable where sunlight is more intense than it is on the earth's surface. Cost mattered little since relatively few cells were required and the space program's budget was otherworldly anyway. Consequently, solar cells developed for space were still far too expensive for widespread use on earth.

The next spurt of interest in solar cells came when electricity prices began soaring in the early seventies. Researchers both in Europe and the United States looked anew at the technology and studied the potential for reducing its costs. Almost overnight, diverse photovoltaics research programs appeared in several countries. To some visionary technologists, solar cells' future as a major electricity source seemed bright.

Most commercial development programs have so far focused on single-crystal silicon cells similar to those developed by Bell Labs. While silicon, the main component of sand, is the second most abundant element on earth, the silicon from which semiconductors are made must have at most one impure atom per billion. One of the purest commercial materials used, it is energy-intensive and expensive to produce.

After purification the silicon is melted and then carefully

drawn from a vat using a technique known as the Czochralski process. The silicon is simultaneously combined with small quantities of another element (usually boron). The resulting crystal, which is about 10 centimeters in diameter and up to one meter long, is then sawed into many thin wafers in a difficult, expensive manner similiar to slicing bologna. Adding to the cost is the waste of about half of the valuable purified silicon in slicing. Each wafer is "doped" with trace elements that form a barrier of electric charge between the two sides of the cell that directs the flow of electrons set free by incoming sunlight.

Metal contacts placed on the front and back of the cell carry the electricity that has been generated to a battery or other device. Groups of photovoltaic cells are wired together in a module that is typically a square meter in size and encapsulated in glass and soft plastic for protection. Each module resembles an ordinary solar collector and has a generating capacity of approximately 100 watts.[5]

Researchers have already greatly reduced the cost of single-crystal silicon cells. From over $600 per peak watt at the beginning of the space program, the cost of solar cells fell to $200 per peak watt in the early sixties and to $50 per peak watt by the early seventies. Today, solar modules cost in the neighborhood of $8 to $15 per peak watt, and the market for photovoltaics for communications installations, small pumps, electrical rust protection for bridges, and other specialized or remote uses is expanding rapidly. Worldwide sales of photovoltaics reached 8,000 kilowatts of capacity in 1982—over ten times the market size in 1977 and four times the 1979 level. This is sufficient generating capacity to supply approximately 1500 modern houses.[6]

Phenomenal technological success aside, the current state of the technology should not be overestimated, nor should the need for continued innovation be dismissed lightly. At $10 per peak watt, solar cells generate power for approximately $1.00

to $2.00 per kilowatt-hour depending on the climate—over ten times the cost of power from conventional sources.[7] Continuing and substantial cost reductions will be needed before photovoltaics can compete economically with electricity from utility grids.

Research Horizons

The future of photovoltaics depends on evolutionary progress in support technologies and further advances in solar cell production processes. So far, industry has concentrated on the technologies that are closest to ready for the market and require relatively little work to meet cost goals. Government, in contrast, has supported work on potentially less expensive technologies that are still a decade or more from commercial readiness. Worldwide, public and private investment in the technology now amounts to approximately $500 million per year, two-thirds of it private money.[8]

The United States has backed the world's most ambitious solar cell development effort. U.S. government spending on photovoltaics—the largest component of the renewable energy research budget—increased steadily after 1973, topping $150 million per year in 1980 and 1981, only to fall to $75 million in 1982. These funds primarily support advanced research on photovoltaic technology, development of low-cost solar arrays, and commercialization programs. As of 1982 the advanced research effort, managed by the Department of Energy and the Solar Energy Research Institute but conducted through university laboratories and private companies, had become the most active part of the program while commercialization efforts have been all but eliminated. In all, these programs have been quite successful—witness the U.S. lead in both advanced technology development and commercial sales.[9]

Until recently, solar cell research in the United States has easily exceeded that of all other nations combined. But the

Reagan administration has reduced the U.S. research program just as other countries are stepping up their efforts. France, Italy, Japan, and West Germany—which have collectively more than doubled their budgets between 1979 and 1982—have the strongest solar cell programs outside of North America. Spending approximately $30 million in 1982, Japan's budget is likely to pass that of the United States in just a few years. More modest photovoltaics research work is under way in Australia, Belgium, Brazil, Canada, China, England, India, Mexico, the Netherlands, the Soviet Union, Spain, and Sweden.[10]

Most of these countries are pursuing two or three promising approaches to making solar cells economical, rather than taking the U.S. approach of developing a whole array of technologies. As a result, some European nations and Japan could soon take the international lead in their specialties.

One of the most important and heavily funded photovoltaic research frontiers is manufacturing single-crystal silicon cells more cheaply. The most conservative approach is to upgrade and automate each step of the current process. At least three techniques now being developed will cut by two-thirds the costs of making high-grade silicon. New methods for growing the crystals and slicing the wafers are also being pursued. Recently developed thousand-bladed saws that cut ultrathin wafers reduce waste significantly. New automated methods of assembling solar cells are also under scrutiny. Simply employing already laboratory-proven processes in more automated factories will cut photovoltaics costs by close to 50 percent in the next few years, while raising efficiency to at least 15 percent.

More radical approaches to cost cutting include bypassing both the crystal growing and slicing stages. Several companies in the United States and one each in Japan and West Germany are "growing" large sheets or "ribbons" of single-crystal or polycrystalline silicon directly from liquified silicon. Complicated, proprietary, and commercially immature as the processes are now, many industry observers expect them to claim a significant share of the solar cell market by the late eighties.[11]

Another solar cell technology with considerable potential is the polycrystalline silicon cell. Sliced from a large silicon ingot that is produced through an inexpensive casting process, these cells can be made from a less pure and less expensive form of silicon. One U.S. company began manufacturing such cells commercially in 1982 and other firms in the United States and West Germany have development efforts under way. Polycrystalline solar cells are still comparatively inefficient, however, so boosting efficiency is a must if this technology is to be successful commercially.[12]

More research attention is being given the so-called "thin film" solar cells that can be made from amorphous silicon, cadmium sulfide, and other inexpensive materials. All thin-film cells require only a small amount of material, which gives them the potential advantage of lower cost. While other researchers take exception, longtime photovoltaics specialist J. Richard Burke claims that "the low-cost pot at the end of the rainbow lies in the use of truly thin-film photovoltaic cells." The hope is that such a material can one day be produced in automated factories for a low cost—much as photographic film is today. In the United States private industry and government have aggressively developed cells made of amorphous silicon, which is a disordered material resembling glass that can conduct current well once hydrogen is added to it.[13]

So far, the highest efficiency that has been achieved for production-line amorphous silicon cells is 3 to 6 percent, and at least 8 to 10 percent is needed for commercial success. To boost cell efficiency, several U.S. and Japanese companies are investing tens of millions of dollars. Already Japanese companies have blazed the way to a commercial market by manufacturing amorphous silicon cells with a modest efficiency and using them in pocket calculators and other low-power devices. By establishing the first commercial market for these cells, the Japanese can employ larger manufacturing plants and thus further lower costs, helping to lay the basis for a vastly larger market in the future.[14]

Other types of thin-film solar cells are made of cadmium sulfide and copper alloys. When they were first produced in the fifties, these cells were so inefficient they were ignored. But interest revived in the seventies when researchers discovered that these materials could be made into solar cells with efficiencies of over 10 percent. Cadmium sulfide now appears to be the leading contender and may enter commercial production in the next few years. Among the other thin-film materials being examined are gallium arsenide, indium phosphide, cadmium telluride, and zinc indium diselenide. While none of these substances can be dismissed entirely, some are outside bets because they contain rare elements or present potential health problems.[15]

Along with solar cell materials, concentrator systems for use with photovoltaics are also being developed. Such devices can increase the amount of solar energy striking a particular cell ten to one thousand times, thus offering the potential of producing relatively cheap solar power even without major advances in basic materials. (The efficiency of most solar cells actually increases in concentrated sunlight as long as the cells are kept cool.) Often mechanical tracking devices are also used to maintain an optimal angle to the sun throughout the day. By using inexpensive Fresnel lens concentrators, large areas can be covered for a reasonable cost. Indeed, the cheapest solar power yet generated comes from some experimental concentrator systems. So far Italy and the United States dominate the concentrator field, but a large commercial market will not develop until the systems become more reliable. One difficulty with solar concentrators is that they work poorly in cloudy or hazy conditions where little focused sunlight is available, which may limit them to sunny climates. Solar cells without concentrators, on the other hand, perform quite well even when it is overcast.[16]

In a nutshell, the goal of most solar cell researchers is to achieve efficiencies of 12 to 15 percent in cells that cost less than a fifth of what they do today. To this end, scientists have developed cost-reduction goals for each component and set tough deadlines for reaching them. In both Japan and the United States government program managers constantly monitor progress and occasionally redirect research to another, more promising aspect of photovoltaics technology. The U.S. Department of Energy price goals established in the late seventies now appear unrealistically ambitious, but substantial cost reduction is nonetheless likely.[17]

Larger manufacturing plants employing more advanced and less expensive processes are scheduled to come on line in the next few years. And intense competition for market shares will tend to push prices down. Conventional crystalline silicon cells, together with ribbon growth and polycrystalline silicon cells, will likely dominate the market for the rest of this decade, though analysts differ as to which of these will be the most successful. Concentrators will probably be widely used in many applications, particularly utility plants. Beyond 1990 amorphous silicon and other thin-film technologies likely will capture the largest share of the market, pushing prices to new lows.

The photovoltaics market will evolve gradually rather than in discrete stages and at each point there should be a range of technologies to choose from—each with its specialized applications. Module prices will probably fall to approximately $3 per watt (1980 dollars) by 1987 and to about $2 per watt by 1990. At that price a total solar electric system will cost between $4 and $8 per watt and generate electricity at a cost of 15¢ to 30¢ per kilowatt-hour (as opposed to over $1 per kilowatt-hour today). This is getting close to standard electricity prices in many parts of the world, including Europe and Japan. Predictions beyond the early nineties are difficult to make since they are dependent on technologies barely beyond the laboratory stage. But further substantial cost reductions are likely since the photovoltaics market will be large enough to attract big investments. Given the rising costs of most sources of electricity, including coal and nuclear power, photovoltaics is likely to be a competitive electricity source in all but a few areas of the world by the mid-nineties.[18]

Building an Industry

The photovoltaics industry is still a young one, with annual sales revenues of about $150 million in 1982. Approximately sixty companies manufacture solar cells today, and over a hundred more build components and support systems. Three U.S. firms had over half of the worldwide market in 1980, but most photovoltaics companies are small and international competition is growing rapidly. Many firms subsist largely on risk capital or government research programs, hoping to begin turning a profit when their product improves. The pressing question for most is how to survive until the cost of photovoltaics can compete with the costs of conventional sources of electricity, thus blowing the solar cell market wide open. Before this harvest, large investments are needed, along with bigger plants and some means of disseminating the technology quickly. Such progress hinges, of course, on the strength of the industry.[19]

Centered in France, Great Britain, Italy, Japan, the United States, and West Germany, the solar cell industry has always stood apart from other renewable energy industries. It is a realm of three-piece suits and carefully crafted investment plans. In Europe and Japan, established electronics giants such as Sanyo, Sharp, and Siemens hold the industry in their hands, but in the United States there is more diversity. Many small companies have sprung up in the U.S., born of risk capital, government research funds, and bright ideas. Solarex, the largest photovoltaics company in the world, was started from scratch by a handful of young American scientists who largely relied on venture capital.[20]

Since the late seventies the solar cell industry has begun to consolidate. Only a few strong companies remain in Europe after a wave of mergers. And in the United States several large corporations have purchased a sizable share of the most competitive solar cell firms. No other renewable energy technology has proved so attractive to large corporations, probably because the potential market for photovoltaics is almost unlimited and because only big firms have sufficient investment capital. Indeed, as of 1982 it cost an estimated $50 million simply to enter the industry.[21] Oil companies in particular have taken a shine to photovoltaics, and the tiny solar cell industry now includes in its ranks such multinational behemoths as Atlantic Richfield, British Petroleum, Exxon, and Shell Oil.

The irony of these developments has not escaped those who first advocated photovoltaics as a decentralized technology. Oil companies now seem eager to get a purchase on every energy source from uranium to solar power, and some watchdogs fear that the oil companies may intend to develop an energy monopoly and impede progress in photovoltaics until the oil wells run dry. Although such fears are understandable, they are probably overblown. The pace of photovoltaics development is unlikely to be affected significantly by the state of the oil market, and in any case there remains sufficient competition in photovoltaics to preclude a monopoly. Indeed, Morris Adelman of the Massachusetts Institute of Technology believes that "the notion that the energy giants, controlling the biggest part of the manufacturing capacity in photovoltaics, could set the price artificially high to protect their other investments, is unrealistic."[22]

The most serious charge against oil companies' involvement in photovoltaics is that they tend to be hidebound and unimaginative and have little experience in this type of industry. Small firms have made a disproportionate share of the world's major industrial breakthroughs, and more small companies would likely speed the development of photovoltaics. Yet too many governments direct most of their research funding to large corporations. True, the sizable investments needed make it likely that large companies will in the long run dominate solar cell manufacturing. But small companies are well equipped to play a pioneering role and later to retail, assemble, and install solar power systems.

Internationally, market competition is a sure thing. Already more than half of the world's solar cells are exported, and each major approach to developing photovoltaics is being explored in more than one country. With exports high and patent protection inherently weak, industry leadership can change hands rapidly. In the white heat of international competition, technological improvements and cost reductions will be spurred.

Initially, its technical prowess and government financial commitments gave the United States a head start in the photovoltaics industry. By the late seventies the U.S. was the undisputed leader in virtually all solar cell technologies. But by focusing on fewer technologies, countries with smaller research budgets are attaining a competitive position. Japan has already moved to the "cutting edge" in amorphous silicon. Joint ventures and international licensing agreements that allow firms in other countries to manufacture U.S.-designed solar cells are also speeding up the diffusion of solar-cell technology.[23]

Since knowledge of photovoltaics technology is already widespread, marketing skills will be as important as cell costs in determining the industry's frontrunners. A particularly competitive market will be that in the Third World. Firms in Europe and Japan will have a natural advantage since they have traditional trading ties and experience selling the diesel pumps, generators, batteries, and other devices with which photovoltaics will be paired. More specifically, French firms have an advantage in West Africa, West German companies in parts of Latin America, and Japanese firms in Southeast Asia. These countries have incorporated solar-cell export drives into development-assistance programs and worked hard to promote the technology. In contrast, the lack of such programs in the United States has some industry leaders concerned that the U.S. will lose its international market lead by the late eighties. The likely heir would be Japan, whose Ministry of International Trade and Industry is devoting increasing funds to photovoltaics and is eager to repeat successes like those earned in the automobile and microelectronics industries.[24]

Still, no one or two companies can dominate this market, and international links between firms will blur the whole question of international leadership. As the world market grows, high transportation costs will also force solar electric systems manufacturers to fabricate at least some components locally. It is possible, for instance, that the silicon may be refined in one country, the cells manufactured in a second, and the panels assembled in a third. Already several developing countries have nascent solar cell industries, assembling components imported from industrial countries as a prelude to manufacturing whole systems domestically. Brazil, China, India, Mexico, and the Philippines are among the Third World nations that are likely to lead the way in photovoltaics.

A Future for Solar Power

Perhaps no other energy technology has the versatility of solar cells. David Morris of the Institute for Local Self-Reliance in the United States observes that "using the same energy source —sunlight—and the same technology, we could have the most decentralized or the most centralized form of electricity generation in history."[25] So far, though, the commercial market for solar cells consists almost entirely of mini-scale electrical systems in rural areas. Most are coupled with batteries and provide only enough power to operate a radio telephone or light a few bulbs. Such systems are crucial, however, in providing reliable communications in Papua New Guinea and lighting rescue cabins in the Swiss Alps.

According to many photovoltaics analysts, the first large use of solar cells will be in the Third World. On farms and in villages there, the power currently supplied by small diesel generators costs several times more than grid electricity. Small solar electric systems could economically power pumps, lighting systems, agricultural equipment, refrigerators, and other important devices. For refrigeration or lighting, batteries or a backup power source will be needed, but for many end uses the device can be left idle when the sun is not shining. As of 1980, photovoltaics is already competitive with diesel generators in rural electricity applications of less than three kilowatts of capacity.[26]

Since 1978 the world's first village solar electric system, with a capacity of 3.5 kilowatts, has been operating on the Papago Indian Reservation in the U.S. Southwest. Since then, several similar systems have been built in Africa and Asia with funds from European and American aid agencies. The largest center of photovoltaics activity is West Africa, where since the late seventies France has been introducing solar-powered pumps and other systems as part of its rural development programs. One innovative effort is to use solar power to run ultra-energy-efficient televisions for educational uses. Another is to provide electricity for refrigeration of medicines at remote health centers.[27]

Within the developing world, interest in solar electricity has risen sharply in recent years. India's government is conducting photovoltaics research, fostering a domestic solar cell industry, and sponsoring solar electric demonstration projects. Pakistan plans to introduce solar electricity in fourteen villages by 1984. In both countries a market for small solar-powered pumps is beginning to emerge.

By 1990 operating experience could combine with technical improvements to make photovoltaics a nearly conventional technology in the Third World. Crucial here will be additional work on battery systems and other support technologies. For villagers, the impact of even small amounts of electricity could be revolutionary. It could mean fresh well water, refrigeration for storing food and medicine, and lights for reading and working at night—modest amenities by industrial country standards but godsends for many of the world's poor.[28]

Somewhat later, solar cells are likely to appear on rooftops in cities and suburbs throughout the world. Like houses with solar water heaters, photovoltaics-equipped houses require a southern exposure and rugged, longlasting materials. Lightweight photovoltaic panels need relatively little structural support, but they need more south-facing roof space than collectors do. (A typical 3-kilowatt residential solar electric system requires 30 square meters of panels.)[29]

Although it will be easier to use photovoltaics on homes specifically designed for their use, it appears that existing suburban communities may be able to get as much as half of the electricity they need from solar cells. There are already approximately 10,000 houses located in areas without power lines that have small (less than 1-kilowatt) direct current photovoltaic systems with batteries that meet essential needs. Providing sufficient power for a typical modern house is more difficult.

To keep rooftop photovoltaic systems from competing with solar water heaters and windows for south-side space and to solve other engineering problems, architects and engineers have designed a few solar electric houses as demonstration projects. Their focus is on making photovoltaic systems easier and cheaper to install and on integrating solar electricity with passive solar architecture and the many other features homebuyers value. One U.S. company has developed a dual-purpose photovoltaic shingle. Another designer is actually using specially-designed solar electric panels as roofing. Although the few solar houses built so far have sold for over $200,000, these homes serve as a proving ground, allowing the refinement of designs and support technologies in preparation for the day when solar cells become cost-competitive.[30]

Solar cells more than any other technology have the potential to decentralize electricity generation. In urban and suburban areas thousands of residential solar systems could be connected to utility lines, doing away with the need for expensive battery storage. Solar homes could draw power from the grid at sunless times and pay for it by selling excess electricity to the utility when sunshine is plentiful. In sunny, dry areas where peak electricity demand for air conditioning occurs when sunlight is most intense, this arrangement could be a boon. Elsewhere, only careful planning will make solar electric houses economical for utilities and consumers.

The prospect for decentralized electricity generation notwithstanding, some utilities see in solar power systems a chance to make centralized generation more versatile. The idea, which many photovoltaics researchers and industry leaders consider practical, is to erect large arrays of solar cells (and perhaps concentrators) in sunny areas and to integrate them with the utility grid. Although solar cells themselves have no economies of scale, photovoltaic systems do, especially the power-conditioning equipment used for utility interconnection. Some researchers believe that centralized solar systems will be the first major use for photovoltaics in industrial countries.

Only a few large photovoltaic systems have been built so far. The largest is a 350-kilowatt system in Saudi Arabia that supplies power for three villages and was funded by the Saudi Arabian and U.S. governments. A larger 1,000-kilowatt grid-connected system is being built with federal and state funds at the Sacramento Municipal Utility District in California. A similar project is under way in Italy. And in 1982 the first contract was signed for an entirely privately financed utility photovoltaic system—between ARCO Solar and Southern California Edison.[31]

The most pie-in-the-sky way of harnessing solar electricity is via the "solar satellite." Several researchers in the United States have proposed placing large arrays of solar cells in stationary orbit around the earth and using microwave transmitters to convey the power to land-based receivers. Since sunlight is more intense outside the atmosphere, it is theoretically possible to reap a great energy harvest in space. But even ardent advocates of this technology admit that it will be decades before launching such vast quantities of materials into orbit is feasible. And skeptics question whether it will ever be economical considering the large amount of energy needed to overcome gravity. More disturbing are the potential health and environmental effects of a high-energy beam aimed at the earth's surface. Microwave radiation causes health problems, and even the earth's atmosphere could be altered. At any rate, no one is banking on solar satellite research at the moment, and many renewable energy advocates believe that the idea gives an aura of science fiction to a technology ready for here-and-now use on earth.[32]

Assessing the worldwide potential for using solar photovoltaic cells takes patience and imagination. Beyond the considerable technical uncertainties are questions about intermediate markets and the industry's strength during the critical mid- to late-eighties, when solar cells will be economically competitive only in areas without conventional sources of electricity. The mid-nineties may be another story, but that will depend on major cost reductions in photovoltaics and on the price of competing electricity sources. In the industrial countries electricity use is likely to grow only slowly in the nineties, but substantial solar cell sales may occur as older power plants are retired. A boom market in the developing countries—particularly those that are industrializing rapidly—is a distinct possibility as well.

Various forecasts of photovoltaics use have been made, all of them based largely on guesswork. The goal of the U.S. photovoltaics program as formulated by Congress in 1978 is to double the manufacture of solar cells each year so as to reach an annual output of 2,000 megawatts of cells by 1988 (250 times the 1982 total). The U.S. Department of Energy subsequently established a goal of obtaining 1 quadrillion Btu's (just over 1 exajoule) of energy from photovoltaics by the year 2000. This would require an installed capacity of over 50,000 megawatts, or about as much capacity as nuclear power has in the United States today. It is now clear that these early goals were high, particularly considering the limited funds the government has devoted to achieving them. In Japan the goals that have been established are more conservative and realistic. The country aims eventually to generate 20 percent of its electricity using solar cells placed mainly on rooftops, but most of this growth is not expected until the 1990s.[33]

Worldwide trends are even more uncertain, but the industry has advanced far enough in the last few years to narrow the range of possibilities. There will likely be at least 1,000 megawatts of solar cells installed by 1990, a large portion of them in developing countries. By the year 2000, the total will probably range between 5,000 and 20,000 megawatts, depending both on the pace of technological improvements and the level of government support. Even the latter figure would provide just 0.4 exajoules of energy, but much more rapid progress seems likely after the turn of the century as the technology matures and many conventional power plants reach retirement age. By mid-century, solar electric systems should be a common rooftop appliance throughout the world and should provide perhaps 20 percent of the world's electricity. This would require a total capacity of around a million megawatts, installed both on rooftops and at centralized power stations. The energy contribution would approach 20 exajoules.[34]

Solar electric systems are clearly among the brightest hopes on the energy scene today. Their potential to provide inexpensive, independent power to people and industries throughout the world is far more important than their gross energy contribution. Thanks to advances being made in this seemingly exotic new technology, the living standards of hundreds of millions of people in developing countries can be significantly raised in the next few decades.

6

Wood Crisis, Wood Renaissance

For most of human history, people have burned wood to cook their food, stay warm, and light their environment. Even today, it remains the world's most widely used renewable energy source. Although deforestation and mounting population pressures are constricting the wood supply, most of the wood burned today is used much as it always has been. Only in the industrial North, where rising oil prices have triggered a revival of wood use for residential and industrial heat, have combustion techniques advanced significantly.

As traditional uses grow, efforts to turn wood into electricity, gas, and methanol are also getting under way. Realizing wood alcohol's potential to power the transportation system will require making wood burning more efficient and phasing out some traditional demands for wood. Indeed, the world's forests can meet rising needs for wood energy only if forest and woodlot management improves in rich and poor nations alike. By the same token, if the health of the forests is neglected, the push to get more energy from wood will backfire, reducing the forests' potential to provide lumber and paper as well as energy.

An Ancient Fuel in Crisis

Approximately 2 billion people rely on biomass energy. While animal wastes, crop residues, and draft animals also supply energy to the world's poor, wood is the principal source of energy for 80 percent of all people in developing areas, and half the world cooks with wood. In Africa fuelwood meets 58 percent of total energy demand. In Ethiopia, Nepal, Sudan, Thailand, and even oil-rich Nigeria, 90 percent of the population depends on wood. Even in larger towns and cities wood is used in the form of charcoal, which is lighter and cheaper to transport than wood and burns smoke-free. In Thailand, for example, almost half the wood used for fuel is first transformed to charcoal.[1]

Dependence on wood reflects a lack of other options. Few in rural areas can afford electricity even if it is available. In the developing world only three out of twenty villages have electricity, while such fossil fuels as kerosene, butane, and propane were pushed out of the reach of many Third World families by the oil price increases of the seventies. According to an expert panel that advised the 1981 U.N. Conference on New and Renewable Sources of Energy, more than 100 million people cannot obtain even the firewood needed to meet minimum needs, and another one billion people need more than they can now get. By century's end over 2 billion people will live in firewood-deficient areas, primarily semiarid regions and highlands. (See Table 6. 1.) Today, the problem appears most acute on the densely populated Indian subcontinent and along the Sahara Desert's edge. In Latin America scarcities of firewood and charcoal plague much of the Caribbean, Central America, and the Andean highlands.[2]

Table 6.1. Fuelwood Shortage in Developing Countries

	1980		*2000*	
Region	*Acute scarcity*	*Deficit*	*Acute scarcity*	*Deficit*
	(millions of people affected)			
Africa	55	146	88	447
Near East & North Africa	*	104	*	268
Asia Pacific	31	645	238	1532
Latin America	15	104	30	523
Total	101	999	356	2770

Source: FAO, Report of the Technical Panel on Fuelwood and Charcoal to the U.N. Conference on New and Renewable Sources of Energy, Nairobi, August 1981.

*Figure is not available.

The fuelwood crisis stems from the practice of ancient traditions in changed circumstances. Although deforestation is as old as recorded history, today's fuelwood crisis has comparatively recent origins. The postwar burst in population growth, the accelerated conversion of forest land into farmland, and the increase in livestock herds have together pressed remaining woodlands inexorably. In short, firewood gathering exacerbates already serious problems of deforestation.[3]

Commercial firewood prices have multiplied almost everywhere over the last decade. In parts of India, West Africa, and Central America, urban families spend one-quarter of their income on wood or charcoal for cooking. When firewood becomes harder to find, people forego their nighttime fire or, worse, their meal. Hard to quantify, the effects of scarcity and high costs of firewood and charcoal are devastating by any measure.

Most fuelwood never enters the marketplace, so a better

indicator of scarcity is the time it takes to find wood. In central Tanzania, providing a family's annual firewood requires between 250 and 300 days of labor. In deforested parts of India, it takes 2 days to gather a week's wood. And in parts of Upper Volta, women spend an average of four and a half hours a day hunting for firewood. Since the burden of firewood collection almost always falls on women and children, critical but unpaid household tasks such as nutrition, sanitation, and education suffer. The costs of this mounting burden show up not in conventional economic indicators, but in indices of infant mortality, disease, and illiteracy.[4]

Fuelwood price rises and supply reductions are also limiting the growth of small-scale industrial enterprises in many Third World countries. Brick baking, tobacco curing, fish drying, and cement making all depend heavily on wood. Although most countries devote only 2 to 15 percent of their fuelwood to such processes, in many these activities represent the fastest-growing use for wood. In some cases critical export industries depend upon wood. Tanzania cures tobacco with wood, and Thailand does the same with rubber. Yet in both countries wood is being cut at an unsustainable rate. Around one fishing center in the Sahel region of Africa, where every year 40,000 tons of fish are dried using 130,000 tons of wood, deforestation extends 100 kilometers.[5]

One way to check these trends in developing countries is to make fuel burning more efficient. The open hearths over which most Third World people cook are only 6 to 8 percent efficient. By comparison, airtight stoves manufactured in the West are 30 to 80 percent efficient. While such stoves are far too expensive for developing-country residents to use, inexpensive improvements over traditional open hearths (such as simple stoves built from locally made bricks) can boost efficiency to 15 percent, effectively halving a household's wood needs. The Lorena stove developed in Guatemala costs between $5 and $15. Molded from mud and sand and fitted with a metal damper and pipe, it is twice as efficient as the traditional stove it displaces. The simple and cheap Junagadh stove developed in India is reportedly 30 percent efficient.[6]

Several social and economic obstacles have kept simple cook stoves from being widely accepted in any Third World country. One is the lighting property of an open fire. Another is its social value. Then too, even though thick smoke from open fires has been called poor people's smog, it also repels insects from the house and roof thatching. A major problem is expense. Many rural families cannot afford even the simplest stove.[7]

To better rural firewood prospects at least a dozen developing countries have started programs to spread simple stoves throughout rural villages. In Senegal an effort sponsored jointly by France and the United States has encouraged over 1,000 villagers to build and use a Lorena-type stove, the Ban ak Sunf. India has also mounted an ambitious effort to build cooking stoves. The key to all such programs will be designing stoves that appeal to the village women who must operate and maintain them. Speed is also essential since new households are forming far more quickly than cook stove use is increasing.[8]

Another approach to conserving firewood is producing charcoal more efficiently. In the most widely used and least efficient method, stacked wood covered with earth is allowed to smolder in the absence of oxygen for several days—a process that wastes 75 to 90 percent of the wood's energy. Switching to kilns made of brick or steel allows the production of charcoal much more efficiently. But steel kilns are prohibitively expensive, so the likeliest replacements in poor countries for highly inefficient earth pits are brick kilns made from locally available clay.[9]

Besides burning wood more efficiently, wood-short countries can make better use of wood cut from lands being pressed into agricultural and industrial use. In Tanzania, for example, tobacco farmers clear one piece of land for crops and then cut wood from another parcel to cure the harvest. Simply storing the wood or making it into charcoal could drastically reduce the amount of wood cut for tobacco production. Two heavy charcoal users, Brazil and Sri Lanka, have learned to make full use of felled trees. All the trees on 65,000 hectares of land scheduled to be flooded when the Tucurui dam is finished in Brazil are being cut for lumber exports and charcoal production before the floodgates close. In Sri Lanka, the Charlanka company will use portable kilns to turn 25 million tons of wood residues that would otherwise be wasted into charcoal for the cement industry, which currently depends on imported petroleum. The Brazilian wood harvest will be a one-time affair, but the Sri Lankans are planning to plant 13,000 hectares of eucalyptus to perpetuate charcoal supplies.[10]

The Return to Wood

Like the developing countries today, Canada, the United States, Europe, and Russia once depended almost exclusively on wood. Augmented by human and animal power and a modest amount of wind and water power, wood formed the energy basis of the New and Old Worlds well into the nineteenth century. Wood was used to cook and heat, and as charcoal, it was used in metal smelting. Forests could not meet the rising demand, and these countries turned to coal.[11]

Since the oil shock of 1973, wood has come into a second age. This revival has been most visible in the residential heating market and the forest products industry in the United States and Canada. Residential firewood use in the United States more than doubled between 1972 and 1981, and the number of homes heated entirely by wood has reached 4.5 million, while another 10 million are partially heated with wood. In Canada some 200,000 homes are heated solely with wood. New England and eastern Canada lead the return to wood-stove use, reflecting in part the region's great dependence on expensive heating oil. (In 1981 half the homes in northern New England were heated at least partially with wood.) Other well-forested regions are taking wood stoves more seriously, though where winters are mild and cheap natural gas or hydroelectric power plentiful the trend is less pronounced.[12]

In heavily forested parts of Europe and the Soviet Union, wood is an important source of energy for residences. In Austria, for example, almost one-third of all homes meet some part of their heating needs with wood. And in the USSR, where fuelwood accounts for 20 percent of the timber harvest, at least one-quarter of the residences are wood-warmed. Because wood use was already high, coal reserves abundant, and natural gas and electricity cheap, wood use in the Soviet Union rose little during the seventies.[13]

The rekindled interest in wood stoves stems partly from recent improvements in stove designs that have been around for a century or more. Although its playful flames and glowing embers may make it more aesthetically appealing than a wood stove, an open hearth lets at least half the warmth of the fire escape up the chimney, so the updraft actually draws cooler air into the room. Airtight or brick stoves radiate far more heat into the surrounding space than fireplaces do. At a cost of from $800 to $2000 per unit, the Finnish or Russian brick stove, which traps hot gases so that the bricks absorb and reradiate more heat, is reportedly 90 percent efficient.[14]

Naturally, the appeal of wood stoves for residential heating depends on how much conventional fuels and the wood itself cost. Compared to a furnace that burns heating oil, a wood stove in a well-forested area can save a household hundreds of dollars a year—more if the members of the household collect and cut their own firewood. The economic advantage of wood stoves is less clear where low-priced natural gas is available. More certain is wood's competitiveness with electricity. In forested regions with electricity prices at or above the U.S. national average of about 6¢ per kilowatt-hour, wood is economic today.[15]

Another factor affecting the use of wood as a residential fuel is ease and convenience of handling. Although chopping, carrying, and loading wood is worthwhile—even invigorating—to some, the convenience of electric, oil, or gas heating cannot be discounted. One way to take the hard labor out of using wood would be to adopt recently developed wood-fired furnaces whose thermostatically controlled feeders automatically convey wood pellets into the fire grates. Still, these problems, along with the difficulty and expense of transporting the wood needed to heat even a medium-sized urban area, mean that wood will probably never be used widely in cities.[16]

The residential wood-burning revival also poses serious health and pollution-control challenges. Proper installation is essential to safe use of wood stoves since hot stoves can cause fires or emit harmful smoke into homes. Paradoxically, burning wood by utilities and industry causes fewer problems than the dispersed use of wood in small stoves because most large wood boilers come equipped with pollution-control systems. Smoke is an especially serious problem in valleys where temperature inversions occur and smoke accumulates. Although recent studies indicate that, except for hydrocarbon particulates, wood burning produces fewer pollutants than fossil fuel combustion does, possible carcinogens have been found in wood stove smoke. In some areas, such as Vail, Colorado, so much smoke from wood stoves has accumulated that their use has been limited by law. Ironically, the more efficient airtight cast-iron stoves now selling so well generally produce more hazardous organic particles than traditional open hearth fires do because the stoves burn more slowly.[17]

Fortunately, a new technology can alleviate the air-pollution hazards of wood stoves. Dow Corning Corporation's "catalytic combuster," a $100 device similar to a catalytic converter on an automobile, burns off a stove's exhaust gases, thus increasing the average stove's efficiency by 20 to 30 percent, enough to pay for itself. Although several major manufacturers plan to sell stoves with the new catalytic combuster, government will have to establish a wood stove tax credit or grant and require the use of combusters to make sure these devices are used widely. By offering such conditional incentives of 25 to 50 percent, governments could avert a growing pollution problem and place wood stoves on an equal footing with heavily subsidized conventional energy sources.[18]

New Uses for Wood

Wood is being put to many new uses, too. A growing number of modern industries are turning to wood, and some utilities are trying it out in electricity generation. Wood gasification is again being used in agriculture, industry, and commerce. The single most important new use for wood may be as methanol, a liquid fuel that could one day edge gasoline out as the preferred transportation fuel.

Rising fuel prices have triggered a renaissance of industrial wood use. Traditionally, industry used wood to generate steam power, make charcoal, and smelt metal. As in homes, wood's use in industry declined during the era of cheap fossil fuels, but has grown dramatically since 1973. In 1966 wood-fired boilers represented a negligible percentage of total industrial boiler sales in the United States. By 1975 they represented 5 percent of the total. As of 1980 more than 2,000 large industrial wood-fired boilers were in use and many thousands more provided energy for smaller operations.[19]

Logically enough, the forest products industry has led industry's return to wood. In the United States and Canada energy-intensive pulp and paper plants consume more petroleum than any other manufacturing industry. In the United States the share of the industry's energy obtained from wood wastes has risen to 50 percent. The largest single U.S. forest products company, Weyerhaeuser, generates two-thirds of its energy from wood and plans to become completely energy self-sufficient by 1990. In Western Europe similar trends are in force. The giant Swedish pulp and paper industry derives 60 percent of its energy from wood scraps and pulp residue. Studies by the Swedish government indicate the industry could become energy self-sufficient and sell excess cogenerated electricity. In Canada this trend has been assisted by the Forest Industry Renewable Energy (FIRE) program, which will spend $288 million between 1979 and 1986 on industrial grants for converting plants to run on wood fuel.[20]

Several other major manufacturing facilities in heavily forested rural areas have also switched from oil to wood. At Dow Chemical Company's new industrial complex in Michigan, a wood boiler will provide 22.5 megawatts of power at less cost than oil or gas. In North Carolina seven brick plants and six textile mills have converted from gas to wood. Cost savings can be dramatic, as a Massachusetts firm discovered when its annual fuel bill went from $720,000 for oil to $270,000 for wood.[21]

Wood's role in industry is expanding partly because new technologies can gather and homogenize abundant wood residues and wastes. Instead of high-quality wood logs, industry can burn the bark, branches, and diseased trees left in the wake of timber and pulp operations. Energy-rich "pulping liquors," which otherwise pose a major disposal problem, can also be an important source of industrial fuel. New truck-sized machines shred trees into standard, matchbox-sized chips and shoot them into waiting vans. About 50 percent water, these heavy chips are expensive to truck long distances, but their use makes sense in well-wooded communities that do not have easy access to oil or coal.[22]

Another alternative is pelletized wood. Made from wood waste bound together under heat and pressure, wood pellets can be used directly in unmodified coal-fired furnaces. Denser and drier than wood chips, they can be transported economically over greater distances. In the United States wood pellets currently cost about as much as coal but contain only half as much heating value. Still, they are economical where there are no railroads to bring coal in cheaply. Pellets are also attractive as an industrial fuel because they give off few pollutants when burned.[23]

An older energy-conversion technology making a comeback is wood gasification. During World War II 700,000 automobiles, mainly in Europe, were powered by wood gasifiers. Unlike the gas fermented from starches and sugars, wood gas is made by heating wood in the presence of only small quantities of air. Although this gas is not energy-rich enough to justify piping long distances, it is well suited for use in gas or oil boilers or in the diesel engines widely used in developing countries. Because burning wood gas is considerably less polluting than burning wood itself, wood gasifiers may become industry's first choice among wood-use technologies. Several firms have begun marketing wood gasifiers that provide energy at roughly the cost of price-controlled natural gas in the United States.[24]

Wood is also being used on a modest scale by utilities to generate electricity. A utility in heavily forested, sparsely populated Vermont recently retrofitted two of its 10-megawatt coal-fired boilers to burn wood chips. The company is also building a 50-megawatt plant that will burn 500,000 tons of wood chips a year to provide electricity for 20,000 homes. For fuel for the furnace, machines will harvest and chip whole trees within a seventy-five-mile radius of the plant. The $76-million facility is expected to generate electricity 20 percent more cheaply than a comparable coal-fired plant can.[25]

By far the most ambitious effort to use wood to generate electricity is taking place in the Philippines. Faced with an oil-import bill that consumes over half the nation's foreign exchange, the Philippine government has embarked on a program to build 300 megawatts of wood-fired power plants in remote areas of the country by 1985. To insure an adequate supply of fuel, the National Electrification Administration provides funding for groups of up to ten rural families to set up plantations of fast-growing leucaena—a strategy that will reverse deforestation as well as provide energy.[26]

Wood's most important new use is likely to be as methanol, a clean-burning liquid fuel that automobiles, trucks, and aircraft can use. Methanol can be produced from the cellulose in wood or grasses, which is vastly more abundant than the sugary and starchy feedstocks used to make the ethanol found in alcoholic beverages and gasohol. Before 1930 virtually all methanol was made from wood. During World War II German cars ran on methanol made from coal, while Brazil's automobile fleet ran on methanol made from wood. Today most high performance racing cars run on methanol. In 1980 almost all the 1.4 billion gallons of methanol produced worldwide were made from natural gas and were used as an industrial chemical rather than as a fuel. In the future methanol may again be produced from coal, which is easier and cheaper to do than producing gasoline from coal—the goal of many synfuel programs.[27]

Methanol is produced from wood through destructive distillation in which wood heated in the presence of a little air decomposes into charcoal, carbon dioxide, and hydrogen. When pressurized in the presence of catalysts these gases become a liquid—methanol. In contrast to ethanol production, methanol production requires little energy from external sources since heat is generated when the feedstock is gasified.[28]

Estimates of methanol production costs vary widely, but the price of the feedstock is critical to all. According to the U.S. Office of Technology Assessment, wood costing $30 a ton can be converted into methanol costing $1.10 a gallon. Where wood or wood-wastes are abundant, technology now for sale can produce a gallon of methanol for between $1 and $1.25. Adding taxes and transportation, methanol would probably retail for between $1.50 and $2.00 a gallon. Since wood alcohol has only about half the energy value of an equivalent amount of gasoline, it is thus cost-competitive with gasoline that costs $3 or $4 a gallon.[29]

Although this technology is widely proved and tested, researchers in Brazil, Canada, the United States, and France are trying to improve significantly the efficiencies and economics of methanol production. Scientists at the U.S. Solar Energy Research Institute (SERI) have doubled the amount of methanol obtained from a given quantity of wood. The SERI gasifier could probably produce methanol for 70¢ to 80¢ a gallon. Researchers in Brazil report other methanol technology improvements that can reduce methanol costs comparably. If pilot-plant experience is duplicated in larger plants, methanol from wood will compete with methanol produced from natural gas.[30]

Methanol has been little used in transportation so far because it blends poorly with gasoline and readily corrodes rubber, plastic, and some metal parts of standard internal combustion engines. Accordingly, it has been necessary to redesign some engine parts, though if mass-produced these methanol-tolerant engines would cost no more than gasoline engines. For now methanol is being used only as a transportation fuel in "captive fleets" such as city buses or company cars that operate in a circumscribed area and fill up at centralized locations. Several extensive on-the-road tests in West Germany, California, and Brazil have demonstrated that methanol-tolerant engines perform at least as well as gasoline-powered ones.[31]

Going beyond these simple modifications of conventional engines for methanol use, engineers are also designing engines particularly suited to methanol. The Ford Motor Company and the U.S. Solar Energy Research Institute have developed a high-compression engine that dissociates methanol into hydrogen and carbon dioxide and achieves a fuel efficiency similar to that of a gasoline engine despite the fact that methanol only has half the energy of gasoline. This new engine could in effect eliminate the cost differential between gasoline and methanol.[32]

The potential for replacing liquid petroleum products with methanol in one heavily forested country, Canada, has been examined in detail. According to a government-sponsored study, Canada could produce over 72 billion liters of methanol in the year 2000, enough to completely replace the 203 million barrels of oil now used for transporation. Although a hybrid natural gas-wood process would be most economical today, cellulose becomes the most economical feedstock if natural gas is priced at parity with oil. According to researchers the principal constraint upon such a strategy is demand related—an abundance of cheap natural gas and ample oil supplies.[33]

Realizing wood's energy potential fully, of course, means locating wood-using systems near wood supplies and keeping system size down accordingly. Indeed, transporting wood beyond fifty to one hundred miles becomes prohibitively expensive, and a plant's size is dictated by the volume of nearby wood —even heavily forested areas can continuously fuel at most a 50-megawatt generator. For wood alcohol, new small-scale units fill an important gap in the technology since the large plants that make methanol from natural gas and coal would require too much wood to be transported too far. International Harvester hopes to market a package methanol plant with an output of 6 million gallons a year, a tenth the size of the typical fossil-fuel methanol plant. Factory assembled and trucked to the site of use, these small plants will not entail high construction costs.[34]

A major constraint to greater wood use for methanol or by industry and utilities is uncertainty about the future price and availability of large supplies of wood. With transportation costs the limiting factor, a sudden surge in local demand could strand large users. As insurance many companies moving to wood fuel are building furnaces capable of burning both coal and wood pellets.[35]

Governments could help the methanol fuels industry emerge quickly. For starters they could purchase fleets of methanol-burning automobiles or offer incentives for large private fleet owners to use methanol. California, for example, has already offered to buy methanol-powered cars from Ford. Such an assured market would give forest products companies the incentive to build relatively small-sized methanol-from-wood plants near existing paper plants and sawmills.[36]

A Growing Resource in Stress

The rising demand for wood energy comes at a time when forests are rapidly being cleared to make way for agricultural land and when demand for timber and pulp is rising. Clearly, new forest-management techniques and policies will have to be devised to meet demand without magnifying environmental stresses. Yet large blocks of virgin forest, the lands replanted for the pulp and timber harvest, and poorly managed or deforested lands, which together make up a quarter of the earth's land surface, each hold surprisingly different potentials for stretching and saving the resource base.

The most economically sound way to increase wood energy use without sacrificing traditional forest products is to remove more logging wastes from commercial forests and—more important—to increase replanting and improve management on small parcels of degraded forest land. In contrast, cutting remote virgin forests or greatly intensifying the harvest from commercial forest lands should be limited both on economic and ecological grounds. Relying on the wrong forests for energy could wreak far-reaching ecological harm.

Virgin forests in remote regions make up the biggest share of the global forest inventory. But their potential as a source of wood energy is small. Tropical rain forests in the Amazon Basin, Central and West Africa, and Southeast Asia are shrinking particularly rapidly as trees are logged for timber. Reforestation prospects in these areas are not bright since the trees themselves, rather than the soil, contain much of the rain forests' nutrients. While dispersed tropical populations face no fuelwood crisis, these remote expanses are being eyed by government energy planners for large-scale energy schemes. Yet caution is the watchword. No more biomass should be removed from most of these lands for energy purposes until other pressures wane and the ecology of tropical rain forests is better understood.[37]

In the northern hemisphere the vast forests covering much of the Soviet Union, northern Europe, Canada, and the United States have actually expanded slightly over the last half century as some farmland returned to forest. Although this resource is vast, much of it is located far from potential markets. Then too, forest regeneration in the thin soils and cold of Siberia, Alaska, and northern Canada can take up to a hundred years, making these forests practically nonrenewable.[38]

Commercial forest lands that supply lumber for contruction and pulp for paper making represent a more likely source of wood energy. The most readily available source of wood energy is the vast quantity of branches, bark, and roots left in the wake of lumber and pulp harvesting. Thus, the rising demand for lumber and pulp could actually increase the amount of wood available for fuel by motivating forest managers to thin slow-growing, diseased, or otherwise unmarketable trees for use in energy conversion. In the United States, the Office of Technology Assessment estimates, wood containing the equivalent of 2.5 percent of U.S. annual energy use is left to rot or is burned at logging sites during lumber and pulp harvests. Were lumber and pulp consumption to double as projected for the year 2000, the amount of wood cut but left unused in the nation's forests would increase by 2.5 to 5 times.[39]

At what environmental costs could these forest residues be removed? And what would the benefits be? In the short-term, clearing the land of dead limbs and branches improves some wildlife habitats, reduces the outbreak of forest fires, and makes tree planting easier. But over the long-term, soil productivity will suffer if the limbs, leaves, and roots that contain most of the forest system's nutrients are removed. Where clear-cutting is practiced, removing logging residues accelerates the erosion of the topsoil upon which all forest life depends.[40]

Removal of logging residues for energy use will clear the air some. Currently, branches, leaves, and stumps of harvested trees are often collected into piles and set on fire. Smoke from these open air fires contributes heavily to air pollution in such diverse locations as Malaysia, Colombia, the northwestern United States, and eastern Canada. Compressed into pellets or gasified, such logging residues could be cleanly and productively burned.

Removing dead trees and periodically clearing the brush could make herbicides largely unnecessary, too. As it is, timber and pulp operations, particularly in the United States and Canada, depend increasingly on the aerial spraying of herbicides to kill species that compete with commercially valuable species for light, soil, and water. What repeated herbicide applications will do to forests, no one can say for sure. But several widely used phenoxy herbicides (such as 2, 4, 5-T, and Silvex) are thought to cause cancer, birth defects, and other health problems in people.[41]

As for productivity, intensifying silviculture on commercial forest lands can expand the supply of lumber, fiber, and fuel. Large pulp and paper companies have begun genetically manipulating trees and practicing short-rotation tree farming to raise output. Scientists at Weyerhaeuser predict that tree production could be doubled if the genetic techniques successfully employed in agriculture are used. In another intensification effort, U.S. Forest Service scientists have increased wood yields of poplar three to five times beyond those of wild timber stands by planting trees close together and harvesting them when they start to interfere with the growth of adjacent trees. Sweden, whose forests are among the world's most intensively managed, has launched a broad investigation of machine-harvestable species that grow rapidly and regenerate without replanting.[42]

Although most advanced tree farming is being done by timber and pulp companies to supply their traditional markets, foresters in several countries are also at work on fast-rotation tree farming for energy. That both groups are at work is important since timber and pulp-oriented silviculture is only partially applicable to energy silviculture: The energy content of plants is seldom maximized in the effort to increase fiber quality and wood strength.[43]

The most important constraint on the general prospects for energy plantations is cost. If lumber and pulp sales are not combined with fuelwood sales, harvesting even fast-growing trees for fuel use is uneconomical. However, if mechanical harvesters can be tailored to a given species and if genetic improvement continues, energy plantations will become more economic. But the calculation may be moot: As lumber and paper grow more expensive, multiple-use silviculture becomes more appealing still.[44]

Another way to increase forest productivity is to plant high-yielding exotic tree species. Indeed, to accelerate forest regrowth, scientists have searched the earth for faster growing, hardier, and more productive tree species. Among the several dozen promising trees located, Eucalyptus is planted most widely throughout the world for fuelwood production. The various species of Eucalyptus—all native to Australia—have adapted to environments as diverse as the cool highlands of the Andes and the moist equatorial lowlands of Amazonia. Its adaptability, drought resistance, rapid growth, and regenerative ability explain its popularity. In Brazil, where annual yields average 12 tons per hectare, Eucalyptus is cultivated for charcoal and for methanol production. There and elsewhere, Eucalyptus cultivation is likely to expand dramatically since manufacturing 300,000 gallons of methanol a day, for example, requires planting 72,000 hectares of Eucalyptus each year for feedstock.[45]

Next in importance among the species with widespread potential is the leucaena tree. Leucaena—with such regional aliases as the Hawaiian giant, koe haole, or ipil-ipil—is a native of Mexico. One of the world's fastest-growing trees, it can grow 20 meters tall in six years. A leucaena plantation can annually yield up to 50 tons of wood per hectare, five times the average of cultivated pines in temperate regions. Leucaena's root nodules also replenish the soil with nitrogen—a boon in agroforestry schemes. In several Southeast Asian countries, it provides shade for coffee and cacao groves. In northern Australia, leucaena is intercropped with pangola grass to make nutritious fodder for cattle.[46]

Tree plantation schemes do entail potentially high ecological costs. The continuous removal of trees chosen for fuel value will probably deplete soil nutrients more rapidly than traditional silviculture combined with residue removal does. In short-cycle energy plantations, cutting takes place every five to ten years (compared to thirty to one hundred years in traditional commercial forests). Then, too, while the nutrient drain is minimal when stems and leaves are left on the ground, in short-rotation energy plantations the younger and more mineral-rich trees are removed.[47]

Monocultural (one-species) forests also tend to need extra pesticides to combat the diseases and insects usually held in check by the more complex ecology of natural forests. Monocultural fuel plantations also fail to provide habitats for the thousands of plants and animals that inhabit natural forests. Simply removing dead wood from forests takes its toll on the likes of owls and woodpeckers, the natural enemies of rodents and harmful insects. Such birds nest in cavities of old or damaged trees that are prime targets for woodchip machines and for firewood scavengers. Southern Brazil's pine tree plantations have been called "ornithological deserts" by Helmut Sick, a leading Brazilian bird authority.[48]

Both ecological and economic factors point toward the superiority of multiple-use, multiple-species forestry over one based on monoculture. Much more research is needed to design forestry practices capable of meeting rising demands for timber, pulp, and energy on a sustainable basis. Until this knowledge is obtained and put to use, energy should be extracted from the world's virgin forests or tree plantations only cautiously.

Reforesting the Earth

For the foreseeable future, the most important wood resource challenge will be to plant trees and better manage forest lands in populated areas. Both where firewood shortages loom and where wood use is rising, trees are being cut but not replanted, used but not cared for. In populated rural areas near markets, soil erosion and flooding are the upshot. The failure to plant and care for trees in these fertile lands stands as a major barrier to the widespread use of wood fuel.

While economic and environmental forces limit the use of remote virgin forests and commercial tree farms, the barriers to greater wood harvests closer to home are social and political. While many wilderness areas are publicly owned and large corporations own most tree farms, ownership of those neglected forests is distributed among millions of people, few of whom see trees as an important resource and fewer of whom have the skills to husband the forest. Where the landless poor rely on wood, those who go to the expense of planting trees have no assurance that they will ever harvest them. New institutions—village woodlots, forest-owner cooperatives, and technical extension services—are the keys to turning these neglected lands into permanent wood fuel resources.

Before the fuelwood problem became widely recognized in the 1970s, many developing countries had forestry agencies that managed forest lands and replaced trees cut down for timber and pulp uses. Traditional forestry of this sort focused on the commercial exploitation of the forests for export, not on the wood needs of the rural poor. Some countries also planted trees around villages where fuelwood needs pressed hardest. But with few exceptions, these measures did not halt the loss of woodlands. Newly planted trees seldom remained in the ground long enough to mature: They were either torn from the ground by desperate villagers and used for cooking or eaten by livestock. Gradually, foresters realized that villagers had to take part in tree-planting efforts if trees were ever to take root. Now traditional forestry practices are being supplemented by "community forestry," which emphasizes village participation in the planting of small woodlots to meet local fuel, forage, and timber needs.[49]

Community forestry breaks with traditional "production forestry" by emphasizing the use of trees for multiple purposes and the integration of tree growing with agriculture. Whereas traditional forestry concentrates on monocultures and closed forests, community forestry tackles chronic shortages of food, fuel, and jobs. Yet this approach is not wholly modern. Intercropping trees with crops is a traditional practice in some parts of the Third World. In Malaysia and Indonesia tall trees valuable for wood are intercropped with coffee, tea, and spice bushes that thrive in shade. In densely populated Java 81 percent of the fuelwood comes from trees planted on the margins of agricultural land. Variants of agro-forestry include the Combretum/rice culture in Southeast Asia, the gum Arabic tree–fallow system in Sudan and Ethiopia, and the coffee/-laurel system used in Central America. Elsewhere, trees planted along field boundaries and irrigation channels break the wind and supply fuelwood. In some traditional agro-forestry systems, trees and food crops contribute essential nitrogen to the soil.[50]

Over the last decade community forestry has made a name and place for itself. Dozens of national governments, international assistance agencies, and appropriate technology groups have started reforestation schemes with differing degrees of local participation, integration with agriculture, and employment of exotic species. Some such initiatives have enjoyed complete success, others total failure. In any event, reversing global deforestation means applying the lessons from these programs on a vastly larger scale.[51]

China and South Korea have most successfully mobilized villagers to plant and care for enough trees to make a difference. Despite admitted false starts and regional setbacks, Chinese officials tell visitors that tree cover in China has grown from 5 percent in 1949 to 12.7 percent in 1978, an increase of 72 million hectares. Outside observers with less information but less reason to exaggerate estimate that between 30 and 60 million hectares have been reforested. Either way, the accomplishment is herculean—a tribute to strong central political support and mass mobilization of village communities.[52]

Smaller but more rapid and thorough has been South Korea's reforestation effort. Before 1973 all attempts at reforestation had failed. Then a new approach emphasizing local participation was launched. Village committees with locally elected leaders were set up and charged with getting private landowners to plant trees on their lands. Since 1976 some 40,000 hectares per year have been planted, and by 1980 one-third of the national land area was covered with young trees.[53]

Village-based tree-planting efforts in India have been less successful. In Gujarat an ambitious reforestation effort has met with only partial success. By 1978 about 6,000 of the state's 17,000 kilometers of roads and canals were lined with new forests planted by hired labor, but the state's attempts to implement the social forestry goals set forth by the Indian government in 1973 have been less rewarding. Efforts to establish wood fuel lots in Africa have fared even more poorly despite the critical shortage of firewood many African nations face. In one World Bank–funded project to plant 500 hectares of trees in Niger, the villagers pulled the seedlings out and allowed uncontrolled grazing in newly planted areas.[54]

Whether village woodlot programs work depends heavily on how well the social structure works. In Korea and China the difference between one villager's wealth and another's usually is small. In most Indian villages caste and economic divisions are great, and cross-caste cooperation is rare.

Land-ownership patterns also affect the success of village woodlots in many areas. Semimigratory Nigerian tribes must leave woodlots unsupervised much of the year. Elsewhere, the nationalization of land has weakened the villagers' sense of responsibility for the soil and their claim to the fruits of their labor. In Tanzania, where the land became state-owned in 1963, farmers do not know if they will reap tree crops eight to ten years hence. In Nepal the government denationalized some forest lands once it became clear that nationalization had contributed to the abandonment of village woodlots. The reluctance of Gujarat's villagers to use woodlots has been attributed to uncertainties about who controls the village commons. The central government directed the commons to be used for woodlots, but villagers fear that the government may authorize some other use before the trees mature.[55]

The importance of social cohesion cannot be ignored in village reforestation projects either. Although attributing the Chinese and South Korean accomplishments to the "Confucian tradition" is simplistic, this explanation does contain an element of truth. Tanzania's woodlot program is modeled after the Korean one, but tribal affiliations in Tanzania hinder common action for such a nontraditional activity as forestry. In many wood-short areas of the Sahel, tribes have only recently, and with great difficulty, given up pastoralism for sedentary agriculture. In such communities social and cultural reorientation—no matter how difficult—may be necessary before woodlots take root.[56]

Rising awareness of the fuelwood crisis in the developing world has motivated a few national governments to act. India's new five-year plan for 1981–1985 commits 1.5 billion rupees (about $165 million) to village energy plantations and biogas units, a rise from almost nothing in the previous five-year plan. These outlays are the first step in reaching India's goal for the year 2000—using biogas and fuelwood to replace all the oil used to power pumps and agricultural machinery, 50 percent of the kerosene used for cooking, and a quarter of the oil and coal used to generate low to medium temperature heat in industry.[57]

The shift in emphasis toward forestry projects aimed at fuelwood production, agro-forestry, and watershed protection has also been marked at the World Bank. Although it budgeted almost nothing in the early 1970s for these activities, the Bank will loan about a billion dollars between 1980 and 1985. Far more is needed to establish adequate village wood fuel schemes, but the World Bank loans will launch critical efforts in various nations and climates.[58]

Increased support and attention notwithstanding, reforestation efforts in the developing world still lag far behind need. Experts who met in Rome in 1981 to advise the United Nations on world energy needs projected that worldwide reforestation efforts would have to increase by a factor of ten. The group estimated that Afghanistan and Ethiopia needed to increase replanting to fifty times current levels; India, fifteen times; and Nigeria, ten times. The Club du Sahel estimates that tree planting must increase in Africa's Sahel by fifty times to meet the needs of people there over the next twenty years. Globally, the group called for spending to double within five years to $1 billion annually, with roughly equal amounts coming from the World Bank, bilateral aid, and developing countries themselves.[59]

In much of the Third World, extensive planting could check widespread and serious ecological problems. Floods and erosion follow deforestation as day the night: Topsoil accumulates as silt and mud in river beds; water overflows banks, inundating cities and fields. In 1981 severe flooding in China's Sichuan province left 753 people dead and 1.5 million homeless—disturbing casualties of deforestation in the Yangtze River basin's upper reaches. The accelerated sedimentation of reservoirs is also drastically shortening the useful lives of dozens of dams in developing countries—seventeen in India alone.[60]

As in developing countries, the most important underutilized part of the wood resource base in the United States and Canada is in the hands of small landowners. Currently, 58 percent of U.S. forestlands is owned by about 3 million small private landholders, few of whom treat their trees as an economically significant resource. Moreover, small private landholdings are concentrated in the East, where potential markets are greatest and growth potential highest. Since the average forest parcel is shrinking as old farms and estates are broken up, few landholders could wrest enough profit from wood sales to justify their cutting, selling, and replanting their trees. Nor are many likely to remain owners long enough to reap the benefits of investing in timber stand improvement. According to a recent forest industry estimate, only one out of nine privately owned acres of nonindustrial forest harvested in the United States is being purposefully regenerated. Instead the rising demand for wood fuel is taking its toll primarily on poorly managed lands. In New England many forest landowners cut the wrong trees so as to make a quick profit—a surefire recipe for a decline in productivity.[61]

In North America the key to maximizing forest productivity may be creating forest cooperatives composed of private landholders who hire a professional forester to manage their lands and to oversee the thinning, cutting, and sale of wood from many contiguous forest parcels. So far, the 165 tree cooperatives operating in the U.S. have dramatically increased the earnings and productivity of previously neglected lands.[62]

Another prototypical initiative with promise is the New England Fuelwood Pilot Project. Under this two-year-old U.S. Department of Agriculture program, landowners receive both technical and financial help in evaluating tree stands, constructing access roads, and marking trees for cutting. For a net cost to government of $1.4 million, the program has brought 20,000 acres under better management and has displaced the need for 20 barrels of fuel oil each year for every acre of forestland treated—some 400,000 barrels overall. Expanding this program to cover all small private land parcels would cost less than $50 million a year, a bargain considering the payoff.[63]

The Wood Energy Prospect

Already high, wood energy use will almost certainly rise over the next two decades. In industrial countries the use of wood is likely to increase by about 50 percent by the year 2000 to approximately 10 exajoules. Fuelwood use in developing countries will increase more slowly since demand is already pressing against supply in many regions: Use will probably increase by no more than one-third over the next twenty years to 38 exajoules. In all, global fuelwood use will reach around 48 exajoules in 2000, compared to 35 exajoules today.[64]

The question more important than demand is supply. How will a 40 percent increase be achieved? With extensive reforestation, fuelwood use could continue upward after 2000, reaching the global potential of over 100 exajoules by mid-century. Without such programs, fuelwood use could plummet after the year 2000 as the resource base begins to erode.

The brighter prospect, of course, is wise management. Properly tended, commercial and small forestlands could probably yield three times as much fuelwood without resorting to short-rotation or to cutting virgin forests. Yet productivity must be raised gradually. A sudden rise in demand for wood for fuel—through crash programs to accelerate wood-to-methanol conversion or a boom in the use of inefficient wood stoves—could trigger widespread deforestation and send timber and pulp prices soaring. And whether technological improvements in tree breeding and a more sophisticated understanding of nutrient cycles will allow further sustainable increases remains to be seen.

These short-term imperatives to plant more trees could in the years ahead be joined by a global environmental one. Increased use of wood for energy could take on added appeal as the search for a tonic to the carbon dioxide (CO_2) released from fossil fuels becomes more urgent. Because trees absorb CO_2 and release oxygen, they are one of the few carbon sinks humans can quickly alter—important since a net increase in the standing stock of wood could slow down the "greenhouse effect." Physicist Freeman Dyson estimates that fast-growing poplar planted over an area the size of North America could absorb enough CO_2 to halve the annual build-up. The greening of Earth may thus emerge as a priority of governments properly fearful of disruptive climatic change.[65]

How this wood is used will be as important as how much of it is used and how it is obtained. A concerted effort is needed to increase the efficiency with which wood is burned, especially in developing countries where the amount of usable energy obtained from wood could be tripled by substituting wood gasifiers, charcoal, and efficient cook stoves for open fires. The reservoir of energy literally going up in smoke in countless open fires is far greater—and more cheaply harnessable—than the entire consumption of fossil fuels in many developing countries. Modernizing, not replacing, wood use is a far more critical national goal for most developing countries than the purchase of more nuclear power plants, oil refineries, or coal mines. Such modernization could take pressure off dwindling forests and give reforestation a chance to catch up with demand. The pull of market forces, the availability of credit for small manufacturers and users, technical extension and demonstration, and mass education campaigns will all be needed to accomplish this transformation of rural wood-burning practices.

By the same logic, wood should eventually be used in its most productive form, probably methanol. Even with improvements in combustion efficiency, direct combustion is likely to become relatively less economical for heating and electricity generation than solar collectors or photovoltaics. And biogas and wood gas could replace wood in cooking and small industry. Of course, the timing of the shift away from direct combustion and toward the greater use of liquid fuels will vary by country and depend on how fast oil prices increase, but will, probably be under way by the 1990s in most countries. Because most developing countries use so little petroleum and use so much wood inefficiently, they could make the shift to methanol-powered transportation systems first. These countries have only small fleets of gasoline-burning automobiles to retrofit, and most of their trucks and buses burn diesel fuel, which makes it easier to adjust them to use methanol.

The impact of such a strategy on the energy picture of one developing country, India, reveals the strategy's potential. According to Amulya Reddy, if biogas digesters were used to meet domestic cooking needs, some 130 million tons of wood currently burned for cooking could be converted into enough methanol and wood gas to power all trucks, buses, and irrigation pumps in India. Since producing wood gas and charcoal involves destructive distillation of wood—the first step in methanol production—these cleaner, more efficient forms of wood use pave the way for methanol. And since two-thirds of the oil India imports is used in trucks and buses, the large amounts of money currently leaving the country to finance oil imports could be diverted to the construction of methanol plants.[66]

In some industrial nations, the prospects for a wood alcohol strategy are also bright. While Europe, Japan, and Australia do not have enough forestland to supply substantial quantities of wood energy, the United States, Canada, and the Soviet Union do. The energy needs of the large automobile fleets in the United States and Canada are gargantuan, but long-term potential is also great. By 2020 these countries could derive as much as 15 exajoules of wood energy in the form of methanol.

This ability to use wood alcohol to supplant increasingly expensive petroleum links the problems of the rural and urban sectors. Indeed, solving urban energy problems will require transforming rural energy-use patterns. Efforts by some Third World countries such as Brazil, Kenya, or the Philippines to ignore the subsistence sector's energy crisis and to produce electricity or liquid fuels from wood could well exacerbate the far more serious rural energy crisis and prove uneconomical to boot. As Gandhi said, the Third World will march into the twentieth century on the back of a transformed rural sector or not at all.

Despite wood's potential to alleviate societies' dependence on scarce petroleum, wood has been passed over in the energy policy debates of the 1970s. In the northern industrial countries the wood renaissance of the late 1970s has gone largely unnoticed by energy policy makers, most of whom omit wood from national energy-use inventories. This myopia is particularly startling in the United States where in 1980 wood supplied more energy than nuclear power, which has received $47 billion in government subsidies. In the developing countries national planners took note of wood only when the mismatch between demand and supply gave rise to widespread hardship. Today it is obvious to those who look that wood should be front-and-center on the energy agendas of many nations throughout the world.[67]

7

Growing Fuels: Energy from Crops and Waste

Among renewable energy sources, fuels from plants—biomass—most defy generalizations. Beyond wood, hundreds of other diverse plant species can be converted to many different energy forms using a variety of technologies. Some biomass energy sources are as old as history, while technologies for using others are emerging at the frontiers of advanced research. The technical feasibility, economics, and environmental impacts of using some biomass resources are common knowledge. A shroud of uncertainty hangs over others.

In one way, though, all biomass sources are alike. None needs expensive manmade collection systems to gather the sun's energy or costly storage systems to compensate for the intermittent nature of solar radiation. Moreover, all are versatile. Biomass feedstocks can be processed into liquid, gaseous, and solid fuels.

Widely used today, biomass energy promises to be even more widely used in the future. In the rural Third World, wood, crop residues, and dung are the major energy sources. Urban refuse containing biomass supplies energy in many cities. And some energy supplies are obtained by converting waste to methane and growing crops especially to produce alcohols.

All energy contained in plants comes from the sun. Plants convert about 2 percent of the energy in light into chemical energy via photosynthesis. In photosynthesis plants absorb atmospheric carbon dioxide, free the oxygen, and build living matter with the carbon. In the most biologically active 1 percent of the earth's land area, plants every year capture and store about 530 exajoules of energy, 50 percent more than annual world energy use. This energy is the foundation of the food chain that sustains life on earth. It can also be tapped for other human uses.[1]

How much of this energy can be harnessed economically and safely can be determined only by research and a close examination of the earth's biomass resources.

The Ethanol Boom

The oil crisis of the 1970s triggered a global scramble to find new sources of liquid transportation fuel for the more than 400 million cars, trucks, and tractors in use worldwide today. Several governments began underwriting ethanol produced from corn and sugar crops. The two largest efforts, those in Brazil and the United States, have concentrated on producing ethanol to blend with gasoline and sell as "gasohol." (Ethanol or ethyl alcohol is found in all alcoholic beverages.) Attempts to use pure ethanol and oil crops on a commercial scale are afoot in several countries. Yet alcohol fuels contribute significantly to the energy supply equation in only a few nations. Their wider use awaits improved economics, more efficient conversion techniques, and cheaper feedstocks.[2]

The use of ethanol as a motor fuel is almost as old as the automobile itself. Fearing an impending oil shortage and hoping to stimulate demand for farm products, automobile pioneer Henry Ford promoted gasohol during the early twentieth century. During the Great Depression the Chemurgy Movement, a group of scientists and farmers trying to buoy depressed agricultural prices by developing industrial markets for crops, also favored alcohol fuels. Use of ethanol as a gasoline extender was widespread in the U. S. Midwest, with more than 250 dealers of so-called "Agrol" in Nebraska alone in 1935. During the 1930s and 1940s ethanol-gasoline blends were used in more than forty countries, but falling prices and abundant petroleum supplies wiped out the nascent alcohol fuels industry after World War II.[3]

The use of ethanol as a gasoline additive particularly appeals to oil-importing nations because it can immediately reduce gasoline consumption. Doing so requires no major retooling of the automobile engine, no new fuel-storage and distribution system. Blending ethanol with gasoline also maximizes the energy value of both fuels since ethanol boosts gasoline's octane level. Mixed with gasoline, a gallon of ethanol provides almost twice the energy that straight alcohol would.

Part of ethanol's appeal is the established character of alcohol-production technology—the foundation of the alcoholic beverage industry. Ethanol (ethyl alcohol) is produced either directly from sugar by fermentation or from starches that are first converted to sugar and then fermented. Ethanol can be derived from three main categories of food crops: sugar crops, such as sugarcane, sugar beets, and sweet sorghum; root crops, mainly cassava (manioc); and all major cereals. The cost of the feedstock is usually the largest cost in ethanol production.[4]

The prospect of substituting alcohol for petroleum has generated intense debate about the "net energy balance" of corn-alcohol production. Some critics of alcohol fuels argue that producing alcohol requires more energy than the alcohol contains. Part of the confusion arises from a failure to distinguish between the energy yield and the liquid fuel yield. The U.S. Department of Energy found that producing 100 Btu's of ethanol from corn required 109 Btu's—44 Btu's to grow the corn and 65 Btu's to produce the alcohol from it. If the energy value of the byproduct, distillers grain (a fermentation residue that can be fed to livestock) is added in, there is a slight net energy gain of 5 percent. If alcohol is produced in an oil-fueled distillery, however, there is no net gain in liquid fuel. But if the distillery is powered by coal, wood, or solar energy, then at least 2.3 gallons of liquid fuel would be produced for every gallon consumed. In short, only properly designed alcohol production based on abundant solid fuels, waste heat or renewable resources will displace liquid fuels from petroleum.[5]

Technological advances could well improve the efficiency and thus the economics of alcohol production. By far the most energy-intensive aspect of alcohol production involves separating alcohol from water through distillation. The prospect of using membranes that permit the passage of alcohol but not water has excited alcohol scientists. Dr. Harry Gregor of Columbia University calculates that by using special plastic membranes producers could bring down the energy cost of recovering pure alcohol from fermentation liquids to about 0.6 percent of the alcohol fuel value. Other investigators are exploring the use of dessicants, solvents, and molecular sieves as alcohol purifiers.[6]

Besides investigating new ways to improve traditional sugar-to-ethanol technology, scientists are refining technologies to convert cellulose into ethanol. In principle, this is easy enough since cellulose is nothing but complex sugars bound together by a substance called lignin. Cracking lignin's hold on these sugars is not easy, however. One very inefficient chemical process, acid hydrolysis, is used in fifty Soviet plants to convert wood chips into sugars that are fed to protein-rich microorganisms that are in turn fed to cattle. Use of acid hydrolysis to produce sugar for fermentation to alcohol is not currently economical, but would become so if conversion technologies were improved. Other cellulose-to-ethanol technologies employing bacteria, viruses, and enzymes are being scrutinized by Brazilian, American, and Canadian scientists.[7]

Among the countries producing ethanol for fuel, Brazil is the unquestioned leader. Forced to import 85 percent of its oil, Brazil launched its alcohol fuels program in 1975. The goal is to attain self-sufficiency in automotive fuel by the century's end. Between 1975 and 1980 alcohol production leapt from 641 million liters to almost 4 billion liters (1.3 billion gallons), and the number of alcohol distilleries jumped from 25 to 300. Aided by government subsidies and cheap loans for car purchasers, the Brazilian subsidiary of Volkswagen produced over 260,000 automobiles designed to run on pure alcohol. In all, over $2 billion in government subsidies have been invested in alcohol production and consumption, much of it raised from taxes on petroleum products. By 1985 Brazil hopes to be producing 10.7 billion liters of fuel alcohol.[8]

Although the primary goal of the Brazilian alcohol fuels plan is to reduce oil imports, the government also hopes that the program will create jobs, reduce the flow of people to the cities, improve income distribution, and promote a more regionally balanced economy. According to World Bank estimates, the program will create about 450,000 jobs between 1980 and 1989. A sore point, however, is that the plantation-style cultivation of sugarcane and the construction of large distilleries may have exacerbated already unconscionable income disparities in many rural areas. (Brazil's richest fifth has thirty-six times more income than the poorest fifth.)[9]

Brazil's alcohol fuels program could also drive up food prices. According to a 1975 World Bank study one-third of all Brazilians have barely adequate nutrition. If Brazil were to achieve self-sufficiency in automotive fuels through sugarcane alcohol production, 2 percent of Brazil (an area half as large as all cultivated cropland in the country) would have to be planted in sugarcane. Producing 10.7 billion liters of alcohol by 1985 will require the equivalent of 10 percent of Brazil's cropland. Already at least 600,000 acres once devoted to rice, wheat, and pastures have been planted in sugarcane.[10]

Brazil's alcohol fuels program also faces economic problems. Originally intended to raise depressed sugar prices and export earnings, the program may in one sense have succeeded. By generating new demand it brought sugar prices up from $200 per ton in 1975 to around $800 in early 1981. At this price, exporting sugar and buying oil makes more economic sense than producing alcohol to displace imported oil.[11]

Another factor at play in Brazil's alcohol fuels venture is pollution. For every liter of alcohol produced, Brazil's distilleries create 13 liters of "swill," a toxic organic pollutant. If Brazil meets its 1985 ethanol goal, distilleries will produce 35 billion gallons of swill, double the amount of sewage Brazil's 126 million people produce. Because substantial investment in treatment facilities may be needed to avoid severe water pollution, many experts doubt that Brazil can reach the ambitious 1985 goal.[12]

Brazil's government has already approved enough distillery projects to produce 8.3 billion liters of alcohol, but will there be enough sugarcane to support production at that level? As of 1980, 2.5 million hectares of sugarcane were under cultivation, but 4.5 million more will be needed to meet the 1985 goal.[13]

The United States also has been driven by heavy dependence on foreign oil to embark upon an ambitious alcohol fuels program. In 1978 every gallon of gasohol containing alcohol from nonpetroleum sources was declared exempt from the 4¢ federal gasoline excise tax. Twenty-two states also partially or wholly exempted gasohol from state gasoline taxes. In some the combination of federal and state tax incentives exceeds $1 per gallon for alcohol used as automotive fuel. In 1980 President Carter announced a goal of producing two billion gallons of ethanol by 1985. Congress then took the initiative a step farther, proclaiming a goal of 10 billion gallons by 1990. If the 1990 goal is met, alcohol will account for just under 10 percent of the 110 billion gallons of gasoline consumed in the U.S. in 1980, a quantum increase from the less than 100 million gallons of alcohol produced in 1979. By late 1980 some 2,500 retail dealers were selling gasohol. In 1974 none did.[14]

The U.S. alcohol fuels program has centered on corn-based ethanol, largely because corn is so abundant and the farm lobby is so powerful. The U.S. farm community sees ethanol fuel as a way to increase demand for corn and, hence, corn prices. Yet corn has other highly valued uses and represents only a tiny share of the U.S. biomass potential. According to U.S. Office of Technology Assessment estimates, ethanol from all grains probably cannot supply more than 6 percent of the biomass energy potentially available in 2000. Yet ethanol from corn has absorbed well over half the federal funds earmarked for energy-to-biomass projects.[15]

Growing interest in alcohol fuels in the U.S. has kindled a lively debate over the relative efficiencies of small on-farm and large central plant production of alcohol. Large plants seem best suited to make the final, highly energy-intensive distillation of 160- or 180-proof hydrous (water-laden) alcohol into the 200-proof waterless variety blended with gasoline. But transporting enough feedstocks for large plants is expensive, and mammoth operations are vulnerable to drought-induced shortages and high prices. Smaller on-farm plants could handle the initial processing, but unfortunately the principal alcohol subsidy—exemption from the federal gasoline excise tax—is available only to farmers who sell their product on the market.

Some 95 percent of fuel ethanol produced in 1980 thus came from six companies, and thousands of on-farm producers got no federal subsidy for their efforts.[16]

If the Brazilian alcohol fuels program threatens the food supply of Brazil's poor, the U.S. program could send the prices of grain and grain-fed meat up all over the world. Since mid-century, the U.S. and Canada have increasingly dominated the world grain market. While distillers grain is a protein-rich cattle feed, the feed market can absorb only so much of this byproduct. Then too, most of the corn's calories are lost in alcohol production. Economist Fred Sanderson of the Brookings Institution predicts that ethanol production above 4 billion gallons a year will drive up corn prices. If gasoline prices in the United States reach $3 a gallon, as they have already in many countries, distillers could afford to pay $6 per bushel for corn without subsidies and credits. At these prices, U.S. corn —a staple of human consumption in some parts of the world and a source of animal feed in many others—would more than double in price.[17]

Despite its immense popularity in the corn belt, grain-based gasohol is unlikely to radically alter the U.S. liquid fuels picture. Long before gasoline price increases make large-scale use of ethanol attractive without federal subsidies, demand for gasoline is likely to plummet due to conservation. For the foreseeable future, investments in reducing fuel use will be cheaper than new fuels. The overriding fact is that current U.S. consumption of liquid transportation fuels is too large to be put on a sustainable basis.[18]

Long-term prospects for the Brazilian gasohol program are considerably better since Brazil can produce more but needs less liquid fuel than the United States does. (See Table 7. 1.) Furthermore, Brazil has substantial quantities of uncultivated land, whereas the United States does not. But in neither country can the prospects for the present goals be described as bright. In both, using diverse feedstocks and improving production efficiency holds the key to longer-term success.

Table 7. 1. Gasohol Prospects

	1980 fuel alcohol production	*1985 goal, alcohol production*	*1980 total gasoline use*	*1985 goal as percentage of 1980 gasoline use*
		(billion gallons)		*(percent)*
Brazil	1.3	5	10	50
United States	.250	2	100	2

Source: Worldwatch Institute from U.S. and Brazilian government documents.

Exploiting a Many-Sided Resource Base

The limited longer-term prospects for the Brazilian and American ethanol-from-sugar and ethanol-from-corn programs have stimulated a thorough search for better energy crops. The ideal one would grow well on marginal land, require little energy or capital for conversion, thrive without fertilizers, and protect the soil from erosion. While research in this area is unsystematic and underfunded, a surprising number of promising plants have been found. Among them are cassava, Jerusalem artichokes, coconuts, and sunflower seeds. Research efforts are also under way to determine whether some crops can be grown on arid lands, in waterways, and in the oceans.

In this rush the environmental stresses from the large-scale cultivation of any one species cannot be ignored. Thus, no single ideal energy crop has been or is likely to be found. Instead, energy farming will have to rely upon a much more diverse plant base than contemporary agriculture does.

With sugar prices rising, Brazil has already begun to use cassava as an alternative feedstock for alcohol fuels. Unlike sugarcane, cassava can be grown on marginally productive land, of which Brazil has plenty, and stored in tropical climates without decaying rapidly. Sugar's advantage over cassava, however, is the ease with which the cane waste (bagasse) can be burned to distill the fermented alcohol. More important, cassava is a staple in the diet of poor Brazilians, so diverting it to energy production could reduce food supplies.[19]

Researchers in several nations have identified plant oils that can substitute directly for petroleum-based diesel fuel. A boon is that such vegetable and palm oils are ready for use without energy-intensive distillation. Simple and inexpensive crushers alone can turn some oil-bearing seeds into fuels for on-farm use.[20]

Brazil's effort to replace gasoline with ethanol has been so successful that the diesel fuel needs of Brazil's large truck fleet now account for one-third of Brazil's oil use. Consequently, Brazil hopes by 1985 to plant 4 million acres with dende palms —the oil of which can be mixed with diesel fuel or used alone in conventional diesel engines with minor modifications—to meet 10 percent of the country's diesel fuel needs. The success of this plan, is far from assured, however, since large-scale cultivation of the dende palm has never been attempted and the trees take five years to mature.[21]

Efforts are under way in North Dakota and South Africa to use sunflower seed oil in diesel engines. South African farmers and the North Dakota State Extension Service have successfully tested sunflower seed oil in farm equipment, and studies indicate that corn farmers could power all their farm equipment with oil grown on 10 percent of the land they cultivate. A selling point is that sunflowers can be grown on poor land with minimal amounts of water. A drawback is price—today sunflower oil costs roughly twice as much as diesel fuel.[22]

The Philippines is successfully substituting coconut oil for diesel fuel. Cocodiesel, as the one-tenth coconut oil mixture is called, burns well in standard diesel engines, though start-up in cooler weather is sometimes a problem. In all, cocodiesel is ideally suited for use in ships, factories, and trucks. Because almost a third of the country's people depend on coconut oil for a living and the world price has dropped, the government is eager to boost coconut oil prices. As with corn in the U.S. and sugar in Brazil, the cocodiesel program is an attempt to boost agricultural income by creating a new market for farm products.[23]

Grasses also hold considerable potential as liquid fuel feedstocks. Although grasses are needed to support meat- and milk-producing animals, they can be grown on marginal soils with few energy inputs and harvested without destroying all ground cover. Like wood, they can be gasified or converted to methanol for use in crop dryers and irrigation pumps. Grasses also add nitrogen to the soil, so a wise strategy would be to intercrop them with nutrient-depleting food crops. According to the U.S. Office of Technology Assessment, an estimated 1.4 to 2.9 exajoules of energy in the near-term and 5.3 exajoules by 2000 could be produced from grasses.[24]

The search for suitable biomass feedstocks has also focused on crops that grow on arid land. Some desert plants contain complex hydrocarbons almost identical to crude oil, and they do not have to be fermented to yield usable energy. Along with the jojoba bean and the copaiba tree, the gopherweed—a variety of milkweed that grows wild in the American Southwest—has attracted the most interest. University of Arizona scientists have found that one acre of gopherweed can annually produce nine barrels of oil, a yield that plant breeders expect to increase to at least twenty barrels per acre at a cost of about $20 per barrel. Jack Johnson, director of arid lands studies at the university, estimates that gopherweed farms three times the size of Arizona's current agricultural acreage would require no more water than farming now takes and could meet all the state's liquid fuel needs. For the millions living in poverty on the world's arid lands, the gopherweed could provide badly needed income and employment.[25]

Another strategy for producing fuels from biomass is to mix new energy crops with traditional food crops. Using land for such multiple purposes—called polyculture—could over the long term reduce food-fuel competition and enhance soil protection. This will mean turning to agro-forestry techniques such as those increasingly employed in developing countries. In the United States a good agro-forestry bet is the honeylocust, a leguminous tree that grows well in various climates and on rocky or easily eroded land. In Alabama one acre of trees has yielded 8,500 pounds of pods each year with a sugar content of up to 39 percent. Other energy crops can be grown under honeylocusts. Another intriguing prospect is the mesquite tree, which produces sugar-rich pods as well as wood in dry regions of Mexico and the U.S. Southwest.[26]

Besides using the output of the world's crop and forest lands for energy, sunlight falling on the earth's waters can be collected for human use by various fast-growing plants. Two extraordinarily prolific aquatic plants, the water hyacinth and ocean kelp, have tantalized researchers with the prospect of turning lakes and oceans into biomass energy plantations. Although aquaculture and mariculture are still infant sciences, the long-term prospects for harvesting aquatic plants for energy are great since these plants do not compete with food crops for fertile land, fresh water, and fertilizers. Water hyacinths can convert polluting sewage wastes into protein-rich biomass even as they generate energy, and kelp and other seaweeds have long been used for food and chemicals in Asia.

Yet, aquatic plant cultivation is no surefire economic proposition. Scientists estimate that a kelp farm covering 46,000 square kilometers—an area the size of Connecticut—would be needed to produce as much methane as the United States now uses, and while estimates of the cost of methane from kelp are still speculative, it is sure to be several times higher than current natural gas prices. Since kelp beds attract and sustain luxuriant fish populations, kelp's energy contribution may be part of integrated ocean aquaculture ventures producing seafood, energy, chemicals, and animal feeds.[27]

A first step in harnessing new biomass energy sources would be inventorying the earth's plant resources and their potential as energy producers. Of the hundreds of thousands of plant species on earth, only a few dozen are cultivated for food or fiber. Why assume the best food crops are also the best energy crops? Surveys of plants suitable for fast-growing firewood cultivation conducted by the U.S. National Academy of Sciences have located several underutilized species. Especially needed are inventories of tropical plants, since rain forest destruction threatens the survival of many unexamined species.[28]

Plant-breeding technologies successfully employed on food crops may also be capable of improving the energy yield of plants. In the United States, for instance, the per-acre yield of corn has been increased from 30 to 100 bushels in the past fifty years. But such yield increases cannot occur without a diverse genetic base, so the need for new energy sources is another reason to preserve the earth's threatened genetic resources.[29]

Recently developed techniques of gene splicing may also increase the energy productivity of plants. Instead of merely selecting and concentrating genetic information found naturally in a given species, recombinant DNA techniques will enable scientists to transfer the genes of one plant species to another, creating an entirely new species or endowing an existing species with new characteristics. Genetic engineering is still a budding science, but it could well revolutionize biomass-energy prospects. Of course caution is required since an error could have such severe environmental consequences.[30]

Agricultural Wastes: The Forgotten Asset

Most people view organic wastes from plants, animals, and humans as a nuisance. But such wastes contain enough energy to alter the energy picture in many agricultural areas. Where firewood is in short supply, animal wastes and crop residues are already being burned extensively to cook food and provide warmth. In India cow dung and crop residues account for 10 percent of the country's total energy supply and 50 percent of rural household energy. On a worldwide basis it is estimated that cow dung and crop residues supply the energy equivalent of 257 million metric tons of coal—2 percent of total world energy use.[31]

This gift of nature has its price. In many areas of the Third World soil quality and the productivity of agriculture are being undermined as more people turn to organic wastes for fuel. When crop and animal wastes are burned, most of their fertilizer value is lost, depriving the soil of nutrients needed to sustain plant life. In Bangladesh, where rice straw is being diverted from cattlefeed to stoves, fewer cattle can be supported so less manure is left on the ground to fertilize the soil. Worldwide, the use of livestock droppings as fuel is estimated to lower annual grain production by some 20 million tons, enough food to minimally nourish 100 million people.[32]

Fortunately, a simple biomass-conversion technology, the biogas digester, opens the way for developing nations to increase the energy value of rural agricultural wastes without incurring heavy costs. Cut down to basics, the biogas digester consists of an airtight pit or container lined with brick or steel. Wastes put into this container are fermented anaerobically (without oxygen) into a methane-rich gas of use in cooking, lighting, and electrical generation. The residue makes an excellent fertilizer, too. If they had biogas digesters, the rural poor would no longer have to confront the Hobson's choice of deciding between today's cooking fuel and tomorrow's soil fertility.[33]

By converting organic wastes into biogas, developing countries could simultaneously meet pressing needs for jobs, fertilizer, and energy. V. V. Bhatt of the World Bank estimates that in India 26,160 biogas digesters could produce as much fertilizer as a large coal-fired fertilizer factory, at roughly the same cost. The biogas digesters would provide 130,750 jobs; the coal-fired plant 1,000. The digesters would yield as much energy as a 250-megawatt power plant, while the coal-fired fertilizer factory would consume enough fuel to run a 35-megawatt plant.[34]

Another benefit of the biogas option is that biogas fermentation can prevent the spread of schistosomiasis and other diseases carried by human wastes since it kills the pathogens wastes contain. Digesters can produce enough valuable gas to defray the costs of latrines and water pipes, and they can reduce the odors that make latrine use unsavory for those who have long used the bush instead. Unfortunately, development-investment decisions are made by agencies with only one goal in mind. Too often, therefore, only one bird is killed with the stone of scarce capital resources.[35]

Biogas technology does have its share of bugs and breakdowns. Fermentation tends to stop in cold weather, and adding insulation so fermentation can go on year-round in cold climates adds to the cost of the systems. Another problem is keeping detergents, pesticides, and air out of the digesters. In all, though, the skills needed to build and operate a digester are considerably less than those needed to operate a diesel pump or a motorcycle.

The most important constraints to greater biogas use are social. Traditional taboos and customs concerning animal and human waste disposal are powerful disincentives. In Islamic countries prohibitions against coming into contact with swine limit the use of abundant animal wastes. In sub-Saharan Africa a taboo against handling wastes in general works against the adoption of biogas technology. In China, by contrast, the long-established practice of collecting "night soil" for fertilizer made introducing biogas units simple.[36]

Another thorn is that unless equity concerns attend technology transfer, biogas units can actually harm the rural poor and exacerbate the ecological problems poverty creates. In India the promotion of household rather than village-sized biogas digesters in rural areas has often worsened the plight of the very poor, who cannot afford single-family units and yet depend upon dung collected from the streets for cooking fuel. When the more affluent animal-owning villagers build "gobar" units, as the Indians call biogas digesters, the dung from their animals becomes a valued resource no longer shared with the poorest of the poor. Denied access to free dung, the destitute forage more firewood, thus worsening deforestation and soil erosion. Community-sized digesters—into which the very poor could put scavenged biomass in return for access to common cooking and washing facilities—would alleviate both problems.[37]

The potential to use biogas digesters in many Third World countries is great, but only China has applied the digesters widely. Chinese leaders began promoting the use of simple biogas digesters to combat rampant deforestation caused by firewood use, declining soil fertility resulting from burning crop residues, and pervasive rural air pollution from cooking fires. So far the Chinese have built 7 million biogas digesters—enough to meet the energy needs of 35 million people. Altogether, Chinese biogas digesters produce the energy equivalent of 22 million tons of hard coal. The government's goal of 70 million digesters by 1985 could be met since 70 percent of China's biogas digesters are located in Sichuan, and many other Chinese provinces have equal or greater potential for the use of biogas. Although the present Chinese leadership has decided that many small-scale rural projects are inefficient, support for biogas still runs high. Despite these pluses, China-watcher Vaclav Smil doubts that the 1985 goal will be met.[38]

China's formula for success has two elements. One is the mobilization of local labor and the use of local materials. The other is an aggressive government effort to transfer technology from universities and laboratories to the rural areas where 80 percent of all Chinese live. Some 200,000 Chinese villagers have attended one-month training courses on the essentials of digester technology and then returned home to supervise construction and teach others.[39]

Use of biogas generators in India has a long, checkered history. Since the late 1940s, the Khadi Village Commission, a government group attempting to implement Gandhi's ideas about village industry, has helped install over 75,000 biogas generators. But India does not have as many pigs as China—a key fact since pig wastes are easier to collect than those of roaming cows—and the Indian digester (made of steel) costs too much for the average Indian villager. Then too, repair and maintenance skills are scarce in Indian villages. Today only about half of the biogas digesters built in India are operating.[40]

Throughout the rest of the Third World interest in biogas technology is growing. The 1980 U.N. Industrial Development Organization Conference on biogas technology held in Beijing drew participants from twenty-seven nations. The Colombo Declaration of the Economic and Social Council for Asia and the Pacific endorses biogas as a priority development technology. India's five-year plan calls for building 500,000 additional digesters. According to the Indian Planning Commission, India generates enough wastes to operate 19 million family-sized units and 560,000 community-sized plants—enough to cut electricity consumption by 44 percent, coal use by 15 percent, and firewood by 79 percent. Small but growing programs are also under way in Nepal and Indonesia, where firewood shortages are particularly acute. Brazil's agricultural extension agency is reportedly redirecting resources so as to disseminate know-how to millions of people on the fringes of the monetary economy.[41]

Opportunities for generating biogas from animal wastes in the industrialized nations are limited primarily to large dairy farms and feedlots where wastes are concentrated and pollution has been a problem. In the United States the Mason-Dixon Dairy Farm annually converts 2.7 million tons of manure from

700 cows into $30,000 worth of gas. In the Philippines Maya Farms (the largest pig farm in Asia) gets all its power from methane generated by 15,000 pigs.[42]

The economics of feedlot conversion depends heavily on how much byproduct protein can be recovered from manure. Therefore, rising protein and natural gas prices will strengthen the already favorable economics in the years ahead. If all the waste from the 13.4 million head of cattle in feedlots in the United States were converted to biogas, enough energy to heat a million homes could be produced annually. This contribution would not substantially alter the national energy picture, but it could help the agricultural system become self-sufficient in energy.[43]

Another potentially significant source of energy from agricultural wastes is in the food-processing industry. Sugar refineries, animal slaughterhouses, canneries, and citrus processors generate mountains and lakes of otherwise troublesome wastes. These wastes can be burned directly, decomposed into biogas, or converted into alcohols. The economics of such waste-to-energy projects depends on such factors as the avoided cost of environmentally sound disposal, the volume of waste, the moisture content of the organic matter, and the market for the energy produced. By U.S. Department of Energy reckoning, four-fifths of U.S. agricultural processing wastes could be economically converted into half a billion gallons of fuel-grade ethanol each year.[44]

Detailed studies of the waste-to-energy potential point to regionally significant energy sources. Feasibility studies conducted by the State of New York Energy Office indicate that 2.5 million gallons of ethanol could be produced from the 1 billion pounds of whey generated by the state's cheese industry each year. Opportunities in the sugar industry are even greater. On the island of Hawaii sugarcane waste provides all energy for irrigation and cane processing, as well as over 40 percent of the electricity the 82,000 islanders use. Studies of the Nicaraguan sugar industry's waste indicate that between 26 and 35 percent of the nation's electricity supply could be provided using standard waste-processing technology.[45]

Energy from Urban Wastes

Not all of the bounty in refuse is found in rural areas. Much of the biological output of the world's forests, farms, and fisheries ends up in city dumps. In fact, an average ton of urban refuse contains about as much energy as 500 pounds of coal. And every year the average American throws away 1,400 pounds of trash, the average West German, 1,000 pounds. (See Table 7.2.)[46]

Table 7. 2. Energy Potential of Urban Waste

Area	*Urban refuse*	*Energy potential*
	(million tons per year)	*(exajoules)*
United States	160	1.9
Western Europe	130	1.3
USSR & Eastern Europe	90	.5
Japan	70	.3
Developing countries	100	1.1

Source: Worldwatch Institute estimates based on U.N. and World Bank sources.

In some urban areas finding environmentally sound disposal methods for voluminous wastes has become a major headache, even a crisis. Burying, burning, or dumping them at sea creates serious land-, air-, and water-pollution problems. Some cities have boxed themselves in with garbage. New York, faced with the problem of disposing of 22,000 tons of refuse a day, has no more land on which to bury it.[47]

Wastes do not have to be wasted. Attempts to derive energy from waste should take a back seat to recycling efforts—which always save more energy. Plastics, paper, and compostable organic wastes should be burned only as a last resort. And of course, separating nonflammable glass and metals from urban waste improves the performance of waste-to-energy plants.[48]

Europe and Japan have done by far the most to utilize the energy potential of urban wastes. In 1977 only 6 of the 262 municipal waste-to-energy plants in operation were located outside of Western Europe and Japan. Munich derives 11.8 percent of its electricity from garbage. Three huge plants in the Paris metropolitan area burn 1.7 million tons of waste per year to produce the energy equivalent of 480,000 barrels of oil. Tiny Luxembourg and Denmark are the world leaders in using urban waste for energy, with over half their total waste converted to heat or electricity. Japan has more plants (85) and more installed capacity than any other country. In Japan and Europe most waste-to-energy plants are cogenerators, serving district heating systems and providing electricity.[49]

The widespread use of urban waste to produce energy in several European nations and Japan predates the oil crisis. The population density of these countries makes converting productive farm and forest land into landfills and waste dumps an unaffordable luxury. Nor can the countries afford surface dumps that leach dangerous chemicals into the water tables and waterways, poisoning fish and fish-eaters. The Germans had produced electricity from a municipal incinerator as early as 1896, but the waste-to-power trend gained momentum in the 1960s with the realization that generating electricity with hot incinerator gases cooled the gas, enabling air-pollution control systems to work effectively.[50]

The United States has the largest potential for turning municipal wastes into energy. By 1990, the U.S. Department of Energy estimates, 200 million tons of solid waste and 15 million tons of sewage solids will be generated each year in the United States—enough to produce more than two exajoules of energy. Yet waste-to-energy technology has found limited application in the United States because open-land dumps have few environmental controls. Municipal waste-to-energy systems are economical only where governments stuck with refuse pay owners of such plants a "tipping fee" roughly equal to the cost of alternative means of disposal. One of the few successful plants in the U.S.—the Revco Plant in Saugus, Massachusetts—employs European technology to produce steam and electricity and depends for half its operating revenue upon a "tipping fee" equal to the cost of environmentally sound disposal.[51]

Still skirting the environmental challenges of landfills, the United States has nevertheless pioneered various advanced technologies designed to convert waste into liquid or gaseous fuel. Unfortunately, applying space-age technology to waste disposal problems does not necessarily solve them. As a rule, the most expensive and complicated plants have failed most miserably, partly because waste containing everything from cans of flammable liquids to discarded motors is hard on complex machinery. In Baltimore, Maryland, a plant opened in 1974 to convert a thousand tons of garbage per day into gas through pyrolysis has never worked more than eighteen days without breaking down. Periodically ravaged by exploding garbage, it runs at about half capacity, millions of dollars in repairs notwithstanding.[52]

In retrospect the failure of the ambitious Baltimore project can be laid to trying to do too many tasks at once and attempting to do with expensive machines what people do better. Instead of employing proven European technology that simply burns a relatively homogeneous waste stream that is usually separated in households and businesses, U.S. engineers hoped to turn a more varied waste stream into commercial-grade fuels.[53]

The barriers to turning urban wastes into an energy source are for the most part institutional. Solid waste disposal costs U.S. towns and cities over $4 billion a year—only schools cost more. Yet most local governments are financially strapped and reluctant to spend money on unfamiliar projects. On the other hand, a private investor would have to have ironclad assurances of the right to the garbage, an ample tipping fee, and a guaranteed market for the steam and power produced. Without those three promises, no such venture could work.[54]

Existing industries that use steam and electricity would appear to be logical builders of waste-to-energy plants. But few are large enough to use all a plant's output, and selling power to utilities is complicated and difficult. Electric utilities have also been reluctant to build waste-to-energy systems, in part because the optimally sized 1,000-ton-per-day plant produces far less power than the plants they use now. In Europe and Japan special municipal authorities have been granted the powers needed to get around these obstacles.[55]

Although most Third World city dwellers are poor and generate little combustible waste, the urban elites of these cities produce nearly as much waste as their Western counterparts do. The waste problem in many Third World cities—in Cairo, for example—is reduced because the poor make a business of recycling things discarded by the rich. Still, extensive underemployment in such cities means that no capital-intensive waste-to-energy plants should be built before recycling opportunities are exhausted. Building plants could worsen the plight of those who derive a living—even a precarious, unsanitary one—by picking through mountains of urban refuse. Indeed, when 30 of the 400 garbage-piled acres of Mexico City's Meyehualco dump caught fire in 1980, planners rejected the idea of building a modern energy-producing incinerator because doing so would have deprived five thousand scavengers of their livelihood. Mexico City's dilemma underscores the general threat to the poor posed by biomass and waste-to-energy systems that give commercial value to wastes that the poor depend upon but do not own.[56]

Although burning wastes in plants to extract energy is more environmentally sound than open dumping and burning, waste-to-energy plants have their share of environmental problems. Urban wastes contain plastics, metal foils and coatings, and chemicals that form noxious gases when burned. One plant in Hempstead, New York, employing simple European waste-to-energy technology had to be closed temporarily when deadly dioxin was discovered in its emissions. Burning wastes at high temperatures and using electrostatic precipitators can reduce harmful emissions, but such technologies are never completely effective. Then, too, residual ash from the plants is typically filled with heavy metals that must be handled as a hazardous waste. Overall, burning waste is more environmentally troublesome than recycling, but less so than dumping.[57]

Energy can also be drawn from urban wastes that have already been buried in landfills. As organic wastes in airless underground cavities decay, they release methane that can be collected by inserting pipes into covered landfills. In the United States fourteen such piped plants, most of them in California, are already operating. The world's first landfill methane-recovery system was built in Palos Verdes, California, in 1975. Currently, it meets the energy needs of 3,500 homes. This energy source is a now-or-never proposition since methane from landfills is lost if not tapped.[58]

Another potentially important source of urban energy is methane from sewage-treatment plants. In Delhi, India, 700 people recently switched from kerosene and charcoal to biogas produced from one of the city's large sewage-processing plants. If all the wastes from the city were thus processed, experts say, 20 percent of the household energy needs in the city could be met. In many urban areas, however, biogas production in sewage-treatment facilities is declining today because mixing industrial wastes with municipal waste kills methane-producing bacteria. The way out, some cities are finding, is to force industries to pretreat wastes.[59]

Despite the many institutional and social problems that plague their use, municipal waste-to-energy plants are well matched to the energy needs of cities. Unlike coal and nuclear plants, far-flung because of pollution and safety factors, waste-to-energy plants can be located near fuel supplies and near customers. By cogenerating electricity and steam for space heating, waste-to-energy plants get twice as much usable energy from fuel as do typical coal or nuclear plants.

Promise and Peril: The Plant Power Prospect

Realizing the energy potential of biomass without sacrificing other values requires sequencing biomass-development efforts carefully. The first step is to put present biomass uses on a sustainable base, simultaneously maximizing existing resources and laying the groundwork for more intensive exploitation in the future. Separating and recycling municipal waste, converting animal and human waste to methane, and brewing alcohol from spoiled crops can turn environmental liabilities into energy assets without further distressing the agriculture resource base. By substituting less-polluting, more efficient end-use technologies for direct combustion, this approach also alleviates environmental pressures. This first phase of biomass use will see industries, farms, and whole regions become less dependent on—even independent of—conventional energy sources.

In poorer rural areas where agricultural wastes are already highly valued and used, this approach is especially apt. In ecologically overtaxed countries biomass shortages and the constraints caused by falling soil fertility bar open-ended development. Yet burning less and returning more to the soil—the ideal option—is not realistic considering how many people depend on such wastes for energy. Thus, particularly where it is warm, the widespread use of biogas digesters makes most sense. China could have at least 100 million digesters in place by 2000, and the rest of the Third World another 200 million. These 300 million digesters could produce 2 to 3 exajoules of energy each year—less than 1 percent of world energy use but a critical 1 percent to a billion poor people with small but pressing energy needs.[60]

The goal for farms should not be to sell energy commercially, but to attain energy self-sufficiency. Although agricultural reform in response to the new energy era has barely begun, large-scale energy plantations modeled after today's agriculture are clearly not the answer. Mixed cropping and agro-forestry schemes that yield food, fiber, forage, and fuel while protecting the soil seem more appropriate—and more likely—to meet the future's multiple demands.

The rush in the 1970s into the large-scale production of liquid fuels based on food-crop monocultures represented a wrong turn. The Brazilian and American ethanol programs caused environmental, equity, and nutritional problems, and even the economics of this all-or-nothing approach remains in question. Relying upon monocultural farming when these practices are themselves being rendered questionable by rising energy costs, soil erosion, and overdependence on synthetic fertilizers simply does not make sense.

The biological energy source receiving most attention—ethanol from sugar or corn—probably will not become a major factor in the energy picture until development programs have been redirected. Only countries with a surplus of quality agricultural land will get large quantities of energy from food crops. Even Brazil will probably find that its land can be put to better use than growing sugar for ethanol. The brightest immediate prospects are a modest output of biogas from large feedlots, ethanol from spoiled crops and wastes, and possibly seed oils for on-farm diesel substitution. Alone these fuels will not color the overall energy picture, but they will help agriculture and food processing begin the switch from fossil fuel use.

Off-farm agricultural processing industries will also move toward energy self-sufficiency by using organic wastes to produce energy. However, as the prices of liquid fuels rise, some companies may sell alcohol and use some other renewable energy source—perhaps direct solar or geothermal heat—to meet their own energy needs.

Municipal solid waste will also make a growing contribution to urban energy supplies. Already enough energy to heat and cool over 2 million buildings is produced annually in this way. With the spread of simple refuse-combustion technology to large cities in North America and the Third World, urban waste's energy contribution could triple. Methane plants that process treated sewage supply domestic cooking energy in some warm countries now, but in colder climates they will probably not contribute much more than the power needed to run sewage-treatment plants. Methane from landfills will provide a valuable local energy supplement for a few decades in some areas. Nowhere for the foreseeable future will advanced waste-to-fuel plants shed their experimental status. Far more important than perfecting these technologies will be putting more sophisticated source-separation and recycling systems into action. In the longer term, directly burning garbage—like directly burning wood—may not be worth the price.

How long the land can provide energy as well as food and fiber will depend on how systematically nutrients from waste streams in cities are returned to the soil. Adding more chemical fertilizers is not enough to check the accelerated depletion of soil nutrients that occurs in waste removal and energy farming. Produced from natural gas, nitrogen fertilizer has become much more expensive in recent years, a fact affecting the economics of energy farming on even the most well-watered and sun-drenched lands. Even where economical, widespread fertilizer use poses environmental problems ranging from the difficult-to-control pollution of water supplies with nitrates to the poorly understood destruction of microorganisms in the soil. Developing countries in particular will be hard-pressed to meet additional fertilizer needs.[61]

Maintaining the land's long-term productivity will probably require returning ash and sludge to the land, a practice seldom followed now. Typically, such wastes are instead buried or dumped in the ocean. One concern is that such wastes contain toxic organic chemicals and heavy metals that can concentrate in plant tissue in health-threatening quantities. Indeed, New York City (which must daily dispose of 8,300 tons of smelly, viscous, black sludge) found its efforts to spread sludge on forest and farmland thwarted because this "goo" contained high concentrations of cadmium, a heavy metal associated with kidney and liver disease. Until such toxic substances are controlled at their source, wastes will continue to be disposal problems instead of tonics to farm and forest.[62]

Only after these steps have been taken can biomass provide energy for other sectors. Only then can biomass-derived fuels move onto the center stage of the world energy scene to help meet liquid fuel needs. Once they are energy self-sufficient themselves, farms may begin "exporting" energy to industry and transportation systems that need high-quality, concentrated fuels. Then, energy plantations on marginal lands and mixed-crop farms will be able to supply small and mid-sized industrial plants with feedstocks for liquid fuel conversion.

Given the enormous environmental impact biomass energy systems can have, environmental planning must occur before investments are sunk. If the usual pattern of choosing a course and suffering the consequences later is followed, human health and the global carrying capacity will decline. Making the right decisions and implementing them effectively will demand a new kind of interdisciplinary decision making. For that to happen, a solid base of information and the political will to stand firm against the narrow goals of entrenched constituencies are needed.

While biomass energy holds varied opportunities and risks for all countries, one generalization holds true: Utilizing biomass resources without paying careful attention to the ecosystems from which they spring, the food and fiber resource systems they can disrupt, and the social systems they are to serve is a recipe for disaster. But if these caveats are heeded, hope for biomass and biomass itself could spring eternal.

8

Rivers of Energy

Whether harnessed by a wooden waterwheel on a tiny stream in Nepal or by a hundred-ton steel dynamo at Aswan on the mighty Nile, all hydropower comes from the ceaseless cycle of evaporation, rainfall, and runoff set in motion by the sun's heat and the earth's pull. By harnessing water returning to the sea, waterwheels and turbines make this natural and endlessly renewable energy useful.

From falling water comes one-quarter of the world's electricity. Among renewable energy sources, only wood makes a larger contribution. No other renewable energy technology is as mature. Yet several times the amount already harnessed remains untapped. Developing this potential will require constructing large dams in the Third World and in the peripheral regions of industrial countries, as well as small dams everywhere. Whole economies could be built around hydropower if the environmental problems, political disputes, and financial uncertainties surrounding its use were overcome.[1]

The Power of Falling Water

Historically, hydropower's use has been shaped more by social and political conditions than by the availability of hydro-technology. In the earliest reference to hydropower, the Greek poet Antipater praised the water-powered gristmill for freeing Greek women from the labor of grinding grain by hand. The Romans had waterwheels, but first slavery and then widespread underemployment removed any incentive to save human labor. Only after war and famine ravaged the disintegrating Roman empire and the Black Plague killed a third of fourteenth-century Europe's population did labor-saving water mills come into wider use. By 1800 tens of thousands were in use throughout the continent.[2]

As primitive hydropower technology spread, so did social dislocation and conflict. Comfortable with traditional hand grinders, small farmers resisted bringing their corn to village mills. Hoping to stimulate the use of water mills where the peasants' grain would be visible, and hence taxable, the French government outlawed hand mills. And in the parts of the New World where slave-holding was not tolerated and labor was scarce, waterwheel technology flourished. By 1800 about 10,000 waterwheels were in use in New England alone.[3]

Hydropower first became a source of electricity during the nineteenth century. Invented in 1820 by the French engineer Benoît Fourneyron, the turbine was to the waterwheel what the propeller was to the side paddle—a submersible, compact, and more efficient energy converter. Turbines were first linked to generators to produce electricity in Wisconsin in 1882, and the development of alternating current by George Westinghouse at Niagara Falls in 1901 made transmitting power over distances economical. During the eight decades since, the technology has been refined but not greatly altered.[4]

The early hydropower facilities, known as run-of-the-river plants, produced little power during the dry season when streams and rivers were low. To obtain continuous power output, large dams with water-storage reservoirs were built. Since the thirties, most hydropower energy has come from major dams set in large rivers. Since the oil shock of 1973, interest in the intermittent power from run-of-the-river dams, many of which were abandoned when petroleum was cheap, has revived.[5]

Rising energy prices have also sparked interest in a largely forgotten hydropower technology that does not depend on dams at all. During the Middle Ages, before dams were common, waterwheels affixed to barges anchored in rivers were widely used. Such floating hydro plants are not ecologically disruptive, and they can tap otherwise inaccessible water flows. Several countries are now trying to modernize this old technique and to assess its costs. If this technology (known as the lift translator) proves economical, the energy potential and environmental soundness of hydropower would increase dramatically.[6]

Since water power was first used to produce electricity, hydro-energy's contribution to the world's electricity supply has risen steadily. In 1980 it accounted for about 25 percent of global electricity and 5 percent of total world energy use. Total world hydro production today is 1,720 billion kilowatt-hours, which is generated at dams with a total capacity of 458,000 megawatts. The world's leading generator of electricity from falling water is the United States (71,000 megawatts of capacity). Next in line are the Soviet Union (47,000) and Canada

(40,000). With a tenth of the planet's potential, China will likely surpass all three over the long run.[7]

If all the energy contained in the water flowing toward the oceans was harnessed, a staggering 73 trillion kilowatt-hours could be produced annually. Yet given technical constraints, probably no more than 19 trillion kilowatt-hours can actually be tapped. But while environmental and economic factors will constrain use of this resource at some point, world hydropower production could still reach between four and six times its present level.[8]

In general, hydropower potential is distributed among the continents in rough proportion to land area. Asia has 28 percent of the world's potential; South America, 20 percent; Africa and North America (including Central America), 16 percent each; the Soviet Union, 11 percent; Europe, 7 percent; and Oceania, 2 percent. Although every continent has hydropower potential, mountainous areas and large river valleys have the most. For instance, India is twenty times as big as Nepal, but Nepal has nearly three times as much hydropower potential.[9]

Much of the world's untapped hydroelectric potential lies far from industrial centers, even far from inhabited areas. Unpopulated parts of Alaska, northern Canada, and Siberia have tremendous hydropower potential. The Amazon, the Congo, the Orinoco, and the rivers snow-fed by the Himalayas all offer sites for large-scale hydroelectric development. Remote reaches of Papua New Guinea, South Africa, Borneo, Tasmania, Norway, the Philippines, Argentina, Guyana, and New Zealand also have many promising dam sites.[10]

Some regions are much farther along than others in developing their water resources. (See Table 8. 1.) Europe, Japan, the United States, the eastern Soviet Union, and southern Canada have done the most to harness this power source. Indeed, Europe has exploited almost 60 percent of its potential. With only a fourth of Asia's resources, it generates nearly twice as much power. Africa has developed only about 5 percent of its potential, half of which comes from just three dams—Kariba in East Africa, Aswan on the Nile, and Akosombo in Ghana.[11]

Table 8. 1. Hydropower Potential and Use, by Region, 1980

Region	*Technically exploitable potential*	*Share of potential exploited*
	(megawatts)	*(percent)*
Asia	610,100	9
South America	431,900	8
Africa	358,300	5
North America	356,400	36
USSR	250,000	12
Europe	163,000	59
Oceania	45,000	15
World	2,200,000	17

Source: World Energy Conference, *Survey of Energy Resources.*

In some areas hydropower is the main source of electricity. More than thirty-five developing and industrial nations already obtain more than two-thirds of their electricity from falling water. In South America 73 percent of the electricity used comes from hydropower, compared to 44 percent in the developing world as a whole. Norway gets 99 percent of its electricity and 50 percent of all its energy from falling water.[12]

Big Opportunities and Big Problems

Few technological changes so dramatically and visibly alter the face of the earth as large dams and artificial lakes. Large modern dams rank among humanity's greatest engineering feats. Egypt's Aswan High Dam, for instance, weighs seventeen times as much as the Great Pyramid of Cheops. The Itiapu Dam, on the Parana between Brazil and Paraguay, will soon generate 12,600 megawatts—as much power as thirteen large nuclear power plants—making it the biggest power complex on earth. The lakes created by such dams number among the planet's largest freshwater bodies. Ghana's Lake Volta, for example, covers 8,500 square kilometers—an area the size of Lebanon.[13]

Even larger dams and lakes, however, are on the drawing board. On the Yangtze River in China, the Three Gorges Dam now under study will probably be the world's biggest dam—capable of generating 25,000 megawatts of power. American, Canadian, and Soviet planners have even grander designs for the giant rivers flowing into the Arctic—the Yukon, the MacKenzie, the Ob, and the Lena. And Egypt is considering harnessing the energy of water now resting in the Mediterranean Sea by channeling it through an artificial canal into the Qattara Depression, an 18,000-square-kilometer sink in the Sahara.[14]

Modern dam building traces back to the establishment of the Tennessee Valley Authority (TVA) in the United States in 1933. Before Franklin Roosevelt created this government body, conflict between private power developers and public power advocates slowed U.S. hydropower development. The creation of the TVA settled the case in favor of the public sector. It also marked the beginning of a basin-wide development program centered around energy production. A unique blend of centralized planning and grass-roots participation, TVA has the power to borrow money, condemn private property, and build dams. It also has a broad mandate to promote rural electrification, control soil erosion, improve navigation, and harness power. TVA spearheaded development in a million square mile area by enlisting the essential help of thousands of small farmers and townspeople and by rewarding them for cooperating. TVA's comprehensive approach to the development of river basins has become the model everywhere.[15]

The decade after World War II was the golden age of large dam construction in the United States, the Soviet Union, and Canada. By the late 1950s the frontier of large dam construction was the Third World and remote regions everywhere. The success of such basin-wide schemes as the TVA and the Soviet development of the Volga and Dnieper Rivers has drawn many Third World leaders to this energy source. So has the prestige value of large dams, which symbolize industrial progress. The generous financing terms and management assistance the industrial world offers make the construction of large dams even more attractive to nations facing chronic capital and technology shortages.[16]

Political cooperation between nations sharing common water resources is a prerequisite to financing and constructing many large-scale hydroelectric dams. (Altogether, some 200 of the world's rivers cross international boundaries.) Where hydropower is most developed—in North America and Europe—nations have successfully devised political mechanisms for cooperative river development and conflict resolution. In North America, for example, the Columbia and St. Lawrence rivers could not have been harnessed had not the U.S. and Canadian governments cooperated closely. In Europe, the Rhine and Danube could not be developed until previously suspicious and oft-warring nations laid aside their differences.[17]

Unresolved conflicts over water rights remain a major barrier to the development of many promising large hydro sites. Long-simmering disputes between India, Nepal, and Bangladesh over Himalayan waters frustrate efforts to harness one of the world's major energy resources. In Canada an old dispute between Newfoundland and Quebec over power pricing has delayed construction of a 2,300-megawatt dam complex on the lower Churchill River. And the hydroelectric and irrigation potential of the Mekong River in Southeast Asia remains untapped because of conflict between Laos, Thailand, Kampuchea, and Vietnam.[18]

Unable to compromise with resource-sharing neighbors, some countries have unilaterally developed the portion of the resource base they control. Yet such a strategy can backfire.

When Egyptian president Nasser initiated the Aswan Dam project in the early 1950s, Great Britain controlled the headwaters of the Nile in the Sudan and central Africa. Thus efficiency and economy were sacrificed so as to place the dam beyond Britain's reach. Today, however, the reservoir is filling with silt far more rapidly than anticipated as soil erodes in impoverished regions of other African countries of the Nile Basin. Should India develop rivers of the subcontinent without Nepal's cooperation, the curse of Egypt will be on it too.[19]

Hydroelectric projects figure prominently in the economies and investment plans of many developing nations. With power from Aswan, Egypt electrified virtually all of its villages and created many new jobs in labor-intensive local industries. Companies attracted by the power of the São Francisco River have brought almost a million new jobs to impoverished northeast Brazil. Venezuela expects to spend tens of billions of dollars over several decades to harness 40,000 megawatts of power from the Caroni River at Guri. And the Philippines, heavily dependent on imported oil, envisions a 45 percent increase in hydroelectric generation in its current five-year energy program.[20]

Yet kilowatt-hours generated is no measure of integrated development. The impacts on agriculture, fisheries, health, employment, and income distribution must all be weighed. Unfortunately, building a large dam in a developing country does not necessarily improve the standard of living for the poor rural majority since the energy-intensive industries that locate near large dams seldom provide many jobs for unskilled local people. A case in point is the $2 billion Asahan aluminum and hydroelectric project in Sumatra, which will employ only 2,100 of the island's estimated 30 million people. Too often, the power not used by nearby industry will be transmitted hundreds of miles to major cities, leaving dozens of villages unlit along the way.[21]

As for the ecological changes wrought by large dams, they

pose both opportunities and dangers. Large dams change a self-regulating ecosystem into one that must be managed. Plopped into a river with no thought to the upstream and downstream impacts, a large dam can bring disaster. Lakes cannot survive some of the abuses rivers can, so traditional ways of life are called into question, especially sanitation practices. Water-borne disease can get out of hand and soil erosion and pollution must be controlled in order to preserve a dam's water-storage and power capacities.[22]

The world over, the silting of reservoirs caused by soil erosion threatens dams. When a reservoir fills with sediment, a dam's ability to store water and generate energy is drastically curtailed. The Sanman Gorge Dam in central China, for example, has lost three-quarters of its 1,000-megawatt power capacity to sediment from the Yellow River. In Nepal deforestation and farming on steep lands threaten to incapacitate the few dams already built on Himalayan rivers. Until the topsoil of Nepal and northern India can be stabilized through reforestation and improved farming practices, both countries' ambitious hydroelectric and irrigation plans will have to be postponed.[23]

A primary motivation for building large dams is to trap water for irrigation. By storing water from rainy seasons and years for use when it is dry, dams mitigate the effects of droughts, increase agricultural productivity, and extend agriculture to dry uncultivated areas. Often the electricity generated at such dams powers pumps that extend irrigation over large areas. Of course, farmland created in this way has a price—the river bottomlands flooded by the dam. Where dams have curtailed the spring floods that once deposited rich silt on the land, artificial fertilizers must be applied to preserve soil fertility, and fertilizer production can consume much of the dam's power output.[24]

On fisheries the impact of large dams is both ambiguous and unpredictable. Gauging impacts is especially difficult in tropical Africa, Asia, and Latin America, where many important but still mysterious fish species live. Where fish species migrate long distances to breed, dams can decimate fish stocks. The rich Columbia River salmon fisheries in North America declined sharply after dams were built on the river—despite well-funded programs to build fish ladders and to restock the river. The unanticipated destruction of the eastern Mediterranean sardine fishery by the Aswan High Dam has been more than counterbalanced by the emergence of a fishing industry on newly created Lake Nasser, but sardine fishermen cannot find the change consoling. Egyptian officials optimistically predict that Lake Nasser will eventually yield 60,000 tons of fish per year. But if experience with other African dams is any guide, production may fall as the lake grows older and becomes more ecologically settled.[25]

On human populations the impacts of large hydropower projects in tropical regions are only too well understood. In warm climates reservoirs and irrigation canals provide ideal breeding grounds for snails that transmit schistosomiasis—a debilitating, sometimes fatal disease that currently afflicts some 200 million people in tropical countries. Better sanitation facilities and improved hygiene could virtually wipe out this and other waterborne diseases, but planners' and governments' best efforts have so far failed to get people near newly created lakes to adopt the sanitation practices that could halt disease.[26]

Another often-neglected cost of large dams is that paid by people whose homes are flooded by the project. Some 80,000 were displaced by Lake Nasser in Egypt and Sudan; 75,000 by Lake Volta in Ghana; 57,000 by Lake Kariba in East Africa; and 50,000 by Lake Kainji in northern Nigeria. China's planned Three Gorges Dam could force some 2 million people to leave home forever. Plans to resettle and reemploy displaced people figure prominently in few dam projects, and most of those made fail for lack of funding. And no amount of government aid can compensate for the permanent loss of one's roots.[27]

Particularly troubling is the threat to native tribes long preserved by isolation. The already beleaguered and shrinking Indian tribes of Amazonia, for example, could be forced by Brazil's ambitious dam development to resettle in a culture harshly and disorientingly modern by their standards. Some native peoples have resisted government resettlement programs, protesting and taking up arms. Holding fast against the Philippine government's plan to place Southeast Asia's largest hydro project on the swift-flowing Chico River, tribes in central Luzon have fought repeatedly with government troops. Still others have won substantial concessions. Native people in the area inundated by Quebec's giant James Bay project delayed construction through the courts and forced the government to grant them $250 million, title to 12,950 square kilometers of land, and preferential employment rights on the project. Isolated tribes in poorer, less developed countries are unlikely to fare so well, although groups such as Survival International have recently emerged to help them.[28]

Dams can also endanger little-known plant and animal species. In Quebec careful environmental monitoring revealed that the new dams and impoundments threatened no species. But many tropical plants or animals with potentially high economic value will be lost forever if dam reservoirs are built because so many tropical species have yet to be named. Even where threatened species have been identified, pressure to destroy their habitats can be irresistible. Over the heated protests of environmentalists, Australia has built a hydroelectric complex in Lake Peddar National Park, flooding habitats of dozens of species found only in Tasmania.[29]

A few hopeful signs indicate that in many countries the dam-building process is now more than a feat of engineering muscle. Many of the hazards of dam construction in the tropics, for example, are better understood now than when the first modern dams were built thirty years ago. The plans for the hydroelectric and irrigation project being built on the Senegal

River by Mauritania, Mali, and Senegal call for extensive ecological monitoring and population-relocation programs. On the other hand, Brazil's "dam it all" approach can be expected to cost the nation plenty later.[30]

While ecological change accompanies any dam project, especially in the tropics, the failure of communities and farmers to make necessary adjustments once the dam is built probably causes more environmental problems than the structures themselves do. Accordingly, even the best-laid plans will not work unless people at the grass-roots level help plan the project and share in its benefits. Unfortunately, the involvement of farmers, owners of small businesses, and local officials—the key to the Tennessee Valley Authority's success—has too often been missing in developing countries. Erosion along the shores of Lake Kariba between Zimbabwe and Zambia, for example, has reached dangerous levels despite efforts by both governments to prevent overgrazing and to preserve a band of trees along the water's edge. Although local farmers and herders know their practices threaten the lake, they cannot afford to forgo short-term production gains.[31]

Tennessee Valley farmers controlled erosion and planted trees because they received cheap loans and cheap electricity in exchange. But poor farmers on the Zambezi are being asked to abandon ecologically destructive practices and offered nothing in return. Often the failure of planners to spread a project's benefits among all affected actually accelerates the impoverishment of marginal groups.

Maintaining Momentum

For the last three decades, large-scale hydropower development in developing countries and in peripheral regions of the industrial countries has occurred primarily because energy-intensive industries need cheap electricity and global-lending institutions have been willing to advance multi-billion dollar loans. Today the failures of the international economic and political systems to adjust to higher prices and higher costs throw such long-term investments into question.[32]

Large dams are extremely expensive, and nearly all are built with borrowed money. Aswan, for example, cost $1.5 billion when it was built in the sixties. Itiapu will cost between $5 and $6 billion, and China's Three Gorges project could cost $12 billion. The U.S. government borrowed to finance TVA in the 1930s; Brazil and Quebec borrowed the needed capital in the 1970s; and China will do so in the 1980s. According to the World Bank, the Third World will need to raise an estimated $100 billion between 1980 and 2000 for hydro plants currently on the drawing boards—a staggering sum considering high interest rates and the financial plights of many developing countries.[33]

Few major dams are likely to be built in developing countries without at least partial World Bank funding because Western banks and lending consortia fear that a Third World country might nationalize a dam once constructed. At the same time, few developing countries are willing to turn over ownership of so important a national resource as a river to foreign private investors.[34]

Capital for hydropower projects is most readily available when sales of power—mainly to energy-intensive industries owned by multinational corporations—guarantee a steady, predictable flow of revenue. In such sparsely inhabited regions as the Amazon Basin, New Guinea, Quebec, and Siberia, the need for power to extract and smelt minerals provides the principal impetus for hydroelectric development. Where prime hydro sites are not close to rich mineral deposits, the main economic force behind large dam construction is the aluminum-smelting industry.[35]

Constructing dams in previously undeveloped areas often leads to a conflict between new users and those who made the dam possible. Dam-owning governments soon see the wisdom in diverting cheap power to employment-intensive activities or to raise prices to fund national development programs. When the availability of cheap power stimulates consumption enough to precipitate shortages, a painful choice emerges: Either governments must raise prices for the heavy energy users, perhaps driving them elsewhere, or it must let smaller consumers alone bear the prohibitive cost of building new coal or nuclear power plants to meet demand. Egypt, the U.S. Pacific Northwest, and Ghana face this painful dilemma today. As the price of electricity produced from fossil and nuclear energy climbs, countries selling hydroelectricity at bargain rates have been increasingly tempted to raise prices to the world average—often ten times the rate they now charge. Yet so far the interests of large electricity consumers accustomed to cheap prices have prevailed.[36]

In the years ahead the key to planning and financing hydropower will be resolving these conflicts and raising prices markedly. Nowhere has the pricing and allocation of hydroelectric power yet changed in response to the oil shocks of the seventies. Between 1970 and 1975, the price of coal quadrupled, the price of uranium increased eightfold, and the price of oil rose tenfold. Owners of reserves of these fuels quickly reaped a massive windfall as prices followed OPEC oil to dizzying new heights. Only the owners of hydroelectric facilities—governments—missed out on the profits. While the prices of oil, coal, and uranium were influenced by the price of imported oil, hydropower's cost continued to reflect the cost of production—the sum of dam-operating costs and the interest on money borrowed long ago, neither of which rose much. As a result, consumer demand and waste of electricity rose. Already underpricing and overdemand are strapping the government-controlled Bonneville Power Authority in the Pacific Northwest. Since electricity prices in the Pacific Northwest are one-eighth those in oil- and nuclear-dependent New York City, Washingtonians and Oregonians use five times as much electricity per capita as New Yorkers do. But meeting this extraordinary price-stimulated demand with new nuclear plants is not working. Nuclear cost overruns already more than equal the cost of all the federal dams on the Columbia River.[37]

For developing countries the cheap sale of hydropower has had even more tragic consequences. Locked into contracts with aluminum companies that were signed before the price revolution of the 1970s, dozens of Third World nations are selling their principal energy resources at a price far below their market value. According to the Center for Development Policy Studies, simply increasing hydroelectric prices to the world average price of electricity would earn fifty-six developing nations over $10 billion annually. Through underselling to Western-owned multinational corporations, these nations are collectively losing about as much each year as the World Bank lends for all development projects.[38]

Ghana exemplifies the problems underpricing hydropower causes. In the sixties this West African nation built a large dam on the Volta River and signed a thirty-year contract to sell power to Kaiser Aluminum Company at three-tenths of a cent per kilowatt-hour (one-twentieth of the current world average price). For Ghana the dam represents a major national investment. It also represents a sacrifice: Waterborne diseases increased once the dam was built, and people from nearby flooded areas had to be relocated. Yet revenues from the dam barely cover interest payments and operating costs, so Ghana gets little net benefit from its principal national energy resource. With the Kaiser smelter taking over 90 percent of Ghana's total electricity production, Ghana faces electricity shortages. It has had to borrow money to build smaller, more expensive dams, and it now imports power from neighboring countries. Attempts to renegotiate contracts with Kaiser have fallen flat, partly because neither the World Bank nor the U.S. government will support Ghana.[39]

Selling hydroelectric power at prices closer to its true market value would make available funds to build other dams and to help consumers cope with higher prices by investing in energy efficiency. In the U.S. Pacific Northwest, for instance, the Bonneville Power Authority could raise nearly a billion dollars a year by selling its hydropower at the national average price of electricity. In developing countries additional electricity revenues could support internal development.[40]

Like dam construction on international rivers, pricing reform demands international solutions. Unilateral action by developing countries to raise prices will undermine investor confidence and jeopardize access to further World Bank loans. Yet inaction has high costs, too, as the debacle in the U.S. Pacific Northwest vividly shows. To avert these problems and boost hydropower development, the World Bank and its principal contributors should encourage a gradual rise in the price of hydroelectricity, one that the aluminum industry could bear without moving its plants closer to centers of demand.[41]

Establishing a realistic hydroelectric-pricing scheme could transform the prospects for hydro development in poorer parts of the world. As the migrating aluminum industry opens up more remote hydropower-rich regions, the price of aluminum could gradually rise to reflect the costs of operating in increasingly difficult terrain. These higher prices would stimulate aluminum recycling. Eventually a smaller aluminum industry located in the most remote regions would reach equilibrium. In the wake of industry's wanderings would be many flourishing and sustainable local economies.

Small-Scale Hydropower for Rural Development

Fortunately, large dams are not the sole hydro development option of developing nations. The power of falling water can also be harnessed at much smaller sites with capacities between 1 kilowatt and 1 megawatt. By constructing small dams, Third World countries can unleash the 5 to 10 percent of their hydropower resources that the World Bank conservatively estimates exists at small sites. Small dams could, in fact, provide roughly as much electricity as Third World countries derive from hydropower today—more, if inexpensive local labor and materials are used.[42]

The economics of building small dams for power production varies widely. The World Bank says that costs hover around $3,500 per kilowatt of installed capacity, but many projects are being built today for between $500 and $1,000 per kilowatt of capacity. Because relatively fixed engineering and site-preparation costs can be spread over a larger power output, larger dams seem to enjoy considerable economies of scale. But small-scale projects look more favorable if the hidden or discounted social costs of large dams are taken into account. In general, developing countries stand to reap more by developing the cheapest small sites available before venturing into additional large dam projects.[43]

Besides generating revenues small hydro plants can reinforce economic development by converting poorer countries' most abundant and least-used resource—labor—into critically needed capital. They can also catch silt-laden storm waters, thus protecting large downstream dams from premature sedimentation.

Among developing nations, China alone has placed high priority on small-scale hydro development. While most developing countries have borrowed money and imported technology to build large dams to run heavy industry, the Chinese have relied on indigenous labor, capital, and technology to build tens of thousands of small hydro facilities. Reports that major cities regularly experience "brown-outs" and that electricity for heavy industry is scarce are true, but China has brought many basic amenities to its vast rural population by building small dams.[44]

Although China was an early user of waterwheels, all but fifty of the nation's hydro facilities were decimated in the strife and economic decay that preceded the communist revolution. After 1949 the government followed the Soviet model of rural electrification by emphasizing large power plants, so the number of small hydro facilities in use actually declined in the late fifties. But with the Cultural Revolution in the mid-sixties came a boom in small-dam construction that China's current leaders continue to promote.[45]

Since 1968 an estimated 90,000 small-scale hydro units with some 6,330 megawatts of generating capacity have been built, mainly in the rainy southern half of the country. Although the average size of the units is a meager 72 kilowatts, small plants account for 40 percent of China's installed hydro capacity. In more than one-quarter of the nation's counties, these small dams are already the main source of electricity, and China expects to add 1,500 megawatts of power annually through 1990 and 2,000 megawatts per year for the ten years following. By the turn of the century, the government hopes small hydro facilities will be providing six times as much energy as they did in 1979.[46]

The Chinese consider small hydro plants just one part of integrated water-management schemes and rural development efforts. Driven by the need to feed and employ a billion people, the government has given highest priority to agricultural water storage, irrigation, flood control, and fishery needs. Chinese villagers have built impoundments and irrigation ditches with simple hand tools and without expensive, heavy earth-moving equipment. Many of the components of hydro plants—turbines, pipes, and gates—have been constructed at small shops by local artisans using local materials and standardized designs. With money earned and saved from agriculture and fishing, communes have upgraded the sites without central government funding. Technical advice from agricultural extension workers has improved dam and plant design and helped lower costs.[47]

Unlike dams that power capital-intensive export industries

in so many developing countries, small dams in China support workshops that turn locally available raw materials into goods used in nearby areas. Hydropowered factories scattered throughout the countryside husk rice, mill grains, make soap, and produce leather and simple metal goods. The power left over is available for lighting, movies, and telecommunications. While the quantities of energy involved are not great, these hydro plants dramatically improve the quality of rural life—and thus halt migration to overcrowded cities—by reducing the backbreaking drudgery of lifting water, sawing wood, and grinding grain by hand. And village-based reforestation, anti-erosion, and schistosomiasis-control programs have enabled the Chinese to avoid the ecological and health problems often connected with hydropower use. The leaders in Beijing claim that their small-scale water development efforts complement rather than displace the need for large dams. By "walking on both legs"—building small dams as well as large—they hope to exploit fully their tremendous water-power potential without incurring high social and ecological costs.[48]

The projects completed outside China confirm the role of small hydro plants in balanced development. In Papua New Guinea, for example, the village schoolmaster in remote Baindoang heard about hydropower on a radio show and asked the national university for help building a small dam. Along with a private group it obliged, and a tiny 7-kilowatt turbine was installed two years later. Celebration and dancing commemorated the coming of power to the village, where it lights the school and store and heats water for communal showers. By mobilizing local labor, the project strengthened the village-level institutions and gave villagers a greater sense of control over their own lives—a far cry from what large hydropower projects do.[49]

Spurred by rising oil prices and such examples, many Third World countries have become interested in small-scale hydropower. Nepal, the most active, recently opened about sixty water-powered mills. These efforts are being assisted by modest but growing international aid programs. The U.S. Agency for International Development has loaned Peru's national power company $9 million for twenty-eight installations ranging in size from 100 to 1,000 kilowatts. France is helping several African nations build small dams, and Swiss groups are helping Nepal set up factories to build small, inexpensive turbines. Unfortunately, the World Bank—which between 1976 and 1980 loaned $1.68 billion for large-scale hydroelectric projects —has spent almost nothing on developing small sites.[50]

Promising first steps, these programs must be expanded considerably to have much impact. By funding badly needed surveys of small-hydropower potential, development assistance groups could document the dimensions of this untapped resource and direct local groups to particularly promising sites. Once specific projects get underway, governments and international agencies can help by providing hydrologists, geologists, and engineers to ensure that dams are built safely and take full advantage of the water flows. International agencies could also take up TVA director S. David Freeman's challenge to establish an international hydropower development corporation to share scarce knowledge and skills. Large lending institutions could help by repackaging capital blocks into smaller parcels and broadening loan criteria to reflect the hidden social costs of large dams and the neglected benefits of small ones.[51]

Developing nations themselves also need to reassess the priority they give large-scale hydro development efforts. Dotting the countryside with small projects may not be as politically gratifying as erecting a few big modern dams, but it would go farther than a "think big" approach toward meeting the needs of the rural populace. Although small- and large-scale hydro projects go hand in hand, integrated village-level water development should precede the construction of large dams as a general rule. Erosion, the spread of waterborne and other diseases, and the other side effects of large projects will be easier to conquer if people have already helped design small dams in their villages. At the same time, rural development will be more balanced and more effective.

Making Better Use of Existing Dams

In the United States, Japan, and several European countries where hydropower resources are well developed, strong public support for free-flowing rivers has brought dam construction to a virtual standstill. In these areas the challenge is not to build new dams, but to preserve both wilderness and ecological values by making better use of existing ones.

Much of the public's opposition to new dams reflects a desire to preserve "white water" recreational opportunities and to keep remote and unaccountable utility companies within bounds. Some also is based on sound ecological principles: Preserving representative river systems in their natural state provides a baseline against which ecological change on other rivers can be measured, as well as sanctuaries for the many species that thrive only in swift-moving waters. Many people also recognize the obligation those who have despoiled so much of the earth have to future generations.[52]

The United States and Sweden have done most to preserve wild rivers with high aesthetic, wilderness, and recreational value. Parts of thirty-seven U.S. rivers—with a combined potential capacity of 12,750 megawatts—are protected from further development under the Wild and Scenic Rivers Act. Another 3,500 megawatts of power potential on the lower Colorado River is going unused because of Grand Canyon National Park constraints. Sweden has permanently banned dams from four undeveloped rivers in its far north.[53]

As the improving economics of hydropower opens the way for exploitation in the years ahead, public officials and citizens should scrutinize dam projects more carefully than they have. In particular, the often-inflated claims of recreational and flood-control benefits must be assessed carefully since power sales alone seldom justify a project. Then, too, the prime agricultural value of bottomland that must be flooded should not be underestimated. In many cases, water conservation and flood-control benefits could be better achieved by reducing water waste and limiting construction in flood plains.[54]

The United States, Europe, Japan, and the Soviet Union all have many small dams (under five megawatts) that represent increasingly viable sources of power as electricity prices rise. France has been so successful in pressing its dams into service that 1,060 micro-hydro stations with a combined capacity of 390 megawatts constitute 1 percent of the nation's total generating capacity. Japan too has aggressively harnessed its abundant water resources with numerous small dams. Recent studies indicate that hilly regions of Wales, Scotland, Spain, Sweden, and Romania all have substantial untapped hydro potential at existing small dams.[55]

The greatest opportunity in the industrial world to take advantage of small dams is in the United States, where many small- and medium-sized dams await renovation. Twenty-one small dams on the Rhône in France produce 3,000 megawatts of power while the comparably sized Ohio in the U.S. produces only 180 megawatts. In all less than 3 percent of U.S. dams produce electricity, even though an estimated 6,000 to 24,000 megawatts is available at small dams alone compared to the present total U.S. hydropower capacity of 64,000 megawatts.[56]

During the last several decades, falling electricity prices and the end of the forty-year life of construction tax concessions led to the abandonment of almost 3,000 dams in the U.S. But the years of neglect are now themselves ending—albeit at a high cost. According to the New England River Basins Commission, the northeastern United States has 1,750 small unused dams that could produce 1,000 megawatts if fully exploited. If these dams were renovated with money borrowed at a 7 percent interest rate, with the understanding that power would be sold at 4.5¢ per kilowatt-hour, 50 percent of their potential could be harnessed economically. If government kept the interest rate at 3 percent and power were sold at 6.7¢ per kilowatt-hour, 80 percent of the potential could be developed.[57]

Both public and private efforts to restore small dams are afoot. In the United States, government encourages the trend in two ways. It grants tax benefits that are twice as high as most industrial investments receive and reduces the regulatory burden on small-dam developers, most of whom are small farmers, small firms, or townships. Ultimately even more important is the Public Utility Regulatory Policies Act of 1978, one section of which requires utilities to buy power from small power producers at fair rates. As a result, applications for permits to produce power—a good measure of hydro development interest if not actual construction—have shot up dramatically from 6 in 1976 to 1,900 in 1981.[58]

Although public attention in the U.S. has recently focused on renovating small, abandoned dams, even more energy is available at medium and large dams that have never been used for power production. Rising power rates have made electricity generation economical at many flood-control and irrigation dams. While estimates of potential vary widely, 44,000 megawatts is probably a conservative figure. Since the federal government owns most of these dams, tapping this potential will require the government either to invest directly or to allow private firms access to the dams.[59]

Opportunities to boost hydropower's contribution to national energy budgets also exist at dams that already generate power. Upgrading the power-generating capacity of dams makes even more sense as the costs of alternative fuels rise and turbine technology advances. At the Grand Coulee Dam on the Columbia River in the United States, for example, one new superefficient generator has been installed and two more may be added. In Switzerland hydroelectric production could be increased by 20 percent if up-to-date turbines and generators were installed on its dams, some of which date back to the 1800s. California could increase power output by raising the giant 700-foot Lake Shasta Dam an extra 200 feet, though buying property along reservoir shorelines and relocating people who have built there would be costly.[60]

In regions where most of the favorable sites have been tapped and where thermal power plants are numerous, hydro facilities can be turned into what are known as "peaking" and "pumped storage" units. Since demand for electricity varies widely over time, sources that can be easily turned off or on are needed to meet demand peaks. Since the water stored behind a dam can be released at any time, hydroelectric plants can become sources of peaking power if additional turbines are installed. Pumped storage facilities further exploit water's flexibility as an energy source by using off-peak power from continuously running coal and nuclear plants to pump water uphill into storage reservoirs. As needed, water is released to run back downhill through the turbines, which recoup two-thirds of the energy used for pumping. Worldwide, some 37,000 megawatts of these energy-storing facilities have been built so far.[61]

Peaking units and pumped storage facilities do have their drawbacks. Pumped storage plants tend to be large, expensive, and difficult to site. Fluctuating water releases erode shore lines, impede navigation, and disrupt fish life. More important, it probably costs less to lower peak demand with conservation and utility load-management techniques than it does to meet peak demand with hydro peaking units.

Where hydroelectric regimes are mature, only institutional inertia stands in the way of a fuller use of hydropower. The Soviet Union, many European nations, and especially the United States could all rehabilitate small dams to acquire needed power. The technology is time-tested, the economic incentive clear.

The Hydropower Prospect

As economically appealing as hydropower is, favorable economics alone is not enough. Political, financial, and environmental obstacles stand in the way. Also key is government's unique catalytic role, since rivers are everywhere publicly owned and since water projects touch upon so many aspects of life. With public resistance to government initiatives mounting and support for development aid declining, only committed and far-sighted political leadership can get world hydropower potential developed.

So great is hydropower's potential that theoretically it could meet all the world's electricity needs, though of course arid lands have virtually no resources. Even quadrupling global hydroelectric production—a realistic goal—would yield roughly as much electricity as the world currently consumes, certainly enough to permit electricity use to grow for many years and to eliminate the need to build most of the coal and nuclear power plants energy planners favored in the wake of the oil-price revolution of the seventies. In some countries and regions hydropower can meet most or all additional electricity needs. Quebec is seriously considering building a fully electrified economy based on water power, while the heavily oil-dependent Central American countries have enough untapped hydro and geothermal resources to become energy self-sufficient. Costa Rica, for example, already gets 35 percent of its energy from hydroelectric plants and 94 percent of its hydro potential remains untapped.[62]

Some nations have enough hydropower to become electricity exporters. Having tapped the swift-flowing headwaters of Europe's rivers in the Alps, Switzerland sells electricity to France and Italy. Nepal and Peru are similarly blessed with abundant hydropower resources, still largely untapped. Nepal could become the Switzerland of Asia, exporting electricity to the Indian subcontinent. Where distance makes transmission of electricity impractical, hydro-rich countries can export such energy-intensive products as aluminum.[63]

The pace of hydropower development efforts varies greatly from nation to nation and continent to continent. (See Figure 8. 1.) North America, the Soviet Union, and Europe all have substantial projects planned or underway. Among the sleeping giants of hydropower—Asia, South America, and Africa—South America, led by Brazil, has come farthest.[64]

Because hydropower plants, especially large ones, take years to plan and construct, short-term projections can be made with some confidence. As of 1980 some 123,000 megawatts of hydro capacity were under construction and another 239,800 megawatts planned. When all these plants are completed by the turn of the century, worldwide hydroelectric output will be roughly double what it is today. But even then, no more than one-third of the power that could feasibly be tapped will have

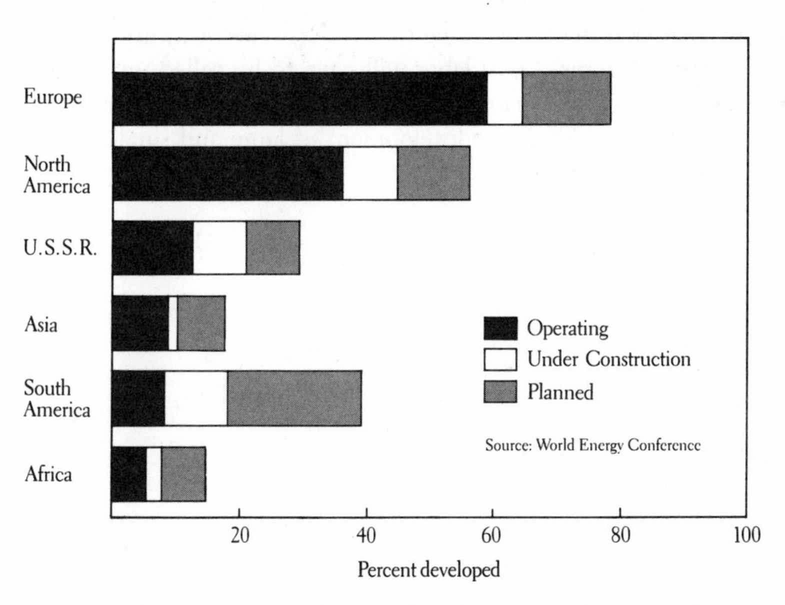

Figure 8.1. Status of Hydropower Development, by Region, 1980.

been brought on line. By the year 2020, the World Energy Conference optimistically projects, hydropower will supply some 8 trillion kilowatt-hours of power, almost six times the present level. But this potential won't materialize unless such economically impoverished but resource-rich countries as Zaire, China, and Nepal attract investment capital and create markets for hydroelectricity.[65]

Financing aside, environmental problems may well pace future hydropower development. In industrial countries the desire to preserve prime agricultural land and unique scenic and recreational resources has already made some large hydro sites off limits. In developing countries environmental catastrophies now unfolding in some regions could damage or destroy the hydropower capacity in others. Unless soil erosion and siltation are checked, the hydropower investments of many Third World countries would be for nought.

Not just dams, but basin-wide development and resource management will have to be the cornerstones of future hydropower programs. Local labor will have to be called upon and rewarded for tree planting and erosion control. And nations will have to take the codevelopment of large and small water projects as a signal rule.

9

Wind Power
A Turning Point

Wind power returns as a breath of fresh air to the world energy scene. Its use is already economical in some regions, and plans for harnessing wind are proliferating in many countries. As technologies and production techniques evolve, wind machines more reliable and less expensive than current models will further widen wind power's use.

Today's wind machines range from simple water-pumping devices made of wood and cloth to large, sleekly contoured electricity-generating turbines with 100-meter blade spans. In Australia and parts of Africa, Asia, and Latin America, wind-driven irrigation pumps are enjoying a renaissance. So too are sail-driven commercial ships in many coastal areas. Small electricity-generating machines are also becoming more popular, particularly in North America and northern Europe. For large wind turbines—sophisticated new machines with computer-based control systems—the energy market will take somewhat longer to open up, but their long-term potential to generate inexpensive electricity appears immense.

The conditions are ripe for wind power. But as with most other renewable energy sources, near-term success is by no means assured. Ambitious government research programs account for much of the progress achieved so far, and decisions by some governments to trim support for wind energy in the early eighties have slowed development. Yet even without government funding wind power development would probably slow rather than stop, so great is the momentum it acquired during the last decade. Indeed, wind power researchers and businessmen are certain that the wind will yield substantial amounts of electricity and direct mechanical power before the turn of the century.

Harnessing the Wind

Wind is born of sunlight, which falls unevenly on different areas of the earth and thus heats the atmosphere unevenly. Since warm air weighs less than cool air and tends to rise, air moves. One large air-circulation system consists of cool polar air being drawn toward the tropics to replace lighter, warmer air that rises and then moves toward the poles. Amid this flow, high and low pressure zones develop naturally and give rise to the persistent trade winds in the tropics, the polar easterlies, and the westerlies that traverse the northern and southern temperate regions. Similarly, coastal winds and such regional turbulence as the Asian monsoons result as cool ocean air flows inland to replace the rising warm air.[1]

Of the solar energy that falls on the earth, only 2 percent becomes wind power.[2] But this small fraction represents far more energy than humanity uses in a year. Of course, most winds occur at high altitudes or over the oceans where they do civilization little good. Even the most ambitious wind-energy schemes would tap only a small fraction of the total resource —comparable to occasionally lifting a bucket of water out of the oceans.

Harnessing the wind's energy is not, obviously, a new idea. Since the dawn of history, sailing ships have transported goods and people, opening up new lands and carrying invading armies to distant shores. Windmills—machines that capture the wind's power to perform mechanical tasks—were developed later, though when and where no one is sure. Windmills first came into wide use in Persia around 200 B.C. These relatively primitive machines were used to grind grain, a practice that later spread throughout the Middle East. Similar devices came on the scene in China at about the same time.[3]

Windmills were introduced in Europe sometime before the twelfth century, apparently by returning crusaders. They found their place first in grain grinding and later in wood sawing, paper making, and agricultural drainage. Europe's windmills were horizontal-axis machines made of wood. Their drive shafts were parallel to the ground, and each machine had four large blades. Gears connected the spinning shaft to a grinding stone or another mechanical device. This design eventually evolved into the Dutch windmill most people think of as the prototype. Sophisticated versions of the Dutch model were found throughout Europe by the fifteenth century. Along with waterwheels, they greatly boosted the productivity of agrarian economies and cleared the way for the industrial revolution. In their heyday in the seventeenth century, windmills numbered about 10,000 in England and 12,000 in Holland.[4]

European industrialists and traders abandoned windmills and sailing ships in the early nineteenth century as coal-fired steam engines became widely used. However, pioneers in Australia and North America held fast to windmills as the only means of obtaining precious irrigation and drinking water. Small horizontal-axis machines with a dozen or more metal blades were developed, and an estimated 6 million water-pumping devices were built in the United States in the late nineteenth century. According to wind-machine expert Peter Fraenkel, the windmill was as important as the Colt revolver in opening the American West to cattle ranching.[5]

An electricity-generating wind machine was developed in Denmark in 1890, opening up a range of new uses for wind power. Not long after, engineers realized that to generate electricity efficiently fewer and thinner blades were needed. The sleek new machines they developed found a wide market in Denmark, the United States, and a few other countries during the twenties and thirties. Most were used to electrify farms.[6]

From the thirties onward, rural electrification sounded the death knell for wind machines in much of the world. New hydroelectric dams and power plants that burned fossil fuels could sell electricity cheaply, partly because they benefited from government subsidies. North American farmers were encouraged by newly formed electric cooperatives to tear down their windmills. A handful of inventors let them stand, however, and even during mid-century when the cost of electricity was low, a few countries launched projects to develop larger, more economical wind turbines. Researchers in Britain, Denmark, France, the Soviet Union, the United States, and West Germany designed wind turbines with over 20-meter long blades and more than 100 kilowatts of generating capacity.[7] Yet the rapid development of nuclear reactors and other decidedly modern energy technologies made even new sophisticated wind machines seem somehow antiquated.

It took the energy shocks of the seventies to spur a wind power revival. Since 1973 dozens of small wind-machine manufacturers have entered the business, and both private companies and national governments have carried out research on larger, more sophisticated turbines.

The engineering elegance of the new machines hints at wind power's still-untapped potential. The blades of a modern wind machine typically occupy only a small space. Yet theoretically they can harness up to 59 percent of the wind passing through the area they sweep. Operating wind machines never approach the ideal but are usually 20 to 30 percent efficient—high compared with other energy-conversion technologies. Given the amount of energy they capture, both the material and energy requirements for the manufacture of wind machines are impressively low. Most wind machines generate as much energy as they take to manufacture in less than 5 years—much quicker than most other conventional or solar technologies.[8]

The amount of energy available in the wind depends on its speed. The amount increases eightfold every time wind speed doubles, so wind at 12 miles per hour contains fully 70 percent more power than wind at 10 miles per hour. A difference of just two miles per hour can, therefore, make or break a wind-energy project. At present, average wind speeds of 12 miles per hour or greater are needed to operate an electricity-producing wind machine economically, though mechanical water-pumping wind machines work fine where winds average only eight miles per hour.[9]

Since wind availability varies greatly by region, each "wind-prospecting" country needs an accurate reading of the size of the resource and its distribution. Initial assessments in North America and Western Europe indicate that in most northern temperate regions there are many areas with sufficient wind to generate electricity economically. Mountain passes and coastlines in these regions appear exceptionally fertile. In both tropical and temperate regions, average wind speeds of 12 miles per hour are fairly common, and many high-potential sites with far greater winds have been pinpointed. And no country is completely windless—an important point considering how many have no coal, oil, or uranium.[10]

A Renaissance for Wind Pumps

The technology that opened the American West in the nineteenth century may turn out to be a lifesaver for the world's semiarid regions during the late twentieth. Diesel pumps are a costly means of drawing up the water so desperately needed for irrigation, livestock watering, and general household use in developing countries, and wind pumps now appear to be a viable alternative. In fact, for drawing water, wind power is a perfect match: When it is windless, water users can simply draw on water pumped into a storage tank on windy days. Storing water is of course far less expensive than storing electricity.

Approximately one million mechanical wind pumps are in use today. Most are located in Argentina, Australia, and the United States, where they mainly provide water for livestock. Since most mechanical wind machines have an energy capacity of less than half a kilowatt, the world's wind pumps supply at best a few hundred thousand kilowatts of power—less than one large thermal power plant.[11] Yet mechanical wind pumps play a crucial role. Imagine, for example, the cost and difficulty of getting coal-fired electricity to isolated ranches in the Australian outback.

Most mechanical wind machines use anywhere from four to twenty blades to capture the wind's energy, which is then transferred by a drive shaft to a pumping mechanism. The most common wind pump in use today is the American multibladed fan-type machine. Heir to the horizontal-axis design that dotted the plains in the nineteenth century, this rugged machine will operate effectively at average wind speeds below ten miles per hour. Most of the machine's parts, including the blades, are made of metal, and the diameter varies from two to several meters. Costs run from around $4,000 to over $10,000 per unit.[12]

Most wind-machine manufacturers are in Argentina, Australia, and the United States, though wind-pump industries can be found in New Zealand, the Philippines, South Africa, and West Germany, too. Although sales plummeted during the fifties and sixties, particularly in the United States, the industry remains strong in Australia and South Africa, where wind pumps are standard equipment on farms and where spare parts and repair services are readily available.[13]

Since the early seventies, the market for wind pumps has again begun to grow. But it remains concentrated in regions where wind machines have been in long use. Modern large farms and deep wells require more pumping capacity than most windmills can supply, so many farmers have been slow to adopt the technology. Wind pumps could be used widely in developing countries, but efforts to import machines for development projects have frequently foundered because the designs were poorly suited to local wind availability, economic needs, or social customs. Many wind pumps go out of commission for want of a few minor spare parts or an oil change. In one project in Zambia, local people eventually dismantled imported windmills piece by piece for use for other purposes.[14]

Solving these problems would both assure a large role for wind pumps and raise rural villagers' living standards. In areas with average wind speeds of at least 10 miles per hour, wind power already can provide pumped water for small-scale uses at approximately half the cost of diesel power. Recent studies found that even in the less windy parts of India, wind pumps are now cheaper to use than diesel pumps.[15]

Research on wind pumps for Third World use has picked up speed in recent years and has been carried out mainly by private nonprofit organizations supported by national governments and international aid agencies. These windmill development projects have relied on materials that are both cheap and locally available, an approach that directly involves and benefits the rural poor. Wind pumps stand as a prime example of what E. F. Schumacher called an "intermediate technology"—one that employs modern engineering and yet is well matched to the needs of the rural poor, providing jobs and creating self-reliant communities.

In particular, the sailwing or Cretan windmill, first developed in the Greek islands but now used for irrigation in several Mediterranean countries and Thailand, lends itself beautifully to local manufacture out of indigenous materials. Improved versions have been built in Colombia, Ethiopia, Gambia, India, and the United States to meet the needs of small farmers. Another innovative design based on traditional windmills is the Savonius rotor, a vertical-axis machine typically made of two oil-drum halves mounted around a perpendicular shaft so as to catch the wind.[16]

Some researchers and government planners are now working out ways to make wind machines an integral part of rural development and to build an indigenous manufacturing capability. Las Gaviotas, a rural development institute in Colombia, has spent six years designing a reliable and inexpensive fan-type windmill that pumps domestic or irrigation water in low winds. A production facility has been built that turns out twenty-five windmills per day, and the national government is helping fund the placement of the wind machines throughout rural Colombia. A similar strategy is used by the London-based Intermediate Technology Development Group (ITDG), which has developed a prototype fan windmill it hopes local industries in many poor countries will one day manufacture. Already, Kijito, a small firm in Kenya, has begun turning out ITDG-designed wind machines.[17]

Another camp of wind-power experts argues that for economy's sake wind-pump users themselves should build the pumps out of local materials rather than waiting for an industry to grow up.[18] In Thailand, where simple wind pumps are widely used by small farmers, this approach has worked. Elsewhere a local commercial market will be needed. Whatever approach is taken, it is sure that domestic manufacturing will go farther toward providing employment and keeping costs low than importing wind technologies will. Programs that train people to install and repair wind machines are also essential for any successful effort to introduce windmills. Agricultural cooperatives and extension services may prove ideal for transferring this know-how.

Clear Sailing

Another use of wind power even older than wind pumps is also being revived. Fishermen and shipping companies looking to reduce fuel costs are adapting sails for their vessels. Using new designs based on modern synthetic materials and computer-assisted control systems, modern mariners are proving that sailing boats can once again serve practical ends. A major study sponsored by the U.S. Maritime Agency concluded in 1981 that a combination sail-and-diesel system is more economical than either used alone. (The power fraction provided by the wind should ideally be between 20 and 30 percent, depending on how high the average winds are.)[19]

The use of wind power appears most feasible on coastal cargo vessels and fishing boats. Many such craft travel in areas with steady winds, and their relatively small size makes it easier to adapt sail technology. The Phoenix, a 20-meter two-masted schooner launched in 1982, provides convenient transportation for passengers and commercial goods around Long Island Sound, with a 20 to 25 percent fuel savings. Another ship, the 60-meter Greek vessel Mini Lace, was chosen as one of the ten outstanding engineering accomplishments of 1980 for its energy-saving sail retrofit. More difficult is harnessing wind power on large vessels, but a Japanese company has already built a sail-assisted 3,000-ton oil tanker and has plans to construct one three times its size.[20]

For developing countries sail-powered boats may have an especially important role to play. The thousands of small fishing and cargo vessels so essential to the economies of developing nations with extensive coastlines eat up large amounts of oil. Sailing vessels could be the answer, as successful reliance on sail power in parts of the Philippines and Sri Lanka shows. If the promise of such examples or the results of feasibility studies are any measure, sails could again become a common sight in the world's commercial fleets. However, as Lloyd Bergeson, a designer of sail-powered ships, noted in 1982, "It took us nearly a century to change from sail to steam, and it will take a while to change back again."[21]

Electricity from Small Wind Machines

Although the wind has been used to generate electricity since before the turn of the century, it has never been a widespread power source. Today change is in the air. Electricity price increases and technology improvements have given the small wind turbine industry a new lease on life.

Before the seventies virtually all wind turbines were used at remote sites with no access to an electricity grid. The machines —small and connected to storage batteries—were designed specifically with that market in mind. Approximately 20,000 direct-current wind turbines of this sort are in use today at fire lookouts, remote airfields, isolated ranches, coastal buoys, and the like. Although the power these wind machines generate costs more than 20¢ per kilowatt-hour, other means of generating electricity in remote areas cost even more.[22]

Today one of the most important technological and economic changes afoot is the development of wind power systems that do not require batteries. Modern technologies convert the direct current produced by a wind turbine into alternating current that can be fed directly into the utility grid. Instead of relying on batteries or going without power when the wind dies down, the user draws electricity from the utility's lines just like other customers do. When winds are high and electricity needs low, excess power automatically enters the electric lines and is sold to other customers.

Most of these recently developed wind turbines have a horizontal axis and two to four blades that rotate at speeds that vary with the wind speed. Most face upwind of the tower with a tail mounted behind the rotor to maintain this position. A few models face downwind, which eliminates the need for a tail, but can cause air turbulence problems. With blades of metal, wood, or fiber glass, the new wind machines are stronger and lighter than older models. Most have blade diameters of 5 meters or less and generating capacities of between 2 and 5 kilowatts (enough to supply the power needed by a typical modern residence in a windy area). Interest is growing in somewhat larger machines of up to 50 kilowatts that could be used by farms or small industries.[23]

These new wind-energy systems are most popular in Denmark and the United States, largely because of high winds, a tradition of wind power use, and favorable utility policies in both countries. In the United States approximately forty manufacturers sold 2,400 small wind machines in 1981. Yet even in these countries the industry is young and subject to normal growing pains. Some firms are barely surviving, selling only a handful of wind turbines a year, and the quality of the machines sold is still uneven.[24]

Wind turbine manufacturers are working hard to resolve these difficulties. They are beginning to replace some "off the shelf" components with those engineered specifically for wind-turbine use, and both private industry and government programs are aimed at increasing rotor efficiency and making transmissions and generators more reliable. Needed still are lightweight, inexpensive, yet rugged blades and lightweight, flexible towers designed specifically for wind turbines.[25]

To break into the mass market wind machines must be reliable and have life spans of at least twenty years. Ned Coffin who heads the Enertech Corporation, a leading U.S. firm, notes: "The key to our business is making a windmill that is idiot proof. It has to be maintenance-free like an icebox." The pace at which costs come down, sales expand, and reliability improves will, of course, be determined in part by large-scale production and assembly techniques. The largest firms in business today produce only a few hundred wind machines per year, which means that each turbine is essentially handmade and that the wind-turbine field is about as far along as the auto industry was before Henry Ford introduced the Model T. Manufacturing wind turbines on an assembly line would be even easier than assembling cars. On a production line several thousand wind machines a year could be turned out at substantially reduced costs, even if the technology were not otherwise improved.[26]

Today a typical household-sized wind energy system of 3 to 5 kilowatts costs between $5,000 and $20,000 and generates electricity for upward of 15¢ per kilowatt hour. At this price wind-generated electricity costs between 50 and 100 percent more than electricity from a central grid, so further cost reductions are clearly needed. Yet wind-turbine researchers believe that technological improvements such as those just described could bring generating costs down to approximately 5¢ to 10¢ per kilowatt hour where the wind averages 12 miles per hour. Then small wind turbines would enjoy a huge market in many areas of the world.[27]

Designs still being investigated could turn out to be both more effective and less expensive than the best conventional machines marketed today. Vertical-axis wind machines resembling miniature merry-go-rounds are already being marketed by one company in Great Britain and another in the United States, though these machines' commercial future depends heavily on further research. Another promising alternative, the sailwing turbine developed at Princeton University, has two curved blades made of wire and cloth. Private industry and several governments are testing a number of other designs that may one day reach the market.[28]

One particularly promising idea has yet to receive the attention it deserves—using wind turbines to heat water for space heating. Recently developed "heat churns" that use mechanical power to heat water are well suited for use with a rotating wind turbine. In windy regions such a device would even now be cheaper than electric resistance heating, and soon it may cost less than fossil-fuel heating in most regions. In Canada and northern Europe, where heating needs are great and winter winds strong, wind-powered churns could be ideal.[29]

Wind Power for Utilities

Quintessentially decentralized, wind may nevertheless power centralized energy systems operated by or for utilities in the coming decades. By clustering large numbers of wind turbines in areas where wind speeds average 14 to 20 miles per hour, "wind farmers" can generate electricity for transmission to industrial and urban areas. Since most areas with such extraordinary winds are only thinly inhabited, wind farms represent the only way the energy potential of these regions can be tapped.

One step in making wind farms a reality is technological. Large turbines appear to have an important long-run advantage for use on wind farms since they are cheaper to build on a per-kilowatt basis and they can more fully exploit a windy site.[30] Since the early seventies engineers in several countries have been working to develop technologically sophisticated turbines that would dwarf those Don Quixote charged at la Mancha.

Typically, a wind machine is considered large if its capacity is 100 kilowatts or more, but several machines capable of generating at least 1,000 to 4,000 kilowatts (1 to 4 megawatts) are in the works. Along with various machines with capacities of 200 to 1,000 kilowatts, five types of multimegawatt wind machines are currently being developed in three countries. The largest of these machines could generate sufficient power for over 1,000 typical U.S. homes or for perhaps twice as many residences in countries where electricity use is lower. Yet it would take a wind farm with 500 of these large turbines to generate as much power as one of the large thermal power plants in use today.[31]

In basic appearance large and small turbines are quite similar. But other differences are great. Large wind machines, essentially an aerospace technology, require meticulous engineering. Their blades are typically as long as a jumbo jet's wings—usually over 50 meters—and the latest computer technology controls the blades' angle and rotational speed. The stress on these blades is enormous, so designing them to hold up in heavy winds has been a world-class engineering challenge for the high-technology firms that dominate the business. In both the United States and Europe engineers who cut their teeth on jet aircraft technology are directing large-turbine research efforts.[32]

The United States has been a pioneer in the development of large wind machines. In 1975 the U.S. National Aeronautics and Space Administration (NASA) began contracting with private firms to develop a series of large horizontal-axis turbines. Under the Department of Energy's supervision this program has resulted in a commercial effort to install thirty-six 3,500-kilowatt turbines at a wind farm in California for $400 million. Designed by Boeing, these breathtaking machines (called Mod-2s) have two narrow blades that describe an arc nearly 100 meters in diameter. On a clear day the turbines can be seen from five miles away.[33]

Plans for other, more advanced but less expensive wind turbines are continuing but have been slowed by the Reagan administration. Meanwhile, however, two U.S. companies—

the Bendix Company and the Hamilton Standard Corporation—have developed large wind machines. These turbines are at the prototype stage, but early performance data indicate that they too could yield power that competes economically with that from conventional power plants. The key to unlocking that potential is to improve the reliability of the large wind machines—particularly their capacity to operate safely in unusually high winds—so that they can become standard utility equipment.[34]

Since the U.S. program was launched, Canada, Denmark, Great Britain, the Netherlands, the Soviet Union, Sweden, and West Germany have begun to develop large turbines. One of the most impressive efforts is taking place in Denmark, where engineers hope a 630-kilowatt machine they have designed will soon be used widely on Denmark's coast. In England Taylor Woodrow Construction, Ltd., a major engineering firm that also builds nuclear power plants, is under government contract to design a 3,000-kilowatt wind turbine that could be mass produced by the late eighties.[35]

Another design, the Darrieus wind turbine, is also coming into its own. The governments of Canada and the United States have separately financed the development of this "upside-down eggbeater." With two or three curved aluminum blades turning a central upright shaft attached to a ground-based transmission and generator, the Darrieus works well in high winds. The blades extend close to the ground where less wind is available, however, and they must withstand varying levels of force as they pass in and out of the "eye" of the wind. It is still uncertain whether Darrieus machines will ever enjoy wide use.[36]

As research on electricity-producing machines continues, utilities are looking for ways to make use of arrays of large wind machines on wind farms. As of 1981, 110 U.S. utilities had wind-energy programs, up from just 50 in 1979. Although most are just small demonstration projects or feasibility studies, several utilities are on the verge of making major commitments to wind power. In Great Britain the Central Electricity Generating Board is seriously studying the nation's potential for wind farm use. In the Netherlands the national electricity association, SEP, is developing a 10-megawatt experimental wind farm and plans to use wind power to generate 7 percent of the country's electricity in the year 2000.[37]

Meanwhile, commercial development of wind farms has begun in the United States. The world's first wind farm began operation in Vermont in 1981, relying on 30-kilowatt turbines developed by U.S. Windpower, Inc. Since then, over twenty wind farm contracts have been signed, including one for a large 80-megawatt project in Hawaii and one for a 125-megawatt project in northern California.[38]

California is clearly the world's pioneer in wind farming. Blessed with mountain passes and other ideal wind sites, California also has a state government keen enough on wind to enact its own wind-energy tax credits, conduct wind resource assessments, and require utilities to buy power from wind farms at a fair price. By the end of 1982 California had 1,000 wind machines with a total capacity of 60 megawatts located at a dozen wind farms, and the industry continues to grow explosively. Most of these wind farms employ small and medium-sized turbines, each with a capacity of between 10 and 100 kilowatts, but large multimegawatt turbines will be used at some projects now in the planning stages. The California Energy Commission's goal is for the state to have 700 megawatts of wind farms by 1987 and 4,000 megawatts by the end of the century.[39]

Much of the early work in developing wind farms in California and elsewhere in the United States is being carried out by small innovative firms formed specifically to tap this power source. Companies such as U.S. Windpower, Inc. and Windfarms Limited have started signing contracts with utility companies to supply wind-generated electricity at the same price as power from newly built conventional plants. The small wind-energy entrepreneurs typically locate their own financing and lease the land on which the machines are constructed. Aided by generous federal and state tax incentives, these firms can invest in new power sources that utilities will not develop on their own. For the utilities, tapping the wind in this manner is of course risk-free, so entrepreneurialism bridges the institutional gap that poses the largest remaining barrier to wind power's widespread use.

The economic verdict on wind farms is now clear. If well-designed wind machines are placed at good wind sites, electricity can already be generated for as little as 10¢ per kilowatt hour. In parts of California, the North American Midwest, northern Europe, and many developing countries where oil-generated electricity is common, wind farms are close to being economically viable now. When wind farms employ later generations of mass-produced wind power technologies, studies in Europe and the United States indicate they will be able to produce electricity that costs between 3¢ and 7¢ per kilowatt hour. By the nineties wind farms will likely have an economic advantage over coal and nuclear power plants in many parts of the world. Until then, what is most needed is more work aimed at increasing these machines' reliability.[40]

Obstacles and Opportunities

Of course, even wind power's economic appeal does not seal its future. The environmental impact of large wind machines as well as the effect of wind power on utility company planning loom as important constraints. Then, too, outdated government policies could impede the spread of wind machines, and few nations have fully charted their wind resources or launched adequate research programs.

As with many energy technologies, the land-use effects of wind machines are key determinants of their acceptability.

Historian and environmentalist Roderick Nash has observed, "Most people do not yet fully realize that obtaining a meaningful amount of power from the wind involves far more than a few picturesque structures surrounded by tulip beds." Indeed, a wind farm with a generating capacity equivalent to that of a 1,000 megawatt power plant would require approximately 82 square kilometers of land. Meeting the California goal of providing 10 percent of the state's generating capacity with wind farms requires placing between 10,000 and 100,000 wind machines on approximately 615 square kilometers (two-tenths of 1 percent of the state's land area). More generally, then, wind-rich countries should be able to get up to half of their electricity from wind machines that will occupy no more than 1 percent of their land.[41]

The most important land-use issue surrounding wind power development is not the total amount of land needed. Instead, it is the potential for ruining scenic wilderness areas or other highly valued land. Clearly, many areas must be kept off limits for wind machines. However, detailed wind assessments show that many good sites exist on land used only for livestock grazing, an activity quite compatible with wind farming.[42] Even large wind turbines are graceful and relatively nondisruptive structures, so dual land use should be possible outside of national parks and other scenic areas. Countries with ample wind should be able to get between 10 and 25 percent of their electricity from the wind without running up against serious land-use constraints.

Noise and safety concerns are another matter. Annoying sounds and inaudible vibrations have been a problem with some experimental wind machines. These problems are avoidable, and wind machine designers are now working to ensure that wind turbines are quiet neighbors. As for safety factors, they will prevent the installation of wind turbines in many densely populated communities. Blade loss is the greatest danger, and even small machines can inflict harm if they fall apart in a heavy wind. Manufacturing and installation codes are clearly needed and some communities have adopted regulations to prevent wind machines from being set up too close to a neighbor's property. Many cities will probably want to ban wind devices entirely, especially considering how paltry wind potential is in most urban areas.[43]

Television interference is another problem, particularly with larger turbines with metal blades. Usually, the effect is quite localized, though some large experimental wind machines have caused video distortion a few kilometers away. Some recently developed wind machines have fiberglass blades, largely solving the problem. Another more expensive remedy is to install cable television in the affected areas. This problem clearly needs more work, and unless resolved, television interference could impede wind power development in some areas.[44]

In U.S. communities where wind turbines have already been erected and public opinion surveys have been carried out, the machines have been well received, so long as there is no noise or television interference.[45] In environmental terms, wind energy is a refreshing contrast to air-polluting coal plants and potentially dangerous nuclear reactors. However, the use of virtually all technologies entails trade-offs, and continued attention to wind power's environmental impact will be essential if wind is to be a major and welcomed energy source. Encouragingly, such concerns are being aired early.

Another critical influence on wind power's future is electric utility policy. The most economical way to use wind generators is to connect them to electricity grids. Yet utility interest in wind power is halfhearted in most regions. Utilities are accustomed to investing only in established, risk-free technologies, and wind machines are only now beginning to meet those criteria. At the same time, many utility managers view wind power as a threat to utilities' monopoly on power production. As a result, some have enacted unnecessarily stringent requirements for wind machine owners who want to interconnect with the electricity grid. Others will pay only low rates for the power produced by wind machines. Even in California government had to pressure utilities to use wind power. True, the use of wind energy presents some new challenges for utility planning, but none is insurmountable. The overriding point is that utilities hard-pressed to finance additional capacity stand to gain by hooking up with small power producers who could save them the initial investment. And soon wind power will be less expensive than coal and nuclear power.

Some skeptics argue that utilities need a more dependable, less intermittent source than wind to boost their capacity. But these critics tend to overlook the unexpected plant shutdowns that make even conventional power plants much less than 100 percent reliable. Nuclear power plants in the United States operate on the average at only about 50 percent of their rated capacity. A single shutdown of such a large plant causes havoc for utility managers who must always have some generating capacity in reserve just in case. Similarly, wind machines sometimes do not operate because of a lack of wind, but this problem can be reduced if thousands of wind machines are spread over a wide area. If developed carefully, wind power can provide reliable electricity and actually add strength to utility grids.[46]

A few progressive utility companies have already begun planning how best to integrate wind power into their electricity grids. One promising strategy is to operate wind turbines in conjunction with hydropower plants. By operating the hydro facilities at full capacity when the wind is not blowing and by slowing them down when the breezes are abundant, utilities can derive maximum benefit from wind machines. The northwestern United States, the James Bay region of eastern Canada, most of Scandinavia, and parts of the Soviet Union all have large hydropower and wind power resources located nearly side by side.[47] Also essential to successful integration of wind power with a utility grid are electricity pricing policies that encourage the use of power when it costs least to produce —that is, when the wind is blowing.

Government support is also needed for the successful development of wind power. The large strides being made in Denmark and California would not have been possible without tax credits. Tax incentives reduce the investment risk and thus stimulate the early development of the wind power industry. Similar subsidies elsewhere in the world could for relatively small sums make the wind power industry strong and independent. Pioneering countries could find themselves with an important new export technology to boot.

Also deserving of government financial support are programs for introducing wind pumps and turbines in the rural Third World. Already, many developing country governments and international aid agencies finance the import of diesel engines and other technologies that are more expensive and less reliable than wind machines. Such agencies could also help individuals and communities buy windmills and help local industries acquire the means to manufacture them.

So far, most government support for wind power has been through research and development programs, mainly on large wind machines. Although many of these programs have encountered technical problems, steady progress since the early seventies has resulted in the development of several large, commercially ready wind machines.

To support development of wind pumps and small turbines, governments have done much less. Some say these machines are simpler and already highly developed, and thus require less assistance. While this is true, it fails to justify the huge disparity in funding levels, especially when one considers the contribution these smaller machines could make. Unfortunately, most governments seem attracted almost exclusively to high-technology utility-oriented research programs. Still, many countries have in recent years begun small-wind-power development projects. Denmark and the United States have helped industry by establishing test centers that allow private companies to test their small wind turbines free of charge—a move that helps consumers too. Cooperation between manufacturers and government on other aspects of the design and use of small wind machines would certainly assist the spread of this technology.[48]

Although government support for wind energy technology expanded throughout the seventies, it has recently stagnated and even fallen. Particularly dramatic are the cuts in the U.S. wind energy program—the world's largest. From a peak of $60 million in 1980 the U.S. wind power R & D budget was cut to $35 million in 1982—ironically, just as the wind power industry was reaching the critical take-off stage. Yet by going beyond traditional basic research and channeling funds into advanced engineering modifications and demonstration projects, government could help companies to commercialize wind machines much sooner than they otherwise would.[49] Unfortunately such programs have been almost eliminated by the Reagan administration's budget cutters. In some cases the reductions are akin to stopping work on a bridge only a few meters short of completion.

Another important task for governments is wind resource assessments. Wind surveys have been carried out haphazardly so far, so knowledge is sketchy. Because the amount of wind available is critical and can vary widely over short distances, governments need to publish general information on the amount of wind in an area as well as lend wind-measuring equipment to individuals or utility companies evaluating a particular location. Such inventories will be essential in mapping out wind-energy development programs, and they could help mobilize business support for these efforts.

In California wind assessments helped energy officials revise their opinion of the state's wind-power potential. Because early estimates were based on data recorded at airports whose locations are chosen in part because they lack wind, California's wind potential had been underestimated. More thorough meteorological calculations revealed a veritable treasure. California's wind prospectors have already found sites for wind farms that could together generate 5,500 megawatts, mainly in windy mountain passes that are relatively unpopulated and yet reasonably close to urban centers. Now "wind prospecting" is a growing business in California, and in some parts of the state early assessments have set off a small land rush.[50]

The international sharing of information on wind availability and of ways to obtain the data could boost wind power development significantly, particularly in poor countries. Now inventorying its own wind resources, the United States will be in a good position to help other nations do the same. The U.N. World Meteorological Organization has also become involved in wind-energy assessments, publishing in 1981 a map showing the general world distribution of wind resources.[51] This agency and relevant professional organizations could adopt standardized assessment procedures and information channels, as agricultural and scientific research centers now do.

Wind's Energy Prospect

While detailed wind data is still scarce, enough information has been collected to assess the wind energy prospect broadly. By almost any account, simple mechanical windmills hold tremendous promise for areas where lifting water is a critical energy need. Since wind pumps can be used effectively where wind speeds average as low as eight miles per hour, they can be used on well over half the earth's land area. Most countries can make at least limited use of mechanical wind pumps, and in such semiarid regions as East Africa, the Indian subcontinent, northern Argentina, northeast Brazil, Mexico, and Peru, they could be a godsend. Only in tropical areas that lack good trade winds is their use out of the question.[52]

How widely wind pumps are used will depend primarily on

efforts to make them available and get them adopted in Third World countries. Success will come first where wind pumps are well matched to current water needs—parts of southern Asia where small-plot agriculture is practiced and parts of Africa where brief but heavy rains could fill reservoirs with water that could be pumped for irrigation during the windy dry months. Agricultural extension services and rural cooperatives can provide the institutional impetus for such projects, while national governments and international aid agencies can provide the financial muscle.

How much energy these wind pumps could supply is difficult to estimate. Compared to major commercial energy sources, the amount is probably not large. But in terms of the number of people whose lives could be improved, the contribution would be tremendous. By the middle of the next century, several hundred million farmers, villagers, and rural poor could be benefiting from wind energy.

Electricity-generating wind machines cannot be used as widely as wind pumps, but their potential is nonetheless large. Preliminary data indicate an abundance of sites for individual wind machines and wind farms throughout the world's northern temperate zone, especially in the plains regions of China, North America, northern Europe, and the Soviet Union. Coastlines also offer good wind power potential. Denmark, buffeted by strong winds near the North Sea, already has about 500 small wind turbines in use—perhaps the largest concentration in the world.[53] In tropical developing countries, wind power generation will probably be most common along Africa's northwest coast, South America's west coast, and on windy islands such as those in the Caribbean and Mediterranean regions where the only source of electricity is expensive diesel generators.

Small wind turbines could be the first technology that allows a significant number of individuals to generate their own power. An extensive survey sponsored by the U.S. Solar Energy

Research Institute in 1980 included a detailed evaluation of the many considerations that affect market potential—wind speed, utility rates, income level, housing density, and the like. The study concluded that 3.8 million homes and hundreds of thousands of farms in the rural United States are well suited to the use of small wind generators.[54] If this study is right, the United States could one day have several million small wind turbines in use at homes and farms, providing 8 percent of the nation's current electricity use.

In the countryside wind can be used for everything from operating milking equipment to running household appliances. Accordingly both the residential and agricultural markets are expected to grow rapidly. The largest market will probably be for relatively small turbines capable of generating perhaps 3 to 5 kilowatts—enough for one household's needs. Intermediate, 10-to-50 kilowatt, machines will also enjoy sales growth as they are put to use in more industries, large farms, and towns. Businesses in particularly windy regions could boost their income by selling power back to the utility.

These sales notwithstanding, the power generated at wind farms may double that produced by individual machines. As California has proven, the wind farm concept is technically feasible and economically appealing. Extensive searches for wind farm sites have begun elsewhere in the United States, and preliminary surveys have been conducted in some Western European countries and the Soviet Union. Many regions with major wind farm potential have been identified, and many more will undoubtedly be found as wind prospecting takes hold.

For coastal nations, one possibility may be placing wind farms offshore. According to an extensive feasibility study carried out in Great Britain by Taylor Woodrow, an engineering firm, platforms similar to those used for oil drilling could be built in the North Sea and a submarine cable could conduct power to a central relay station onshore. Even with all the extra costs entailed in working at sea, the study concluded, offshore wind farms may soon be economically competitive with British nuclear reactors.[55]

Together, wind farms and independent wind machines should one day provide 20 to 30 percent of the electricity many countries need. Even where winds are not high and the overall electricity supply systems are not very compatible with wind power, it can make some contribution. Overall, it seems reasonable to expect that if electricity generation worldwide increases another 50 percent the wind can one day provide 12 percent of total generating capacity—that is 350 gigawatts—or slightly less than 10 percent of actual power generation—that is 900,000 gigawatt hours. In other words, wind power should be able to provide 10 exajoules of primary energy, or about half as much as hydropower does today. This would require millions of wind machines and perhaps half of 1 percent of the world's land.[56]

Surrounded by uncertainties, the pace of wind power development is difficult to predict firmly. One "if" is the wind turbine industry: Wind machines must become more reliable and be mass produced if they are to be used widely. Public and private investments over the next five to ten years may well lay the groundwork for rapid expansion by the late eighties according to industry observers. Also critical now are programs to carry out wind assessments, modify utility policies, and ensure the environmental acceptability of wind power. Under favorable but less than ideal conditions, wind power could provide a few thousand megawatts of generating capacity by 1990 and as much as 20,000 megawatts by the century's end.

Once the initial market breakthrough is made, the wind power field could unfold rapidly. Among other things, increased investments and the pioneering work being carried out in Denmark, California, and other areas are erasing some of the credibility problems that originally plagued wind energy. Some engineers and technocrats who earlier steered clear of "unconventional" technologies are now enthused about wind power, as are an increasing number of rural development planners, utility executives, and consumers. If recent technical achievements are backed up by effective industry and government policies, wind power could reach the all-important turning point.

10

Geothermal Energy
The Powering Inferno

One major renewable source of energy does not come from sunlight. Geothermal energy comes directly from the earth's vast subsurface storehouse of heat. Like the sun's energy, that heat is the product of gravitational collisions, atomic reactions, and radioactive decay. Just as the sun will eventually cool, so, too, will the earth. But meanwhile—for millennia—it can supply immense amounts of energy.[1]

By no stretch of the imagination is geothermal heat today oil's equal, or even wood's. The twenty countries that use geothermal energy for purposes besides bathing cull approximately 0.5 exajoules of energy each year, 60 percent of it in the form of direct heat and the rest as electricity. Although not yet a major component of the global energy budget, this is enough direct heat to meet the needs of over 2 million houses in a cold climate and sufficient electricity for 1.5 million houses.[2]

Geothermal energy use is rising rapidly. Where resources are abundant and accessible, geothermal power is already an energy bargain, usually less expensive than electricity generated by coal and nuclear power plants. If technological advances proceed apace, the few countries that have already committed themselves to geothermal development will be joined by dozens more.

Subterranean Fires

All geothermal heat comes from magma—the molten rock that underlies the earth's roughly 40-kilometer-thick crust. While temperatures typically increase only 25°C with each kilometer of depth, temperatures as high as 360°C are in some areas found close enough to the surface—2 kilometers—to be reached with current drilling technology. These anomalous "hot spots" are also home for volcanoes, geysers, and hot springs.[3]

Most geothermal activity occurs where two plates of the earth's crust meet, allowing the cauldron of fire to reach close to the surface. As a result, the world's geothermal riches include the area where the mid-Atlantic ridge bisects Iceland, areas around the Mediterranean, the Rift Valley in East Africa, and the "Ring of Fire" that extends around the Pacific Basin. Yet even outside these areas, which comprise about 10 percent of the world's land mass, are abundant lower-temperature geothermal deposits.[4]

The world's geothermal resources fall into four broad classes, each of which has unique problems and possibilities. *Hydrothermal reservoirs* are found where permeable, water-bearing rock sits atop very hot, impermeable rock. There, water touching the heat source rises to the surface, cools slightly, and flows outward and down to be heated again. Over time this water reaches an equilibrium temperature, ranging from only slightly above that of groundwater to far above boiling. Lower-temperature hydrothermal reservoirs are widespread, whereas resources with water (or steam) over 150°C are limited to a few regions. A few hydrothermal deposits have sufficient temperature and pressure to yield dry steam, the most valued geothermal resource.[5]

While hydrothermal reservoirs are the primary source of geothermal energy today, the three other types of geothermal energy also have long-run potential. *"Geopressured" reservoirs* are formed when plant matter trapped in sedimentary basins decomposes and produces methane—the main component of natural gas. As the overlying sediments exert increasing force, the pressure and heat build. Such reservoirs are not as widespread as conventional hydrothermal reservoirs, but in the U.S. Gulf Coast and a few other areas they are abundant.[6]

Hot dry rock and *magma* are the ultimate geothermal resources, but using either poses difficult problems. Utilizing the hot dry rock found throughout the world will require developing a novel heat-extraction technology since there is no naturally circulating water present. As for magma itself, tapping its heat will probably be confined to volcanoes and other geophysical anomalies that bring magma close to the surface. Technologies for using the hellish temperatures found at such sites have yet to be developed.

One spur to developing geothermal energy is the extent of the resource. The earth contains substantially more energy within it than humanity has used so far. Yet exactly how much energy can be tapped and where remain largely unanswered questions. Only modest geothermal resource surveys have been carried out until now. What these crude estimates do make clear is that geothermal energy is much more abundant than was once believed and is sufficient to allow a vast expansion in its use.[7]

The Earth's Energy in Harness

Making use of the earth's heat is not a new idea. For millennia people have flocked to hot springs. Two thousand years ago both the Romans and the Japanese were relaxing in elaborate geothermal hot baths. By the ninth century Icelanders were using geothermal heat for cooking. In the Middle Ages several towns scattered around Europe distributed naturally hot water to heat houses.[8]

While geothermal technology has advanced far in recent decades, the simplest uses remain among the most popular. Japan's 1,500 hot springs resorts are visited by 100 million people each year, for example, and require no drilling, little piping, and a minimal investment. Yet it would take five large conventional power plants to heat these baths were nature less obliging. In parts of Mexico people wash clothes with naturally hot water. Some Thais and Guatemalans use it to boil vegetables and tea.[9]

In the Philippines and Kenya some crops are dried with low-temperature geothermal heat. In Idaho an aquaculture facility that uses geothermal water has found that the fish grow 25 percent faster and seldom succumb to disease. The largest agricultural application is greenhouse heating. Hungary already has 70 hectares of geothermal greenhouses in use, while Italy is saving $600,000 of fuel oil a year by using several such greenhouses.[10]

In scattered applications geothermal heat has also found a place in industry. In northern Iceland a mineral-processing plant uses geothermal energy to remove the moisture from siliceous earth. In New Zealand the Tasman Pulp and Paper Company relocated its mills during the 1950s to be near geothermal energy sources. Saving 30 percent on energy, the company pockets an extra $1.3 million annually now. At Brady's Hot Springs, Nevada, an onion-dehydration plant using geothermal energy is saving $300,000 per year, enough to motivate the company's managers to expand the original plant and construct an additional one.[11]

Heating homes is the widest application of geothermal energy today. Since the turn of the century, the residents of Klamath Falls, Oregon, have drilled more than 400 wells to tap the 40° to 110°C water beneath their houses for space- and domestic water–heating. Household wells there have heat exchangers that transfer heat from the briny subterranean reservoir to the pure water circulating to the house. Using a heat exchanger conserves the resource, minimizes corrosion, and skirts the problem of waste-water disposal. These systems cost from $5,000 to $10,000, and the one-time investment can in many cases be shared among several households so that the life-cycle costs compare favorably with conventional heating options.[12]

Further boosting the economies of shared systems, many communities have turned to geothermal district heating. The most impressive example is Iceland, whose immense geothermal resources provide 75 percent of the population with heat. In Reykjavik, the capital, nearly all of the city's 112,000 people use heat from two geothermal fields under the city and from another 15 kilometers away. Visitors to this frigid city in the 1930s recall the pall of coal and wood smoke that engulfed it in winter. Today the air is clear, and home heating costs 75 percent less than it would using fuel oil.[13]

Where human settlements sit astride geothermal resources, low-temperature district heating is an unbeatable bargain.[14] Careful planning and major investments by a local government or special heating district are needed, but large fuel savings justify both. A few such systems are already in place in France, Hungary, the Soviet Union, and the United States, as well as in Iceland.

To boost the efficiency and, thus, the economics of geothermal heating, some systems feature heat pumps. These electrical devices send a refrigerant, usually freon, through a series of chambers in which heat is extracted from one medium (such as naturally hot water) and transferred to another (such as air). If a heat pump is used, geothermal water at relatively low temperatures and even ordinary groundwater can be employed for heating. Some 50,000 groundwater heat pumps are in use in the United States, and the National Water Well Association predicts that rural use will expand rapidly. This is bound to happen, especially if groundwater heat pumps can be adapted to provide cooling in the summer—an intriguing idea still being tested.[15]

Even more popular than the direct use of geothermal heat today is its use in electricity generation. Small wonder. Moving hot water is expensive because insulated pipelines are expensive. Indeed, a 60-kilometer pipeline in Iceland is the world's longest. If converted to electricity, however, the energy in more distant geothermal sources can be put to use in cities and factories. Today approximately one hundred geothermal power plants of from 0.5 to 120 megawatts of capacity are operating in fourteen countries. (See Table 10. 1.) Including a few commercial plants and many experimental ones, the total generating capacity is approximately 2,500 megawatts and rising rapidly.[16]

The simplest technology for generating electricity is the dry steam system used at steam-only reservoirs. Only four such systems are in operation: two commercial complexes in Italy and the United States and two smaller systems in Indonesia and Japan. Electricity generation at these rare but prime sites is mainly a matter of piping the steam to a standard turbine. The largest complex is one at the Geysers in northern California that as of 1982 used 200 wells and 17 separate power plants to provide 1000 megawatts of generating capacity for the Pacific Gas & Electric Company and other utilities. These plants are among the most reliable and least expensive sources of electricity in the state.[17]

More common are geothermal reservoirs that contain both

Table 10. 1. Worldwide Geothermal Capacity in 1981 and Plans for the Year 2000

Country	*1981*	*2000*
	(megawatts)	
United States	932	5,824
Philippines	446	1,225+
Italy	440	800
New Zealand	203	382+
Mexico	180	4,000
Japan	168	3,668+
El Salvador	95	535
Iceland	41	68+
Kenya	15	30+
Soviet Union	11	310+
Azores	3	3
Indonesia	2	92
China	2	2
Turkey	0.5	150
Costa Rica	0	380+
Nicaragua	0	100
Ethiopia	0	50
Chile	0	15+
France	0	15+
Total	2,538.5	17,649+

Source: DiPippo, "Geothermal Power Plants: Worldwide Survey," and United Nations Conference on New and Renewable Sources of Energy, "Report of the Technical Panel on Geothermal Energy."

steam and water. Plants that tap this less ideal form of geothermal energy are found in at least ten countries, though many such projects are still experimental. One of the most successful facilities is the one in Wairakei, New Zealand. In nearly continuous operation since the mid-sixties, this reliable 190-megawatt plant has nonetheless run into problems. Electricity generation at the site declined during its early years, apparently because water was being extracted faster than it was being replaced. Generation has stabilized in the last decade, however.[18]

Wherever high-temperature geothermal water or steam is available, electricity generation looks to be an attractive proposition. Such areas are rarer than those suited for direct use, but electricity's versatility makes their rapid development a near certainty.

Technological Frontiers

The cost of harnessing geothermal energy would drop precipitously if three technical challenges were met. One is improving the means of locating and drilling for geothermal resources. Another is finding ways to use more abundant, lower-temperature resources for electricity generation. The last is overcoming the corrosion and pollution problems associated with the use of mineral-laden geothermal water.

The presence of hot springs and the like made finding most of the geothermal reservoirs now in use easy. Yet though many identified reservoirs have not yet been developed, attention has already turned to means of finding now hidden geothermal resources. Undoubtedly, some industries and urban areas sit atop geothermal resources that contain cheap energy they desperately need. But which cities and which industries? Most exploration and drilling relies on techniques similar to those used in natural gas development, and so petroleum companies are heavily involved in many geothermal projects. Their geologists have developed sophisticated remote-sensing techniques that are being adapted to geothermal energy prospecting. Chance still enters in, but such techniques can pinpoint the most promising drilling sites and reduce the number of "dry" wells drilled.

At present, drilling wells accounts for more than half the cost of some geothermal projects. A deep geothermal well can cost several hundred thousand dollars, twice the cost of the average oil well. When petroleum drilling techniques are adapted to the unique conditions of geothermal reservoirs, costs should fall, though by how much is hard to predict. Fred Hartley, president of the Union Oil Company of California, optimistically contends that the cost of geothermal wells can be halved.[19] Also needed are simple and accurate means of estimating short- and long-term production from geothermal wells so that planners and investors can make sound decisions.

Then, too, the pace of geothermal development will pick up when generating plants can make use of the more abundant geothermal resources that contain both steam and water. The conventional "separated steam" design employed in several countries uses the naturally available steam alone to run the turbines. In more efficient "double flash" plants in use in Iceland, Japan, the Philippines, and New Zealand, hot water brought to the surface is directed to a vessel where the pressure is reduced and additional steam is generated. These plants have met with some minor corrosion problems, but for steam and water over 200°C, double flash plants are likely soon to be the most widely used technology.[20]

A recent geothermal innovation that allows efficient electricity generation using lower-temperature water between 150°C and 200°C is the binary cycle plant. At these mainly experimental facilities, geothermal water is circulated in a closed loop and run through heat exchangers that transfer the geothermal heat to a secondary working fluid with a low boiling point. Since the moderately heated working fluid vaporizes and runs a turbine, relatively low-temperature geothermal water can be used. Testing in pilot plants in China, Japan, and the United States indicates that more research and operating experience is needed, but also that for the long run the binary plant appears to be a most promising design.[21]

Another way to use geothermal heat efficiently is to employ the same resource for both electricity generation and direct thermal uses—in effect geothermal cogeneration. Waters discharged from geothermal generating plants can be hot enough to use for residential heating or industrial processes. In two Japanese plants discharged geothermal water is distributed to households for space heating, cooking, and bathing.[22]

Impurities are a common problem at many geothermal energy projects. Picked up from subterranean rock by the hot circulating water, such nuisance materials as salts and silicates give rise to scaling and corrosion. While the geothermal water at Reykjavik is pure enough to drink, mineral concentrations have forced other plants to close. Moreover, the materials that corrode or scale the inside of a geothermal system often become pollution outside it. Hydrogen sulfide, a noxious gas that smells like rotten eggs, is the worst culprit. Found at almost all geothermal sites, occasionally it is concentrated enough to cause lung paralysis, nausea, and other health problems. At Larderello, Italy, emissions of hydrogen sulfide that are seventy times the U.S. Environmental Protection Agency's suggested standard have been detected. Pollution-control devices developed for use on coal gas can remove approximately 90 percent of the hydrogen sulfide, but so far only the Geysers plant in California and a few others use them. At less than 10 percent of the systems' cost, expense is no excuse for this lapse; geothermal plants need not become major polluters.[23]

Mercury, arsenic, and other potentially dangerous substances are found dissolved in geothermal water. Unfortunately, many plants simply discharge the toxic water they use into nearby streams and lakes. The river into which the Wairakei plant in New Zealand discharges its water has arsenic concentrations two to five times as high as those permitted in U.S. drinking water.[24] These problems could grow severe if no action is taken, but fortunately most of the dissolved substances can be kept out of water supplies by chemical removal or by reinjecting the geothermal water back into its subterranean reservoir.

Reinjection could also lessen other problems. One is subsidence. In some areas where large amounts of geothermal water are withdrawn from the earth, the land has started to sink. At Wairakei the ground level drops 20 to 60 centimeters per year. Like other subterranean processes, this one is poorly understood. Even so it can probably be forestalled by reinjecting the geothermal water to maintain the underground pressure balance. Reinjection is already used at many projects, but techniques need to be made more effective and affordable. One danger is that reinjected geothermal water could contaminate groundwater, a hazard that must be avoided.[25]

Another approach to avoiding subsidence and groundwater contamination is to place heat exchangers in the geothermal reservoir so that water does not have to be extracted at all—in many ways the cleanest, most elegant solution. A U.S. manufacturer, the Sperry Corporation, is developing a heat exchanger that it claims will generate electricity as efficiently as a binary plant although using geothermal water that is significantly cooler.[26] These systems remain at an early stage of development, however, and whether they will live up to expectations is uncertain.

Geothermal Horizons

To date only hydrothermal deposits—geothermal reservoirs containing steam, hot water, or both—have been exploited commercially. But alongside the heat in "geopressured" deposits of methane-saturated water, hot dry rock, and magma, even the substantial amount of energy in hydrothermal reservoirs seems paltry.

In 1975 the U.S. Department of Energy began assessing "geopressured" methane reserves at the site of the largest known reservoir along the Gulf of Mexico. There wells are being drilled to depths of 3 to 6 kilometers to determine whether the volume, temperature, and methane concentration are high enough to make the resource commercially exploitable. Early signs are not encouraging. Reservoir pools appear to be smaller and more expensive to reach than originally anticipated, so industry is losing interest. Even if the economic picture were to brighten, "geopressured" deposits entail serious pollution and subsidence problems that would be hard to resolve since the high pressure makes reinjection difficult.[27]

Hot dry rock is a much more common geothermal resource. It is widely distributed around the world. If a circulating fluid can be introduced into fractured rock, naturally occurring hydrothermal systems can be mimicked. Researchers in both England and New Mexico have demonstrated the feasibility of extracting usable energy from hot dry rock. (Hydraulic fracturing techniques were used at Fenton Hill in New Mexico and explosives at Cornwall in England.) But making this an economical source of energy requires considerably more research. Finding sufficient water to use hot dry rock could also be a constraint in many parts of the world.[28]

The ultimate technological challenge for geothermal engineers is to use molten rock directly. Although most of this magma is inaccessible, volcanoes sometimes bring molten rock with temperatures over 1000°C close to the surface. Several years ago the Soviet Union announced a plan to build a 5,000-megawatt power plant using magma at the Avachinski Volcano on the Kamchatka Peninsula. Construction has yet to begin, however, and many geothermal experts consider the idea unworkable. The only example of actual use of lava's extraordinary heat is in Iceland. On the island of Heimaey, a volcanic eruption that occurred in 1973 and forced the evacuation of a town of 5,000 people has provided a lava pool that the returning townspeople have tapped for district heat. In general, however, materials and equipment must be improved before it makes sense to use volcanic heat directly.[29]

Hot-Water Institutions

Like all renewable energy sources, geothermal energy cannot flourish until various institutions make accommodation. But in the case of geothermal development, some institutional expertise can be borrowed from the petroleum and utility industries. In particular, geothermal resource surveys similar to oil and gas surveys can be borrowed. As with petroleum, government's role here is to conduct the broad preliminary assessments that indicate whether and where the private sector should carry out more detailed surveys. Most countries with major geothermal development programs, including the Philippines and the United States, have begun such surveys, though few are as extensive as they might be.[30]

The legal status of geothermal resources also remains uncertain and potentially bothersome. In many countries the government owns all mineral resources, including geothermal deposits, found beneath the earth's surface, so it must participate in their development. In the United States, on the other hand, the law varies by state, and many landowners are unsure of their geothermal energy development rights. In most market economies it makes sense to follow the petroleum model, giving the private sector primary responsibility for developing geothermal energy, but standardizing leasing procedures, and charging the industry royalty fees if the state owns the resource.[31]

As geothermal resource policies are developed, environmental considerations must be woven into them. A major geothermal development can turn an area of tens of square kilometers into a giant construction site covered with piping, wells, and power stations. Some of the most valuable areas for geothermal development are even more highly valued for their aesthetic qualities. Some hot springs and geysers are held to be national treasures and are found in national parks entirely off limits to developers.

For privately owned geothermal resources, too, it is wise to develop geothermal energy carefully and to assess the potential environmental consequences. In the United States, for example, some fear that geothermal development just outside Yellowstone Park could irreparably damage the spectacular geysers within the park. And the Japan Hot Springs Association has formally opposed the government's geothermal plans on the grounds that they could damage Japan's many hot springs resorts. To minimize the environmental impact of geothermal development, governments can limit plant size, regulate pollution levels, and ensure that adjoining uses for the land are compatible with geothermal development. Although such industrial and zoning requirements may at first seem constricting, they ultimately work to the benefit of geothermal developers. Given a firm set of guidelines at the outset, they can avoid most legal uncertainties and disputes thereafter.[32]

Even where access to an economical geothermal resource is undisputed, financial considerations can stall development. In the early stages of a project, risk is high since expensive exploratory wells must be drilled with no guarantee of success, and few utilities, local governments, or small companies can afford such high risks. It is no surprise, then, that national governments, oil companies, and venture capital firms are financing most geothermal exploration. In the Philippines, for instance, a subsidiary of the Union Oil Company of California has signed a contract with the Philippines government and is the principal geothermal developer. In developing countries financial constraints are particularly acute, but such international financial institutions as the World Bank and the regional development banks have begun to support geothermal projects there.[33]

An approach to risk sharing taken in France, Iceland, and the United States is for the government to reimburse some proportion of the cost of exploratory wells. In Iceland an Energy Fund provides loans to cover 60 percent of exploration and drilling costs. If the well is successful, the loan is repaid at normal bank rates using the proceeds from the project. If it is dry, the loan becomes a cash grant and the project is dropped. These incentives have been successful, but they must be carefully designed and their use monitored so that they do not encourage frivolous projects with little chance of success. Ideally, the private developer should venture some capital and assume a reasonable risk, while government should receive royalty payments for successful projects.[34]

Beyond the initial exploration, even establishing commercial facilities at identified geothermal sites involves financial uncertainties since some of the technologies are so new. Few lenders can supply large blocks of capital at reasonable interest rates for experimental technologies, so some form of government incentive will in most cases be needed initially. In the United States the government grants a geothermal tax credit of 15 percent that can be added to a standard investment credit of 10 percent and a depletion allowance similar to that permitted for petroleum development. So far these incentives have stirred up only slight interest. To cultivate more, government could make tax breaks more generous, though direct subsidies may be a more effective and equitable alternative.[35]

Municipal governments have a vital role to play in some forms of geothermal development. Most district heating systems are owned and operated by municipal governments and regulated as public utilities, so they have guaranteed markets and access to capital at relatively low interest rates. The city government of Reykjavik has developed a successful geothermal heating system, providing a model that other geothermally rich cities may want to copy.[36]

For electricity generation using geothermal energy, electric utilities are obviously the key institution in most countries. With their access to low-interest capital, utilities can manage such large investments with relative ease. Moreover, geothermal plants should have special appeal since they are extremely reliable, operating at a higher proportion of rated capacity than either coal or nuclear power plants. At the Geysers in California, the Pacific Gas & Electric Company has financed and owns most of the power plants built so far, though utilities without such plum sites to exploit have naturally been slower to get involved. While reluctance to take on projects perceived as too risky for customers and stockholders is forgivable, paranoia about new technologies that could widen the options for electricity generation and lack of imagination are not. Utility managers are beginning to wake up to the benefits of geothermal energy, but in many regions pressure from governments and citizens groups seems to be the necessary nudge.[37]

The Geothermal Prospect

Expanding more than 10 percent per year since the mid-seventies, geothermal energy use appears likely to increase five- to tenfold by the end of the century.[38] Direct heat and electricity generation from geothermal sources are likely to share in this growth, though the industrial countries will place more emphasis on direct heat and the Third World more on electricity generation. Naturally, early development efforts will stay concentrated in those countries with abundant and easily accessible resources. Gradually, however, the use of geothermal energy will expand, particularly as technologies for using less accessible or lower-grade sources are developed.

Estimates of how quickly the direct use of geothermal heat will grow vary widely. Now less than 0.3 exajoules per year (enough to heat 2 million typical buildings in a northern climate), geothermal heat use could by the turn of the century amount to between 1 and 3 exajoules, depending on how many countries shape and act on firm plans. Iceland expects 82 percent of the country's homes to be using geothermal heat within three to five years. France, which has low-temperature geothermal resources underlying two-thirds of its land area, aims to have a half-million geothermally heated homes by 1990 and to supply one-tenth of the country's low-temperature heat using geothermal energy by the year 2000.[39]

Canada, China, Japan, the Soviet Union, and the United States could also expand the direct use of geothermal heat dramatically. China's national exploratory program has paid off: Geothermal resources that a few years ago seemed negligible today appear abundant indeed. Approximately 2,300 hot spots have been identified, and geothermal experts believe China alone may harness more than 0.1 exajoules per year in direct geothermal heat by 1990. In the Soviet Union, much of which is underlain by low-temperature geothermal deposits, several district heating projects are underway. In the United States no big direct-use projects are yet on the drawing board, and government support is anything but solid. Nevertheless, U.S. geothermal heat use in the year 2000 could range from 0.1 to 1 exajoules—enough to meet 1 to 10 percent of U.S. residential space heating needs.[40]

Geothermal electricity development has also been erupting in recent years. National plans for the year 2000 add up to over 17,000 megawatts, nearly seven times the current level. One-third of this total or 5,800 megawatts will be in the United States. Surprisingly, though, many of the new plants will be built outside of the four industrial nations that have thus far pioneered in geothermal electricity—Italy, Japan, New Zealand, and the United States. El Salvador in some years already generates one-third of its power supply using geothermal energy, and Mexico plans to build 600 megawatts of geothermal capacity by the mid-eighties. Other developing countries with noteworthy geothermal programs include Chile, Costa Rica, Indonesia, and Turkey.[41]

Second only to the United States' geothermal power efforts are the Philippines'. Although the country has less than 500 megawatts of geothermal generating capacity today, it plans to have over 1,200 megawatts by 1989. Eventually, geothermal energy will rival hydropower as the country's largest electricity source. Earmarking $347 million for this program over the next several years, the Philippines is launching major exploration efforts and technology-development programs. Short on fossil fuels and eager to build industries and "electrify" villages, the government will also construct an undersea transmission cable to transport geothermal electricity to a neighboring island. Few other developing countries are investing comparable amounts of time or talent in sophisticated new energy technologies.[42]

By any sound reckoning, geothermal energy use will be substantial in the year 2000. But it will figure much more prominently in some national and regional energy economies than in others. Such countries as Iceland and the Philippines will draw heavily on their rich geothermal endowment, but overall, geothermal sources will furnish no more than 1 to 2 percent of the total world energy supply until the technology for tapping them improves and some industries relocate to geothermally rich regions. Still, such constraints do not mean that geothermal power cannot gradually become another strong link in a diverse global energy system.

II

Working Together Renewable Energy's Potential

Solar water heaters are found atop more than 5 million houses that relied on fossil fuels a decade ago. Wind machines that in 1973 merely summoned up memories of a bygone era are rapidly becoming a standard power source for utilities. Community forestry projects, underway in no more than a smattering of nations in the early seventies, are now found in more than fifty nations. This is certain progress—the vanguard of the renewable energy development effort.

But what about the future? Can the various renewable sources of energy together provide sufficient energy for modern societies? And if so, how long will it take and what will the transitional period be like?

Such questions can be addressed only by stepping back from assessments of the individual renewable energy technologies and taking a wider view of renewable energy's prospects. The simple technical or economic potential of an energy source means little unless coupled with an understanding of the needs the resource is to meet. Household cooking, aluminum smelting, and automobile manufacturing, for example, each use energy in a different, uniquely evolving way. The main question is how in each of the major end uses renewable energy sources can interact with conventional fuels, other renewable sources, and with efficiency improvements to provide economical and safe energy.

The broad picture that emerges here is one of diversity. Tomorrow's various needs, resource availabilities, and evolving technologies will combine in different ways in different countries so that no two national energy systems will be exactly alike. Even within nations, energy systems will rely on six or eight major sources rather than on three or four, as most do today. Renewable energy's future has the potential to be much more than the sum of its parts. By working together in innovative and productive ways, renewable energy technologies can form a strong base to support societies.

Rebuilding

Today roughly one-quarter of global energy use goes to heat, cool, and light buildings, and another 5 percent is used in water heaters and other appliances. Two-thirds of this total comes directly or indirectly from oil and natural gas, premium fuels that could be put to better use in automobiles, petrochemical production, and industries.[1] The immediacy of the problems has begotten many solutions and already energy conservation and renewable technologies are influencing the shapes of the world's buildings. Substantial optimism is warranted since many of the least complicated, most economical renewable energy technologies will play their largest role in buildings.

Increased efficiency is the first step in reducing energy bills in buildings virtually everywhere. In small residential buildings such measures as adding insulation and employing improved furnaces and air conditioners can reduce energy needs by 25 to 50 percent at a minimal cost. In larger apartment and commercial buildings, combining such simple conservation measures with computer-controlled energy-management systems can result in similar improvements. Even more encouraging, architects and engineers now know how to build new buildings that use 75 to 90 percent less fuel than conventional buildings do. Statistics for the industrial countries show reductions in energy use in existing buildings averaging 10 percent or more in the last decade alone.[2]

As the energy needs of buildings decline, supplying the remaining needs with renewable resources becomes easier. Fewer solar collectors or wind machines are needed to supply sufficient heat or electricity to an efficient building, for instance, and many renewable energy technologies that would not be economically viable in a conventional house are so in a "low energy" one. Still, cost remains paramount in determining renewable energy use in buildings. Large capital outlays are beyond the pale for most building owners even if the new technology will pay for itself in fuel savings in a few years. Ease of maintenance is also critical since few people want to spend much time fiddling with a faulty energy device.

By all of these criteria, passive solar design shines brightly. Energy efficiency and solar design complement and reinforce each other, and once conservation has reduced heating needs to a certain point, passive solar design becomes even more cost-effective than further conservation measures. Climate-sensitive buildings are both inexpensive and uncomplicated, factors that have already found them a following in the middle-income housing market in some countries. The simple practicality of climate-sensitive design virtually guarantees that one day it will be employed in varying forms throughout the world. Based on current trends there could be as many as 100 million such buildings by the year 2000.[3]

Climate sensitive designs will be coupled with other renewable energy technologies. Solar collectors, the original vanguard renewable energy technology, are, for instance, the most economical means of heating water in many regions. Since they can easily be added to existing houses, solar collectors have the potential to catch on rapidly, as is seen in Japan where 11 percent of homes are already using the devices. By the end of the century, solar collectors should be a thoroughly conventional household appliance, with 50 to 100 million gracing the world's roofs.

Buildings can also become their own power stations, although the eventual popularity of such systems is difficult to calculate. Photovoltaic panels mounted on rooftops and small wind turbines in the backyard have the potential soon to be economical means of electricity generation under the right conditions. Wind turbines will likely be restricted mainly to rural areas, but rooftop solar cells could become a common suburban and even central city technology. Such systems have the potential to give individuals a measure of energy independence that is unheard of in the modern world, transforming "consumers" into "producers."

Other renewable energy sources have an important but limited role to play in buildings. Contrary to forecasts made in the mid-seventies, residential wood use will grow during the next two decades, especially in such forest-rich regions as North America, Scandinavia, and the Soviet Union. Fuelwood reliance is inconvenient and expensive in many cities and suburbs, however, and residential use there will be limited by a lack of ready access to wood supplies and by increasing competition from industrial users. In the aggregate, household reliance on fuelwood in industrial countries will probably double and could triple by the end of the century, supplying 10 to 20 percent of residential energy in most nations.

In cities with cold climates district heating using municipal solid waste, wood, geothermal energy, or solar ponds will be an important supplementary source of heat. Already many European cities make use of district heating, solving waste disposal problems at the same time. District heating is efficient and inexpensive, and it can make use of first one conventional or renewable fuel and then another as the relative prices of energy sources shift. While other renewable energy technologies encourage individual building owners to work independently, this one will push them to cooperate.

Local adaptation is obviously essential to successful use of renewable energy in buildings. In relatively mild climates a climate-sensitive design combined with solar collectors for hot water could provide most of the energy needed. In a large northern city a superinsulated townhouse or apartment building could feature rooftop solar collectors and derive most of its space heat from a garbage-fired central heating plant. In a humid, tropical region a climate-sensitive design might be assisted by a solar powered air conditioner.

Virtually all regions have the potential to power their buildings with renewable energy. It will be up to local communities and individuals to overcome the institutional and financial barriers that are the largest impediments to a transformation of the world's buildings. Davis, California, is a model. Its comprehensive building code, innovative developers, and enthusiastic citizens have encouraged a solar energy and conservation revolution and have begun to wean the town of fossil fuels.[4] Dozens of other cities are in the process of writing their own versions of the Davis success story, and today this is one of the most exciting frontiers in renewable energy.

A Fresh Start for Industry

The world's industries currently account for approximately one-third of global energy use, though the percentage varies widely by country. Industry's energy requirements are highest in nations that produce aluminum, cement, chemicals, or steel. They are lowest where agriculture or light industry dominates the economy. Here the line between developed and developing countries breaks down. Japan, the Soviet Union, Brazil, India, and the Philippines all use a large share of their energy in industry.[5]

In many countries the productivity of energy has taken its place alongside the productivity of labor as a key measure of industrial achievement. In the United States, for example, the more energy-intensive industries together spent 55 percent of their research and development funds—or over $5 billion—on reducing fuel requirements in 1980. Across Japan, Europe, and North America industrial energy use has leveled off, even fallen, while the output of industrial products continues to increase. In Japan the steel industry has cut energy consumption per unit of production by 12 percent.[6]

All signs point to continuing energy-efficiency improvements and a gradual shift of emphasis toward more fundamental changes that require larger investments or more sophisticated technologies. As the record-breaking recession of the early eighties ends, many companies will be introducing new energy efficient technologies at a rapid pace.

Just as industry has realized the potential of energy conservation, so too has it begun to evaluate renewable energy's role in increasing profits. As renewable energy investments become profitable, they will multiply, though on an application by application and use by use basis. The catch is that renewable energy technologies cost more on average than does conservation and the commensurate risk is higher, so businesses are slower to respond. One impetus for renewable energy investment is the major changes in plant equipment dictated by the need to improve efficiency or generally upgrade technologies. Renewable energy equipment can be added at the same time for a relatively modest cost. New cogenerators or conventional boilers can easily be designed to run on biomass fuels as well as coal or natural gas, giving a plant manager welcome flexibility.

Today wood is the most rapidly growing renewable energy source in industry, mainly as a substitute for fuel oil in industrial boilers. In fact, the wood products industry is fast approaching energy self-sufficiency, while various other industries located in forested areas are also turning to wood. Already Brazil relies heavily on charcoal for smelting steel and half of all new industrial boilers sold in the United States are wood-fired. In the future gasified wood is likely to find a place in industries that require a clean, steady energy supply—such as brick and textile production. Wood's industrial future is so bright that in some northern temperate nations and in heavily forested countries in the developing world wood could overtake coal as the fastest growing industrial fuel.

Roughly half of industrial energy use goes to produce direct heat, and in the United States more than one-third of this heat is low-temperature—less than 180°C—and fully 80 percent is at temperatures below 600°C.[7] To reach these temperatures, simple solar collector systems are appropriate at the lower end of the spectrum and solar concentrators and solar ponds at the upper end. Geothermal energy could figure importantly here, too. Most industrial solar and geothermal systems are unlikely to begin making a major contribution for at least a decade, but progress could accelerate rapidly thereafter. Eventually, many industries will probably begin to relocate to take advantage of solar and geothermal energy.

Industry employs energy in more diverse ways than does any other sector. Besides heat and the electricity needed in electrolysis and mechanical systems, specialized energy requirements include metallurgical coal for the steel industry and petroleum feedstocks for petrochemical production. Some of these needs will be very difficult to meet with anything but fossil fuels, but this should not preclude rapid progress toward meeting larger, more flexible energy needs with renewable sources.

Renewable Energy on the Farm

Renewable energy could give agriculture a new lease on life. Farming has grown increasingly energy-intensive in recent years, with oil-fueled equipment now performing many tasks once done by people or animals. Heavy use of fertilizers, pesticides, and irrigation are maladapted legacies of the era of cheap fuel. Although agriculture accounts on average for only 3.5 percent of the commercial energy used in industrial countries and 5 percent in developing countries, nearly all the energy it uses is in the form of highly valued liquid fuels and electricity.[8]

Several factors bode well for renewable energy's use in agriculture. Most farms have ample land for solar collectors, wind machines, and other devices. Most farms use energy in forms well-matched to some of the renewable resources. And many farmers are comfortable handling a variety of technologies and adapting new devices to their needs—witness the quick spread of wind pumps throughout rural North America in the late nineteenth century.[9]

Producing biological fuels, including ethanol, methanol, and biogas, is a logical first step for farmers. Agricultural wastes are widely available for fuel production, and they could be supplemented by special energy crops or forest materials. Some farms may use a small share of their land to grow crops such as sorghum, Jerusalem artichokes, or sunflower seeds that can provide fuel for their tractors. Fast-growing trees could be another popular energy crop. It may also make sense for farmers' cooperatives to build bio-fuel plants: Farmers could contribute feedstock wastes and draw out a proportional amount of fuel, selling the rest.

Direct use of solar energy is one of the most attractive options on many farms. Low-temperature solar heat can be used for drying crops and heating farm buildings. In the United States more hogs than people live in solar heated "homes," and solar milking parlors are popular as well. Solar grain dryers are also being used by North American farmers today. Where the crops are not too moisture-laden and the grain can be dried gradually, these dryers have performed well. Most of these systems are still built on the farm, but most likely commercial systems will be developed soon, particularly if encouraged by government programs or farmers' cooperatives. Meantime, only innovative farmers who are good with their hands have solar grain dryers or heaters.[10]

Traditionally, wind power has been widely used for water pumping. Today wind pumps suitable for small farms and livestock grazing are enjoying a renaissance. Large irrigated farms, however, can probably make better use of pumps powered by photovoltaics—systems that for now remain experimental and costly but nevertheless have the best chance ultimately of meeting modern irrigation's high energy demands.[11]

Farms may also be in a good position to generate their own electricity in the near future. Surveys indicate that there is ample wind available for electricity generation in most temperate farming regions and ample sunlight for electricity generation on most farms everywhere. As these technologies are perfected, energy farming and crop farming could increasingly go hand in hand. By the nineties farms could be adding strength and diversity to utility grids.

In agriculture renewable energy is a good fit. Much of the energy used on farms is needed in summer and autumn when sunlight is abundant, and the forms of energy needed are in some cases those most readily available. Over the long run most agriculture should become energy self-sufficient. In the interim, however, technologies will have to be refined and energy efficiency improved.

Energy for the Rural Poor

The energy problems of modern industry or agriculture pale beside those of the world's poor. For the roughly 2 billion people in developing countries that rely mainly on fuelwood and agricultural wastes to meet energy needs, choices are constrained by shortages of traditional fuels and of resources to pay for new ones. The world's poor thus confront energy problems in immediate human terms—as a daily scramble to find fuel to cook the family's meal or heat its home. As the World Bank noted gloomily in 1981, "The crisis in traditional energy supplies is a quiet one, but it poses a clear danger in the lives of much of the population of the developing world."[12]

The world's rural peasants and villagers use only a tiny share of the world's energy, and small additional amounts could provide large benefits. Yet many energy programs introduced in developing countries are grim parodies of those in industrial nations. The emphasis is on large power plants and imported fuels that can aid in industrialization, but that do not touch the lives of the poor majority. Many nations have begun to right this imbalance in recent years, an overdue development given further impetus by the United Nations Conference on New and Renewable Sources of Energy held in Nairobi in 1981. Representatives of both industrial and developing countries emphasized the overriding importance of rural energy solutions. Unfortunately, financial commitments here continue to lag behind rhetorical ones.[13]

Shortages and abuses of fuelwood and other biological energy sources are the crux of the rural energy problem, and no conventional or renewable fuel can be substituted quickly and at a reasonable cost for a large share of these traditional fuels.

In the next decade or two only wiser management and more efficient use of biomass fuels can save the day. Establishing large forestry programs and introducing efficient wood stoves pose no stupifying technical challenges, but the massive efforts needed will require an unprecedented mobilization of human and financial resources.

In that sense bringing renewable fuels into use on a sustainable basis in the Third World is even more difficult than developing renewable energy sources from scratch in industrial nations. Eventually, village wood lots and privately owned fuelwood plantations must be established so that natural forests are not plundered out of desperation. The new fuelwood supplies can be shared among community members and become the basis for new village industries. Erik Eckholm, an American researcher who has studied community forestry programs, observes that "the process of creative community action that successful village forestry requires is the essence of what real development is all about."[14]

Other pressing rural energy needs also require attention—crop drying, water pumping, mechanical power for agriculture, heating, and refrigeration among them. A steadily increasing stream of research and demonstration projects in the last decade have been aimed at evaluating the potential for renewable resources to meet these needs. The results have been mixed. While a few of the ideas that once generated excitement can now be written off, most of the difficulties encountered indicate not that the technologies must be scrapped, but that small changes are needed, particularly in the way they are introduced. John Ashworth, a U.S. energy expert who has visited many rural development projects, notes that there is a growing awareness that "new technologies must undergo adaptation in order to be compatible with local cultural practices, local needs for technology, and the structure of the greater society."[15]

Among the most promising "new" renewable sources of energy for rural areas is biogas. Ideal for cooking and lighting, as well as for electricity generation, biogas digesters allow people to use the energy in biological wastes without sacrificing the valuable fertilizer they contain. The key is to develop inexpensive and easy-to-build community-sized biogas digesters. That way all families (some offering only their labor) can participate, not just livestock-owners. Of little use in extremely cold or arid regions, biogas digesters could nonetheless be used in a sizable share of rural communities.

An important rural energy need is for electricity, small amounts of which greatly improve living standards by providing power for agricultural equipment, refrigeration, and lighting. The 1960s dream of extending central electric grids into the "heart of darkness" appeared to fade in the face of the prohibitive costs of building so many power plants and electric lines. Today it is evident that if the rural poor are to have electricity anytime soon, small decentralized systems will have to provide much of it.

Now most out-of-the-way places that have electricity are served by diesel generators, typically run just a few hours a day to supply power for agricultural equipment and for a few lights in the evening. But diesel generators are expensive and—a bigger problem—unreliable. They require regular maintenance and an occasional complete overhaul. Since there are so few trained mechanics in rural villages, broken-down diesel generators are a common sight throughout the Third World today. Then, too, fuel supplies are by no means guaranteed in remote villages served by pocked, mud-washed roads.[16]

Many diesel generators in rural use could be replaced by more reliable renewable energy technologies that would generate electricity for the same cost or less. The smaller the needs to be served, the more the advantage shifts to renewable energy technologies since economies of scale are larger for diesels. Small-scale hydropower and biogas-fueled generators have already proven effective and economical. Wind power can also provide electricity where wind is ample. Over the long-run, easy-to-maintain solar electric systems will probably be the most popular way to generate precious electricity for village and agricultural use.[17]

For irrigation, livestock watering, and domestic water supplies, wind pumps and solar pumps are the best bets. While they are an established technology, mechanical wind pumps are still being adapted to developing countries' needs. Still, evidence indicates that in many areas, wind pumps can be both cheaper and more reliable than diesel pumps, particularly in small-scale use. Solar pumps are less technically mature, but their potential in windless areas looks great.

Other promising renewable energy technologies are still at the trial-and-error stage. The initially cool reception to solar cookers might change if solar ovens with enough storage capacity to work in the evening hours were developed. Solar refrigeration could be a big help in preserving medicine and food where electricity is not available, though more work is needed on this technology. Many other good ideas are on the drawing boards, awaiting application or an engineering twist.

It is a popular notion today that the rural poor should lead the way to reliance on renewable energy. They do, it is true, already rely heavily on renewable energy, but the difficulties the Third World faces in using renewable energy on a sustainable, economically productive basis are nonetheless substantial. Developing countries' renewable resources are currently eroding at a frightening rate, and they often lack the technical expertise or financial resources needed to develop or adapt new energy technologies. However, working toward some realistic goals—more efficient use of biomass energy and gradual introduction of other renewable energy sources—could greatly improve the energy situation in rural areas in the near future while longer-term solutions are developed. So far only China has taken a truly comprehensive approach to solving rural energy problems. Although cultural differences will prevent other countries from copying the Chinese model exactly, that nation's successful use of biogas, small-scale hydropower, and community forestry despite minimum financial resources gives an idea of the potential.

Transportation Dilemmas

Providing inexpensive alternative fuels for automobiles, trucks, and aircraft is the problem within the energy problem. Transportation vehicles use 20 to 40 percent of the oil in most nations or over 2 billion liters of liquid fuel a day.[18] In many developing countries, dependence on automobiles and trucks is nearly complete since the capital investment needed to build rail systems is prohibitive. Because oil-derived fuels pack a lot of energy and are easy to transport, finding good substitutes will be difficult.

Conservation and fuel-efficiency have begun to make a dent in the transportation energy problem. In 1980 new cars sold in the United States (which uses $14 million worth of gasoline each hour) were 50 percent more efficient on the average than they were in the early seventies.[19] Less dramatic shifts are occurring elsewhere. These changes in new car fuel efficiency translate only slowly into reduced gasoline consumption since many old cars remain on the road. But global gasoline consumption has already declined from its peak in the late seventies, and further reductions can be banked upon. In industrial countries gasoline use per vehicle will probably fall an addition 30 to 50 percent by the year 2000, though some of this decline could be negated by increases in the number of cars on the road in the developing countries.

Among the alternatives to gasoline, synthetic fuels derived from coal or oil shale have received the most attention. During the seventies it was frequently predicted that future gasoline price hikes would assure synthetic fuels' economic viability.

But as price rises occurred, "synfuels" continued to remain out of reach. Today cost estimates for synthetic fuels plants are rising faster than gasoline prices, and both industry and government are abandoning major projects after spending hundreds of millions of dollars on them. No convincing evidence indicates that synfuels will ever be anything but a minor and expensive replacement for gasoline.[20]

Electric cars are also being considered as an alternative to gasoline-powered vehicles. Since electricity can be derived from many types of energy, renewable resources included, it seems in some ways to be a good power source for tomorrow's automobile. But batteries developed so far are expensive and inconvenient: They must be recharged every hundred miles and replaced after a few hundred rechargings. Battery research continues in government and private laboratories, but a breakthrough that would put electric cars on the commercial market in large numbers before the year 2000 is unlikely. The next generation may, however, see electric vehicles widely used in commercial fleets and later in privately-owned cars. Whether electric batteries can ever largely replace gasoline is not yet known.[21]

Hydrogen is a more problematical transportation fuel. It can be produced from a range of conventional and renewable energy resources that are first converted to electricity, an inefficient and expensive process. However, a new technique using iron oxide holds out the potential of cheaply separating hydrogen from water using sunlight directly. Hydrogen is a clean-burning fuel, but because it is a gas at normal temperature and pressure, hydrogen must be cooled and liquified or chemically converted before it can be stored in a fuel tank—a minor technical problem but a major expense. In all, hydrogen is unlikely to hit the road during the next twenty years, but it could very well become a popular automotive fuel after the turn of the century.[22]

The renewable fuel most acclaimed as an alternative to gasoline is, of course, ethanol—a form of alcohol obtained from sugarcane, cereals, and many other crops. But the very fertile agricultural land needed to produce these fuel crops is itself under increasing pressure, and such alternatives as cassava and sweet sorghum cannot measure up economically to sugarcane or corn. For alcohol to become a major transportation fuel, new means of producing it must be found.

Wood alcohol—or methanol—is the alternative fuel with the most potential. Methanol can be produced from a wide range of energy resources, including wood, biological wastes, coal, and natural gas. Already used extensively as an industrial chemical, methanol can be used in slightly modified internal combustion engines that could be built for about the same cost as gasoline-powered ones. Essential to extensive use of methanol is finding inexpensive ways to make it from various energy crops—a search that is already paying off.

One of the most encouraging things about methanol is that it might serve very well as a transitional fuel. The gradual shift from natural gas to coal, wood, and waste products as feedstock could go almost unnoticed by drivers. Different nations might produce methanol from different feedstock materials, and some could even want to export surplus methanol, making it a common energy currency.

Frank von Hippel, a senior research physicist at the Princeton University Center for Energy and Environmental Studies, observes that "if you can economically increase efficiency to, say, 60 miles per gallon, then you can easily absorb the demand with biomass."[23] Indeed, improved fuel efficiency together with greater use of public transportation are essential if we are to maintain mobility while gradually switching to methanol and perhaps electricity and hydrogen in future years. Cars will undoubtedly be among the last users of oil, however, and it will be many decades before the transition is complete.

Sustainable Electricity

Like liquid fuels, electricity is a particularly precious form of energy. Used extensively in industry and buildings, it has helped raise living standards the world over. In industrial countries close to a third of national energy use goes for electricity generation. Nearly half of this comes from coal, a quarter from oil and gas, and 8 percent from nuclear power. Renewable energy in the form of hydropower provides the remaining fifth.[24]

During the postwar period, government and industry in most nations vigorously promoted electricity use. The hard sell paid off, largely because electricity is so versatile and because technological improvements and more efficient power plants pushed prices down. New plants in Western Europe and the United States averaged 150 megawatts of capacity in 1950 and 400 megawatts in 1978. Growth in electricity use became predictable, rising by 5 percent or more each year regardless of economic ups and downs.[25]

In the seventies the electricity picture began to change drastically. Rising capital costs caused in part by the need to limit the social risks that large power plants pose combined with high interest rates to boost electricity costs. Rising oil prices caused additional increases. As a result, electricity prices kept pace with or outstripped inflation in many nations. In response consumers cut back, and slow economic growth acted as an additional brake. The rate of growth in electricity use has fallen from 6 percent to 2 to 3 percent in the United States and by similar amounts in Europe. Unexpected conservation, in turn, upset the high-growth assumptions on which utility planning has been based, and utilities found themselves with expensive but unneeded coal and nuclear power plants planned or partially built. Many of these plants have been canceled since the mid-seventies, and some utilities are adapting to the new era by investing heavily in load management and conservation.

Many utility planners now recognize that conservation can provide for consumers' needs at a lower cost than can new power plants.[26]

For utilities, the newer sources of renewable energy are not yet as sure a bet as energy conservation. But "renewables" are climbing steadily on utility agendas in many regions. Geothermal plants, wind turbines, photovoltaics, and solar ponds are beginning to compete economically with conventional power plants now under construction. Some utilities are using these new small-scale technologies to cope with uncertain trends in electricity use since the capital expenditures are modest and they can add or delete generating units relatively quickly as electricity use trends vacillate. The world leader in this endeavor is California, which will get most of its additional generating capacity in the late eighties and nineties from cogeneration, geothermal power, wind power, and solar power. Coal and nuclear power plants originally planned for that period have all been canceled.[27]

One of the chief concerns surrounding reliance on solar power, wind power, and hydropower for electricity generation is the problem of power interruptions caused by hourly, daily, or seasonal weather changes. While it is true that such fluctuations will place limits on the use of some renewable energy sources, much of the problem can be alleviated through careful planning. Luckily the problem does not even arise if the renewable energy source contributes only a small proportion of the overall generating capacity on an electric grid. And as renewable energy becomes a major power source, the various generating technologies can be balanced to ensure uninterrupted power. Often wind is available when sunlight is not and vice versa. By interconnecting dispersed areas with different climates (and renewable energy resources), interruptions will be fewer. Hydropower has unique potential as a balancing agent since water can be stored and power output increased or decreased as other energy sources on the grid wax and wane.[28]

In many regions electricity systems may be self-sufficient. But in others long-distance trade in electricity may be necessary to enable regions to sell surplus renewable power and buy it from others as needed. Modern computer-control systems can regulate the flow of the electricity and ensure that the lowest-cost power is used at any given time. They can also automatically alter the price of the electricity according to how it is produced and when it is used.

Outside of California only a few utilities have wholeheartedly embraced conservation and renewable energy so far, but already some impressive plans have been put together. The U.S. state of Hawaii, which today gets most of its electricity from oil-fired power plants, plans to get between 79 and 94 percent of it from indigenous renewable resources by 2005. Large numbers of wind turbines, geothermal plants, and ocean thermal plants are projected to be in place by then. The Philippines has a similarly ambitious program under way, although its emphasis is on wood-fired "dendrothermal" plants, hydropower dams, and geothermal plants. In the U.S. Pacific Northwest and in New England several utilities have ambitious programs to develop renewable energy sources. Wind power developments are on the planning tables of several utilities in northern Europe, including Denmark, the Netherlands, and Sweden.[29]

The key to making widespread use of electricity generated from renewable resources is making the change gradually so that technical and institutional impediments can be ironed out. By means of productive conservation, utilities will be able to buy time to experiment with new ways of generating and distributing electricity. Renewable energy technologies can be brought into action as they become economically competitive, first with oil- and gas-fired plants and then with new coal and nuclear power plants.[30] By thus reducing costs and risks, utilities can begin paving the way for a sustainable electricity system. In the more distant future enormous diversity is likely.

Most electricity grids will be "fed" by combinations of central generating projects (such as wind or solar "farms") and decentralized plants located at houses and industries.

Special note should be taken of the electricity problems developing countries face. Since most Third World nations today use only small amounts of power, they will have to add substantial generating capacity simply to maintain modest economic growth.[31] Conservation will help, but it will still be necessary to develop many new power sources in the near future. Most Third World electricity planners cannot afford to await the outcome of experiments with new technologies. For them, the best tack is to use their unexploited hydropower resources in the interim. If they lack hydropower and must expand coal- and oil-powered plants, cogeneration is the ticket. Meanwhile, officials in developing countries must begin evaluating their nations' renewable energy resources and considering the adaptation of some of the new electricity-generating technologies being pioneered in the industrial world.

Adding Up the Numbers

Technology assessments and end-use analyses alike make it clear that renewable resources' contribution to the global energy budget will grow steadily. If current trends continue and governments adopt moderately supportive policies, renewable energy use is likely to increase by at least 75 percent by the year 2000, rising from the 1980 level of 63.5 exajoules to between 113 and 135 exajoules. (See Table 11. 1.)[32] Renewable energy's share of world energy use would thus rise from the current 18 percent to around 26 percent.

Such numbers inevitably obscure as much as they reveal since the type of energy produced is as important as the quantity, and the efficiency with which energy is used varies enormously. Today renewable energy actually supplies less *useful* energy than these figures indicate. But in the future overall

Table 11. 1. World Use of Renewable Energy, 1980, 2000, and Potential

Source	1980	2000	Long-term potential
		(exajoules)	
Solar energy: passive design	<0.1	3.5–7	20–30
Solar energy: residential collectors	<0.1	1.7	5–8
Solar energy: industrial collectors	<0.1	2.9	10–20
Solar energy: solar ponds	<0.1	2–4	10–30+
Wood	35	48	100+
Crop residues	6.5	7	—
Animal dung	2	2	—
Biogas: small digesters	0.1	2–3	4–8
Biogas: feedlots	<0.1	0.2	5+
Urban sewage and solid waste	0.3	1.5	15+
Methanol from wood	<0.1	1.5–3.0	20–30+
Energy crops	0.1	0.6–1.5	15–20+
Hydropower	19.2	38–48	90+
Wind power	<0.1	1–2	10+
Solar photovoltaics	<0.1	0.1–0.4	20+
Geothermal energy	0.3	1–3	10–20+
Total	63.5	113–135	334–406+

+ indicates that technical advances could allow the long-term potential to be much higher; similarly, a range is given where technical uncertainties make a single estimate impossible.

< means less than.

Source: Worldwatch Institute

efficiency is likely to increase along with the adoption of renewable resources.

During the next two decades, the traditional sources of renewable energy will continue to be the most abundant. Fuelwood use will probably rise by at least a third, providing an additional 13 exajoules each year. Hydropower will grow even more rapidly, more than doubling by the end of the century and providing the equivalent of an additional 19 to 29 exajoules. In both cases the main constraint will be the environmental chaos now caused by uncontrolled forest clearing—which damages watersheds and makes fuelwood scarcer.

The short-term prospects of the other renewable energy sources are less certain. Even the mature and economical technologies run up against market obstacles: consumer unfamiliarity and the lack of a ready means of distribution. Yet some studies make projections for the year 2000 as though such constraints did not exist—a surefire recipe for inflated expectations and subsequent disappointment.

Some renewable energy sources are likely to break these initial barriers relatively soon, however. Passive solar design, already commercially established in a few nations, could supply 7 exajoules of energy by the end of the century, and solar collectors 4 exajoules. Wind power should be contributing 1 to 2 exajoules by the end of the century. For geothermal power, the figure is 1 to 3 exajoules.

For other energy sources, developments over the next twenty years will probably be slower. The immediate prospects for solar photovoltaic systems are uncertain because it is not known how fast costs will be brought down. With major technical improvements in the next five to ten years, solar electricity could provide as much as 0.4 exajoules of energy by the year 2000, but its contribution could well come later, too. Solar ponds have immense potential. But the technology has not been extensively used yet, and industries and cities will have to make adjustments to use the ponds effectively. For the same reason, energy form liquid biological fuels and urban wastes looks to be fairly limited during the next twenty years.

Considering these projections along with the outlook for conventional energy resources reveals that renewable energy will provide close to half of the additional energy the world will be using by the century's end. (See Table 11. 2.) Coal, natural gas, and nuclear power will also become more important, of course. Yet with world energy use as a whole growing more slowly, increasing by little more than 1 percent per year over the next twenty years, the renewable energy share of the global energy budget should reach 26 percent.[33]

Table 11. 2. World Energy Supplies, 1980 and 2000

Sources	*1980* *Amount*	*1980* *Share*	*2000* *Amount*	*2000* *Share*	*Change 1980–2000*
	(exajoules)	*(percent)*	*(exajoules)*	*(percent)*	*(percent)*
Oil	133	38	113	26	−15
Natural gas	61	18	79	18	+30
Coal	82	24	105	24	+28
Nuclear power	8	2	23	5	+190
Renewable energy	63	18	113	26	+75
Total	347	100	433	100	+24

Source: Worldwatch Institute.

At first glance, these numbers do not appear impressive. That is because many of these energy sources are starting from a base of nearly zero, while the established, conventional sources of energy have already acquired great momentum. Yet this early progress lays the groundwork for major leaps forward in the future. Twenty years is quite a short time horizon for assessing the future of new energy sources with the potential to support humanity for millennia.

What about the more distant future? It is obviously impossible to make firm predictions about the world's energy systems fifty or a hundred years from now, but the long-term prospect for renewable energy is undeniably promising. Given enough time for technological developments and institutional adaptation, some currently insignificant energy sources could flourish: solar ponds, wind power, photovoltaics, and methanol from biomass. With proper management, renewable energy sources could easily supply over 300 exajoules—as much energy as the world uses today—before running up against resource constraints.

Many changes will obviously be needed before the world can hope to rely entirely on renewable energy. On the average, energy will have to be used perhaps three times as efficiently as it is now. New supply networks more appropriate to renewable fuels will have to be developed. And many of the world's energy institutions will have to be restructured. Each of these changes is beginning to occur, and none is as economically and environmentally overwhelming as the conventional energy path now seems.

Energy efficiency and the use of renewable energy sources are now central to the world's energy future. Even in the next twenty years they will provide a cushion that allows most nations to limit the use of coal and forego nuclear power development altogether. Synthetic fuels and advanced nuclear technologies once intended to be major energy sources in the twenty-first century should also be reevaluated. By all accounts, these years should witness a major flowering of renewable energy, greatly limiting the need for more hazardous energy sources.

12

Institutions for the Transition

Institutions and politics—not resource limits or technological immaturity—most constrain greater use of renewable energy. Greatly expanding use of renewable energy is a prudent step to meet widely shared goals rather than a radical redirection of social values. Renewable energy does not need special favoritism. Rather, renewable energy—along with energy conservation—is a logical centerpiece of sound energy policy.

Unfortunately, most energy policy is myopic, focusing on the maximization of energy output with little reference to the critical values—jobs, equity, the environment, and national security—that are affected by energy investments. This tunnel vision reinforces the already immense power of the institutions that provide and benefit from conventional energy sources. A sound energy policy is one that seeks to serve society rather than to harness society for the production of ever vaster quantities of energy. Such a policy puts high priority on conservation and renewable energy.

By and large, the institutions that have grown up around conventional energy sources are inappropriate to renewable energy. While exploitation of fossil and nuclear energy hinges increasingly on the management of complex and far-flung institutions, tapping renewable energy requires the transfer of decision-making power, technical skills, and financial resources to individuals, local governments and the marketplace. To achieve these ends, research and development programs must be redirected, technical extension programs expanded, financial transfer mechanisms fashioned, and utilities opened to market forces. The moving force for these changes must be new coalitions of energy consumers, farmers, homeowners, workers, businesspeople, and environmentalists cognizant that their traditional agendas will be met or lost in the crucible of energy policy.

A New R&D Agenda

Tapping renewable energy is first a question of creating new technologies—the task of research and development. The last decade has witnessed a great increase in funding for renewable energy research and development, most of which has been productively spent. The new and exciting avenues for further research that these advances open up in turn give added weight to the case for a more balanced allocation of R&D monies among renewable, fossil, and nuclear energy. Carefully channeled into neglected areas, R&D funding increases can be expected to yield high payoffs in the years ahead. The challenge is to give programs more flexibility and direction and to diversify them.

During the 1950s and 1960s, energy research was synonymous with research on nuclear power. Hopes for an unlimited and cheap supply of atomic energy led scientists to neglect the study of synthetic fuels, photovoltaics, and other energy sources. More sobering, no one looked critically at the rosy claims for atomic power until a large number of plants had been built. Had research agendas been flexible, funds would have been shifted into other sources as signs of trouble arose. Instead, the nuclear industry had by the seventies become large and entrenched enough to bend the research agenda toward the atom with little reference to the economic potential of nuclear power plants or the strength of the alternatives.[1]

Energy research and development grew much more diverse after 1973, when budgets for energy research of all kinds surged upward. Spending increases for renewable energy have been particularly dramatic. Starting with less than 1 percent of energy R&D funds in 1973, government and government-stimulated expenditures for renewable energy in the Western industrial countries and Japan rose to 7 percent of the total research budget in 1977 and then to 13 percent in 1981. In absolute terms spending rose from about $20 million in 1973 to almost $1.3 billion in 1981. Private industry in those countries spent an additional $1.5 billion dollars on renewable energy R&D in 1981. In both absolute and per capita terms the United States (followed by France) spent the most on renewable energy R&D in the 1970s. Japan is catching up rapidly, however.[2]

Increases aside, the lion's share of energy R&D funds still goes to fossil and nuclear energy. (See Figure 12. 1.) While budgets have expanded, few countries have weighed the relative potential of each new energy technology or based forward-looking programs on realistic assessments of energy needs. In 1981 nuclear and fossil fuel R&D still absorbed 75 percent of the Western industrial countries' energy research budgets. Billions of dollars are being spent annually on advanced nuclear reactor systems that cannot be economic until well into the next century, if ever. Meanwhile, many promising avenues of renewable energy research that could make a difference within the next decade or two are left unexplored for lack of funds.[3]

The great leap forward in the 1970s came because energy supplies had by then become a matters of crisis. But many programs born of crisis die with the passing of immediate peril.

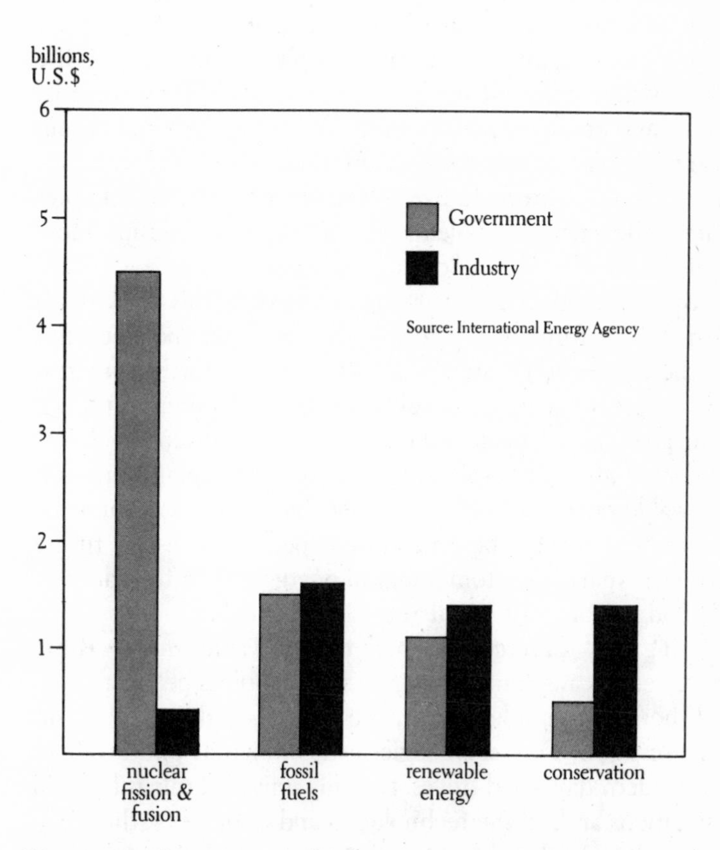

Figure 12.1. Government and Industry R&D Expenditures in 17 Industrial Countries, 1981.

Today the need is to look beyond the immediate crises to a gradual but purposeful transition to reliance on renewable energy during the next several decades. While increasing spending for all energy technologies during the 1970s spared governments the need to make painful, politically divisive choices, spending for energy generally and renewables in particular in the 1980s is in danger of being cut back in the face of economic hard times. Much as the last hired is the first fired, cutbacks in energy spending are falling disproportionately on the newer, more promising technologies. In the United States, for instance, the Reagan administration is trying to turn back the clock by drastically cutting government support for renewable energy and energy efficiency while increasing spending on nuclear power. Yet unless governments do an about-face, committing themselves firmly to balanced and well-funded R&D programs, the energy problems of the 1970s will erupt more virulently in the late 1980s and the 1990s.[4]

As the technical and economic reviews in this book make clear, the performance of renewable energy technologies during the 1970s and the immediate prospects for further progress warrant greatly increased R&D allocations. A minimum short-term goal should be to spend one-third of all energy R&D funds—in absolute levels, twice the present expenditure—on renewable energy. If overall energy R&D budgets cannot be raised, both nuclear power's current performance and future prospects make the atom a logical energy source to reduce in the budgets of most countries.

Using the resources devoted to renewable energy R&D effectively means confronting the sticky choices between long- and short-term, public and private interests and applied versus basic research. For technologies as dissimilar as photovoltaics and modern-day wood stoves, the only universal rule is to build institutions around the technologies and resources rather than forcing R&D efforts to fit into preconceived or established organizations.[5]

In most countries renewable energy research is conducted along the same lines as other research ventures. In the United States and France, the two largest spenders for solar energy research, there are large central research facilities—the Solar Energy Research Institute (SERI) and the Commissariat a l'Energie Solaire (COMES). But they have a better record for accumulating scientific knowledge than for adapting technologies to user needs. In Japan the New Energy Development Organization (NEDO) has the major responsibility and is noted for its close cooperation with Japanese industry. Research in China, on the other hand, has been made part of a wider effort to diffuse technologies into the countryside. There, advanced scientific work has been somewhat neglected. In developing countries where research funds are scant most resources have been spent adapting imported technologies to local conditions and keeping abreast of Western developments.[6]

While no research setup is applicable in every country, continuity is vital to every nation's success. Assembling high-quality research teams and conducting sophisticated research requires time and institutional stability. In the highly politicized, crisis-buffeted 1970s, such continuity was lacking. In the United States frequent reorganizations, disruptively short budget cycles, shifts in program goals, and political meddling have reached epidemic proportions and have seriously compromised the large U.S. investment in renewable energy research. Needed in the U.S. and elsewhere are performance goals and long-term budget commitments.[7]

Increased spending for renewable energy R&D should be directed toward establishing a sensible balance between the various technologies. Thus far research priorities have paralleled those dominating the overall energy R&D agenda. (See Table 12. 1.) Technologies that produce electricity, (solar thermal electric devices, large wind machines, and photovoltaics) are favored over those yielding liquid fuels or direct heat (biomass-conversion technologies or passive solar design). A bias toward high-technology, capital-intensive approaches is also evident: Solar thermal electric and ocean thermal electric technologies, which have only long-term and geographically limited potential, rank higher than methanol-conversion technology and solar ponds. The primary recipients of increased spending should be biomass energy, direct use of solar energy, and small-scale applications of all renewable energy technologies.[8]

Table 12. 1. Breakdown of Government R&D Expenditures for Renewable Energy in Industrial Countries, 1979 and 1981

Technology	*1979*	*1981*
	(millions of dollars)	
Solar heating & cooling	196	146
Photovoltaics	143	203
Thermal electric	116	156
Wind	85	140
Ocean thermal energy conversion (OTEC)	57	54
Biomass	64	106
Geothermal	178	243
Total	839	1048

Source: International Energy Agency, 1982.

Given its present contribution and long-term potential, biomass energy R&D has been inadequately funded almost everywhere. Combustion, fermentation, gasification, and distillation techniques have been in use for a long time, but performance and efficiency could be vastly improved through modest investments in chemical and engineering research.

Energy crop research also deserves increased priority. Because research into alternative feedstocks has been so neglected, new efforts to use biomass fuels in the United States and Brazil have been based on food crops. Many little-used food crops, coppicing trees, arid-land plants, aquatic plants, and agro-forestry combinations cry out for examination. Cultivating promising candidates for energy cropping on pilot farms in various climates over extended periods alone can answer basic questions about yield, water requirements, and soil impacts. Investigations of alternative feedstocks take researchers into agriculture and basic plant biology—sciences alien to most energy planners but essential to sound biomass energy use. Hence, it makes sense to conduct such investigations in agricultural research centers expanded with the support of national governments and international organizations.[9]

The second critical research gap is in support for small-scale and community-sized systems—small wind pumps, on-farm ethanol stills, climate-sensitive design for tract homes, solar ponds, and biogas digesters. Since many small firms, some of them struggling, are marketing these systems, government programs that do not include industry as a partner can be counterproductive. An especially efficient approach to research is to channel funds into industrial product improvement. When the U.S. Department of Energy tried developing methane-from-landfill technology already on the market, potential users became concerned about the product's lack of commercial maturity and they withheld purchases while government researchers duplicated the systems. A better way to assist small companies is through programs like that of the Rocky Flats Wind Center, where government scientists purchase small machines, check their performance in field conditions, and help companies improve the machines.[10]

Among high-technology applications, the most deserving of funding is photovoltaics. While the kind of intensive, well-financed private-sector interest that exists in Japan, Europe, and the United States lessens the need for government aid, government should still carry out basic research and ensure that all promising leads are pursued, especially those related to neglected technologies or high-cost demonstrations. In the

United States, for example, both large wind machines and alternative technology for hydropower facilities fall into this category.[11]

Of course not all renewable energy advances will come from energy-technology development programs. Photovoltaic technology benefited from advances made in semiconductors, while large wind machines now feature strong, lightweight synthetic materials and alloys pioneered by the aerospace industry. Further advances in materials science are the key to overcoming the corrosion problems plaguing active solar collectors and heat exchangers for geothermal and solar gradient ponds. Biomass conversion could be made more efficient if it employed catalysts used in the chemical and refining industries. Advances in plastic film technology could revolutionize the economics of active solar collectors by reducing weight and cost. These and other opportunities underscore the need for fundamental research in materials science as well as for adequate basic science budgets.[12]

Research on technologies of special importance to developing countries is especially urgent. Only a few of the largest Third World nations—such as Brazil, China, and India—can afford to mount sizable research programs. The rest make do with sometimes maladapted technologies pioneered in industrial countries. Only limited funds are available to work the kinks out of biogas technologies, fuelwood crops, wind pumps, and passive solar designs for humid, tropical climates. One way to make sure this important work gets done is to first identify priority technologies that need work and then to mount new R&D efforts within developing countries. Another is to establish a small network of international research laboratories modeled after existing agricultural research centers. One of these interconnected centers could work on fuelwood, another on other biomass research, a third on solar energy technologies, and a fourth on small hydropower, wind power, and miscellaneous technologies. Funds for such programs could come from the United Nations, the World Bank, the industrial countries, the wealthy oil-exporting nations, and the Third World countries themselves. The new laboratories would stimulate important new research efforts in the developing countries and encourage the international exchange of information among Third World scientists.[13]

A final high-payoff area of research is resource assessment—studies of geothermal gradients, water flow in smaller rivers, wind availability, biomass inventories, and the like. By drawing attention to unused or overlooked energy sources, resource assessments can catalyze energy development. But because the benefits of resource assessments are diffuse and difficult for private firms to turn to profit, governments must take the lead in making such surveys. No small incentive is that such resource assessments probably represent the most cost-effective investment in renewable energy governments can make.[14]

Using Vernacular Technologies

Unlike conventional, mostly standardized systems, renewable energy technologies must be engineered to withstand diverse environmental stresses and handling by people with little technical training. While not all renewable energy systems are what Ivan Illich calls "vernacular technologies"—small-scale and dispersed machines and tools—many are. The concept is important since where engineering and human elements have been given proper emphasis, technological adaptation has been rapid and negative side effects minimal. Where they have not, the opposite holds true.[15]

To fully meet the unique engineering challenges posed by the use of renewable energy, technology development programs must focus more on durability and simplicity, less on high performance—the touchstone of fossil fuel and nuclear engineering. A new institutional focus is needed, too, one centered on extension services, consumer education, and technical training systems rather than on laboratories alone.

Renewable energy technologies are unique among energy sources in the degree to which they must be fine-tuned to local climatic conditions. While a nuclear power plant or oil refinery would be essentially the same whether built in Central Africa or Scandinavia, a single solar collector design would simply not work in both regions. Variations in dust, temperature, sunlight, humidity, wind, and rain impose different design requirements. Since renewable energy systems are designed to tap the energies of the climate, they cannot easily be sheltered from the wearing effects of the weather. Rain, dust, wind, and sunlight can rust, pit, topple, and crack systems. Renewable energy technologies must be durable enough to withstand the extremes of the weather as well as the incessant variation of day-to-day natural energy flows. Some of the biggest problems of renewable energy systems come from their inability to withstand extreme storms. Small hydro facilities are often damaged by severe floods that occur every decade or so; some types of active solar heaters can be ruined by an unseasonal frost; wind machines can be destroyed by extremely high gusts; and hurricanes could sink OTEC plants. Adapting renewable energy systems to climatic extremes is an economic imperative because they must often operate over a period of one to three decades with only minor maintenance so as to justify their initial costs.[16]

Often high efficiency must be sacrificed to improve the durability and to lower the cost of the technology. During the early 1970s, when active solar heaters were reinvented in the laboratories of many industrial countries, scientists stressed high performance only to discover that less efficient but more durable models worked better on real rooftops. This lesson was brought home on a grand scale in the U.S. government's Solar Heating and Cooling Demonstration program. Between 1974 and 1978 the U.S. government funded the installation of several dozen different solar water and space heater designs. Most of the complicated, ultraefficient systems soon broke down, while the simple but hardy models—most little changed from the solar collectors used in California and Florida half a century before—worked.[17]

The failure to take climatic and ecological variations into account when transferring hydropower technology from one region to another has engendered many problems. When Soviet planners helped China design the Sanman Gorge Dam during the 1950s, they assumed sedimentation rates typical of Soviet rivers. The Yellow River, it turns out, silts up much more rapidly, and three-quarters of the dam's power and water-storage capacity were lost within a decade of the project's completion. Similarly, many active solar collectors and photovoltaic arrays designed by European firms have failed in Sahelian Africa, where dust storms impair their performance. In response to these challenges, corporations and development groups are adapting these technologies to local climatic conditions. Within developing countries, the need is to enhance local or regional technical capability to assess and modify renewable energy devices.[18]

Every bit as essential as how well a renewable energy technology fits into the physical environment is how well it fits in with the habits, needs, and skills of the people who must use it. The classic case of mismatch between users' customs and abstract technological promise is the solar cooker, which has run up against a solid wall of user resistance. Yet users' preferences cannot be considered a dead crust of habit that must be broken before "development" can be undertaken. Instead they embody long-proven practices that are best built upon rather than scrapped. Naturally enough, technologies that mesh with rather than disrupt traditional patterns of living are the most likely to come into widespread use.[19]

Like any other technology, renewable energy systems also require alert and trained operators. Houses burn down when wood stoves are used cavalierly; biogas generators have to be periodically purged of impurities; and pipes in solar water heaters freeze and burst when owners forget to drain out the water or add antifreeze, just as power plants shut down when somebody closes the wrong valve. Still, renewable energy technologies do differ from conventional energy technologies in one regard: Large numbers of nonexpert users must be able to master their maintenance and use. Living in a passive solar house, using biogas, or burning wood is different from switching on a heater supplied with nuclear electricity because the user is the producer and the operator as well as the consumer.

In many cases, high user involvement has been both a blessing and a curse. Much of the pioneering use of small renewable energy systems in the United States and Western Europe has come from backyard tinkerers and "do-it-yourselfers" who make up a small percentage of most communities. However, what these people see as opportunities to become more self-reliant and exercise their technical skills, many people view as chores. Consequently, the widespread use of small-scale energy devices depends on making their operation as simple as possible. Automatic governors for small hydropower facilities, thermostatically controlled drains for solar water heaters, automatic pellet loaders for wood furnaces, and light-sensitive shade controls for passive solar houses are important steps toward accelerating mass diffusion of renewable energy technologies. Here, as in environmental adaptation, success may involve small sacrifices of performance and efficiency.[20]

Education, information, and extension programs are the key to the effective use of even simple-to-operate systems. Basic skills and basic knowledge can be taught in elementary and secondary schools much as courses in mechanical arts, home economics, and driver's education are today. Low-cost or free energy audits for households and loans for feasibility studies have already helped American and European consumers and small resource owners assess their energy-investment opportunities. The industrial and agricultural extension programs that have played such an important role in spreading skills to small businessmen and farmers could also be retooled to help adults become familiar with renewable energy's technical side.

To appreciate the importance of teaching maintenance skills to users, consider how biogas digesters have been introduced in China and India. In China installers and operators from every village that was to receive a digester were trained. In India the program was almost exclusively hardware oriented, focusing on building as many generators as possible and relying on outside technical expertise. Today China has twenty-five times as many digesters as India and fully half of India's digesters are in disrepair. In energy, as in health care, the Chinese have emphasized raising the entire rural population's technical competence rather than refining the skills of a technical elite. Other countries, particularly the Western industrial countries, need to follow this example and balance their elite-oriented approach to technical education to ensure that all their citizens acquire the minimal technological literacy necessary to function in a world of increasingly dispersed energy systems.[21]

Simplifying the operating requirements of a technology can, of course, make the system itself more complex—which makes installation and repair networks all the more important. Just as the automatic transmission simultaneously simplified operation and complicated repair of the automobile, so too the automatic control systems of increasing importance in renewable energy systems will make quality installation and repair more critical. According to the U.S. Consumer Product Safety Commission, improperly installed or operated wood stoves of the newer and more efficient but hotter-burning variety are causing an estimated 1,300 house fires a year. And the improper installation of some solar hot water heaters has undermined the economics of the systems and tainted the public's perception of the new industry. As with skills like auto repair, plumbing, and home building, installer certification and training standards set by industry and unions and supported by government are sorely needed.[22]

In developing countries the lack of adequate installation, maintenance, and repair networks and facilities greatly impedes the use of renewable energy technologies. Many otherwise promising projects to install wind pumps, biogas digesters, and other devices have failed once technical experts who set up the projects have departed. To remedy this problem governments need to start training programs within villages (especially economically marginal communities) that teach people to operate and repair their own devices or to establish small village-level industries to manufacture, install, and repair the new technologies. China has taken the latter approach in small hydropower development, and the Intermediate Technology Development Group is attempting a similar strategy with wind pumps in several countries.[23]

Many of the undesirable side effects associated with dispersed energy systems—deforestation, air pollution from wood burning, and agricultural soil erosion—can best be tackled by designing control systems into the technology and including environmental protection in user-education programs. It simply is not feasible to monitor smoke emissions from each of millions of wood stoves, wood gasifiers, ethanol stills, and methanol distilleries. Vastly more efficient and effective is regulating the manufacture of the technology. Governments could require wood stove manufacturers to equip their models with catalytic combusters much as some now require the auto industry to equip vehicles with pollution-control features. Still, active owner involvement remains vital. For just as a $300 automotive catalytic converter can be ruined if a consumer unthinkingly fills up with leaded gasoline, so too a wood stove combuster can be impaired if painted wood or metal foils are tossed into the fire. Realistically, trying to design an idiot-proof technology will probably remain an elusive goal rather than a reality.[24]

Seed Money: Financing the Transition

New technologies and the support structures to make them work create energy opportunities. But people cannot exploit these opportunities unless they have both financial incentives and access to capital. Taxes on conventional fuels can both motivate people to use new sources of energy and raise the revenue to finance them. Gradually raising energy prices and making financing assistance more equitable could speed an orderly transition to renewable energy and help the poor, often the victims of energy policy. Building equity into energy programs is more than a matter of social justice; it is essential to making a successful transition to renewable energy. While some countries have recognized the link between large but temporary oil revenues and the possibility of building more permanent energy systems, few have successfully channeled capital resources to those best able to exploit renewable energy. Grants, loans, and tax breaks are needed to put capital into the hands of consumers and businesses.

The greatest disincentives to the widespread use of renewable energy technologies are government controls that keep the price of fossil fuels below their true or replacement costs. Price controls encourage energy waste and put renewable energy technologies at a competitive disadvantage. Indeed, in parts of the United States that rely on price-controlled natural gas, things have hardly changed since the sixties. But in oil-dependent northern New England, where prices for heating fuel rose from 20¢ a gallon in 1972 to over $1.00 a gallon in 1980, oil consumption has fallen by 12 percent annually and some 55 percent of all households have turned to wood as their principal source of heating fuel. Use of wood and solar energy for residential and commercial heating would undoubtedly have increased even more rapidly had not 45,000 oil consumers shifted to price-regulated natural gas.[25]

In other industrial countries energy-pricing structures take a variety of forms. Consumers in Western Europe and Japan pay prices close to the cost of obtaining additional supplies of energy today. In both regions expensive imported oil is an energy staple so the incentive to restrain oil consumption through taxes is strong. In contrast, Canada and the United States have done the least to bring energy prices to replacement cost levels. Long accustomed to cheap energy, both must now make a painful choice between a purposeful phase-in of higher prices or more sudden price shocks. While the two nations have recently decontrolled oil prices and are slowly decontrolling natural gas prices, those moves need to be supplemented by further natural gas price increases and additional taxes on gasoline.[26]

The Soviet Union produces virtually all the oil and gas it uses, providing insulation from the world market. The government also sets prices far below prevailing world levels. Low prices do not, however, necessarily stimulate additional consumption because central planners may not allocate resources to the sectors that would take advantage of low prices. The Soviet Union has been spared the problems of an oil-dependent transportation sector primarily because planners have not given high priority to building automobiles. Then, too, during the 1950s and 1960s the Soviets stuck with coal while the rest of the industrial world switched to oil and gas. At the time these policies were viewed by Western observers as archaic, but partly as a result, Soviet oil and gas reserves are high and there is relatively little oil-dependent capital stock today. Still, even this stability in the energy economy has had its price. Soviet industries that rely on oil have had little incentive to conserve or use new energy sources.[27]

The practice of holding fossil fuel prices below replacement cost is also widespread in oil-exporting countries. All the Middle Eastern members of OPEC, as well as Venezuela, Nigeria, and Mexico, sell petroleum products to domestic consumers for a fraction of their export value. Gasoline in Saudi Arabia, for example, sells for 25¢ a gallon, and the roads are clogged with large American-made automobiles. While such cheap energy has been a powerful stimulus to rapid internal economic growth, it is also locking these countries into petroleum-based economies.[28]

In many developing countries without significant hydrocarbon resources, governments subsidize the price of imported kerosene to shelter the poor who depend on kerosene to meet basic needs. According to Indian analyst Amulya Reddy, the same distorting impact of price controls operating in New England or the Soviet Union is at work in India. Three-quarters of India's 116 million households depend wholly on kerosene for lighting, and controlling kerosene's price even for the sake of the majority has had a serious unintended effect on transportation patterns. Since kerosene and diesel fuel are interchangeable, the government had to extend price controls to cover diesel fuel as well. As a result, the number of diesel-using trucks rose rapidly and railroad use declined, even though railroads are several times more efficient movers of goods and use domestically mined coal. Spurred by increases in trucking, India's oil imports have continued to rise. In 1980 oil imports consumed 80 percent of export earnings. Meantime, investments in biogas, wood gasifiers, or other domestic renewable energy sources that could directly replace kerosene go a-begging.[29]

Behind such seemingly backward policies is a well-placed fear of harming the poor. Since energy costs account for a disproportionate share of poor people's budgets, higher prices often mean doing without. Small wonder riots broke out in Cairo and other major Third World cities in the late seventies when kerosene subsidies were reduced. For policymakers the challenge is to balance energy goals and equity goals—a crucial task since simply decontrolling the prices of widely used fuels deprives many consumers of the means to invest in energy efficiency and renewable energy at the precise instant they have the most incentive to do so.[30]

The problem of the poor has put energy goals on a collision course with the desire for equity in many countries. This conflict has been particularly unfortunate because from the standpoint of society as a whole the greatest opportunities for using renewable energy and improving energy efficiency are among the poor who have the least ability to invest—whether it be in home insulation in Appalachia, a new wind pump in East Africa, or a more fuel efficient tractor in Mexico.

The key to steering between the Scylla of price controls and the Charybdis of inequality is to raise fuel prices through fuel taxes and to recycle the revenues for consumer and business investments in energy alternatives and efficiency. This approach gives consumers the incentive and the capital they need to invest. On a global basis the shift to replacement cost pricing through taxes will yield an enormous windfall—the several hundred billion dollar annual difference between the cost of extracting, transporting, and refining oil and its market value. This windfall could be the world's operating budget for the energy transition. Prudently reinvested in energy efficiency and renewable energy, this treasure is the bridge to a sound energy system. If squandered on unproductive subsidies to declining industries, defense buildups, corporate buying sprees, or luxury consumption by the elite, this treasure will be irretrievably lost, making the transition to renewable energy much more difficult.

For each country the optimal way to spend its oil revenues to promote sustainable energy systems will differ. In the United States the most productive use of the revenues of a "windfall" severance tax on decontrolled oil and gas is to fund housing weatherization for low-income people and to provide tax credits and loans for a variety of renewable energy sources. For developing countries without significant oil reserves, government subsidies from imported petroleum products should be redirected to the purchase of domestic biogas digesters and efficient cook stoves, the development of village woodlots, and the creation of domestic industries that manufacture renewable energy technologies. Sri Lanka has already stopped subsidizing the purchase of imported fuel and begun instead to fund tree planting and charcoal production.[31]

Recycling the money from the oil windfall into alternative energy systems will be easiest where oil reserves are substantial. The state of Alaska, for example, has already set aside $5 billion in oil revenues from the North Slope for an ambitious hydroelectric development program. Venezuela has followed a similar path, devoting most of the funds it has set aside to carrying out its oil-financed five-year energy plan to hydroelectric projects. Using a small part of its vast oil revenues, Saudi Arabia has funded various photovoltaic and solar pond investigations in the hope of making sunlight a source of permanent wealth.[32]

Western industrial countries with declining oil and gas reserves have not fared well at keeping alive the link between temporary oil revenues and an alternative energy future. The United Kingdom has used the revenues from its North Sea oil and gas primarily to meet general government operating needs. The Netherlands, faced with the decline of the natural gas fields that have been a major source of postwar wealth, has yet to institute replacement cost pricing to extend supplies or finance alternatives. Even the largest oil producer (and consumer) in the West, the United States, has allowed oil owners to reap most of the benefits from oil decontrol. An attempt in 1979 to fund low-income energy assistance, alternative fuels, mass transit, and the Solar Energy and Energy Conservation Bank with proceeds from the windfall profits tax has largely lost momentum.[33]

Financing the transition to renewable energy will be most difficult in developing countries that lack petroleum. International experts meeting under the auspices of the North-South Roundtable in May 1981 noted that "the energy crisis in developing countries is not a crisis of scarcity of energy resources but a scarcity of finance." The World Bank estimates that developing nations need to invest $60 billion to $80 billion in energy during the eighties. Approximately one-quarter of this would be for renewable energy—initially, fuelwood and hydropower projects. Yet competing needs to raise agricultural productivity, build industries, and establish modern sanitation, health, and educational services press these countries hard.[34]

Most developing countries will have to rely upon external investment and aid supplemented with internally generated resources and set their investment priorities carefully. They will also have to depend heavily upon the international private banks. Such government-supported institutions as the World Bank and the regional development banks can have an all-important leveraging effect, encouraging private investment as well as providing loans for projects of scant interest to the private sector. The World Bank significantly increased its energy lending in the late seventies and early eighties, but resurrecting the proposal for an energy affiliate to the World Bank —a move proposed and then blocked by the United States— is the only way to accelerate adequately the flow of public as well as private investment dollars. A successful example is Brazil, which has financed large renewable energy investments with foreign loans that supplement revenues from taxes levied on imported petroleum.[35]

Oil exporting nations have begun helping the poorer Third World countries retool their energy systems. Mexico and Venezuela, for example, rebate 30 percent of the oil payments of ten Latin American countries in the form of loans, with the interest rate set at 2 percent if the funds are invested to promote greater energy self-sufficiency. To take such financial recycling one step farther, Thomas Hoffman and Brian Johnson of the International Institute for Environment and Development have proposed setting up international mechanisms to channel oil revenues into high-payoff investments in poorer countries.[36]

Renewables often represent a good investment choice for capital-poor developing countries because the equipment needed to harness this energy can be manufactured domestically, employing their most abundant and underused resource —human labor. Unlike coal or nuclear power, which require imported equipment, many renewable energy technologies such as small dams, efficient wood stoves, fuel lots, and biogas digesters can be produced by people who would otherwise be idle. This is the secret of China's rural energy successes. Off-season agricultural workers in China have built a basic infrastructure that would have been unaffordable if financed with borrowed funds from abroad.[37]

Putting capital resources into the hands of those best able to use renewable energy has been achieved only rarely. Direct grants and cheap credit (low-interest loans or interest subsidies) are the least cumbersome and most equitable ways to transfer capital. Unfortunately, the subsidies for renewable energy technologies most widely available now are tax credits and loan guarantees for a fairly restricted set of technologies. As such, they represent an important first step to reversing longstanding discrimination against renewable energy, but incentives aimed primarily at the affluent are ultimately limited in their effect.

These drawbacks notwithstanding, tax credits and exemptions have had some positive effects. In the United States a 40 percent tax credit on the first $10,000 spent on renewable energy equipment has been a major stimulus in the growth of the renewable energy market, particularly that for solar water heaters. Brazil has exempted solar equipment and 100 percent alcohol-fueled automobiles from the national value-added tax, a subsidy of approximately 35 percent. Many states and cities in the United States have also exempted renewable energy equipment from property and sales taxes. But do large tax breaks add up to boondoggles? A recent study of California's generous solar tax credits found that gas brought from the North Slope of Alaska to California would still enjoy greater tax subsidies than solar water heating. A more serious problem is the bias of renewable energy tax credits that apply to active solar collectors but not passive design, to ethanol plants but not wood stoves, and to alternative fuels but not fuel-efficient automobiles.[38]

The single biggest drawback to the incentives approach is equity: The many who do not pay income taxes cannot benefit. Where credits are funded through taxes on fossil fuels, the poor suffer twice—first from higher prices and then from denial of access to aid. A minimum-equity goal should be making tax credits to individuals refundable so that the poor can receive direct grants. Also equitable and effective are loans and interest-rate subsidies. Since most renewable energy systems entail high initial costs but no fuel costs thereafter, lowering the cost of borrowed money can be a powerful investment incentive. For just such reasons the U.S. government's Solar Energy and Conservation Bank—which would have made available $1.2 billion over a four-year period for interest-rate subsidies on loans to households and small businesses—was created. When its way is cleared of political obstacles, the Bank could be an important aid to low-income homeowners who cannot obtain commercial loans on any terms and to older industrial cities with badly deteriorated building stock.[39]

The least common but potentially most effective and equitable means of financing the transition to renewable energy is the direct grant. Grants foster equity because not only those well-off enough to pay taxes benefit. Then, too, under the grants system the consumer gets the money immediately instead of at tax return time. In Japan the solar collector industry has benefitted greatly from government grants to consumers equal to 30 percent of the cost of the system.[40]

Another worthy model for direct grant programs is the Canadian Home Insulation Program (CHIP) pioneered first in Nova Scotia and then expanded to the entire country in 1979. During the program's first two years in Nova Scotia, 15 percent of the households took advantage of $800 cash grants for home weatherization. Canadian energy analysts attribute the program's success to its simplicity and to an effective media campaign aimed at potential participants. Similarly, Canada's Forest Industry Renewable Energy (FIRE) program has channeled over a quarter of a billion dollars into cost-shared projects, primarily in wood burning. Room for applying similar programs elsewhere in the world is great indeed.[41]

Opening Up the Grid

No institution is more important to the fate of renewable energy—or more in need of redirection—than the electric utility. The last decade has been a particularly painful one for utility managers. Slowing rates of growth in electricity use, rising environmental conflicts, and confusion about the potential of new electricity-generating technologies have called into serious question practices and expectations inherited from an earlier era of rapid growth and rising centralization. Yet nowhere more than among electricity producers is conservation the logical institutional forerunner of renewable energy.[42]

With a few notable exceptions the utilities have so far resisted change. But such entrenchment has been at the sacrifice of profits as well as broader social concerns. In serious need of reform are pricing, financing, competitive access, and structure. More specifically, the price consumers pay for electricity must more accurately reflect the social costs of producing electricity; utilities must help finance energy efficiency and renewable energy; nonutility power producers must be encouraged to operate; and today's giant utilities must be restructured into a greater number of smaller, more workable entities.

Sweeping in their implications, these reforms are neither

untested nor incompatible with sound business practice. Where these changes have been made over the last decade, the record reveals success. Kept on their present course, utilities could be major obstacles to a renewable energy future; redirected, utilities could be powerful instruments of change.

As for the major source of consumer problems—price setting—the rising cost of generating electricity from new power plants has made traditional patterns largely obsolete. Logically enough, prices were set to encourage more consumption as the cost of generating electricity fell steadily between the beginning of the century and the early seventies. But as power became more expensive to generate, rate structures were slow to change. While the overall cost of electricity delivered to the consumer has risen in the last decade, most rates still reflect the *average* cost of producing electricity, which is a mixture of older cheap power and more expensive power from newer sources. As a result, consumers buy at a price considerably below what it will cost the utility to produce additional power, and thus consumers use large amounts of electricity.[43]

One solution to this squeeze—simply decontrolling the price of electricity—has too much consumer opposition to work. But establishing "life-line rates" for the minimal amount of electricity needed for each household to meet basic needs and then letting the rest of electricity sold reflect the true cost of production is one means of encouraging conservation and the use of alternatives without harming consumers. With life-line rates, consumers can get at an extremely low price enough electricity to power household appliances and lighting. Tried experimentally in California, life-line rates give consumers reason to reduce electricity use but keep bills fair and affordable.[44]

To restrain demand and encourage consumers to use alternatives without debilitating price increases some utilities have devised ingenious demand-management schemes. As early as 1966 utilities in West Germany began charging consumers less for power supplied during periods of slack demand. Over ten years this simple change altered consumption patterns so dramatically that plans for several large new power plants could be canceled. Other utilities have achieved similar results by "interrupting" power to large customers during peak demand periods. Detroit Edison, for example, can shut off over 200,000 water heaters from its headquarters. Consumers who volunteer to be interruptible customers receive a 35 percent reduction in rates. A growing number of utilities are also trying to influence electricity demand by offering consumers free energy audits to diagnose energy-saving opportunities or by giving away thermal blankets to wrap around uninsulated electric water heaters. As utilities become more deeply involved in structuring electricity demand, they will be well equipped to benefit from and shape new technologies that displace electricity but rely on utilities for backup. Extending these demand-management schemes should be a high priority for consumers, utility regulators, and utility managers alike. All clearly benefit.[45]

Exceptions aside, the trend toward utility demand management is most advanced in Japan, Western Europe, and oil-reliant parts of the United States. Soviet utilities and U.S. utilities with extensive coal, nuclear, or hydropower capacity have been slower to adopt new pricing practices, and in the Third World few utilities manage demand. Yet electricity costs in developing countries are high, and recent analyses of power systems in urban industrial regions of China, India, and Brazil indicate that instituting more rational pricing systems along with investing in more efficient transmission networks offer higher rates of return than building new power plants does.[46]

Utilities are in a near-ideal position to make financial assistance available to their customers to improve energy efficiency and use renewable energy. They have access to large amounts of capital at low interest rates and, unlike commercial lenders, they are accustomed to making investments with long-term payoffs. Utilities also have established relationships with electricity consumers. Expanding these ties is much easier and cheaper than building new institutions from the ground up. Many utilities, most of them in the United States, have begun making low-interest loans, offering cash rebates, or renting systems to their customers. In many cases customers make loan payments through monthly electric bills. So compelling is the financial logic of these arrangements that many utilities could be spending half their capital funds on consumer financing by 1990.[47]

The prospect of extensive utility involvement in renewable energy and conservation has many alternative energy advocates worried. Their fear is that utilities will drive up the costs and reduce competition. Such dangers are real. The key to skirting them is regulatory supervision, not cutting utilities out of the transition. Regulations should enable consumers to choose their own contractors and systems and should protect consumers who choose not to take advantage of the utility's plan, making sure they do not end up subsidizing those who do.[48]

As utilities take on new roles they will have to shed and share others. The most important change will be in electricity production itself. All utilities, whether in capitalist or socialist nations, operate as legal monopolies. They have been granted the sole legal right to buy, sell, produce, and distribute electricity in a given area. These monopoly concessions have been granted on the assumption that electricity production and marketing would be less efficient and effective were several companies competing to provide service. Once that rationale made sense. The pioneers of electrification in the early twentieth century quickly learned that setting up competing parallel networks of distribution wires and transformer stations for the same customers inevitably raised consumers' costs and lowered producers' profits. Charters for the exclusive right to service a particular area were thus granted by municipal governments. Along with that right came a legal monopoly on power generation, even though production of electricity was never really a natural monopoly. The appeal of building ever larger plants obscured this distinction until recently when large scale lost some of its economic luster.[49]

Competition will revive in the electricity industry if utilities are required to purchase power from independent power producers. In Europe utilities commonly purchase cogenerated electricity from industry and municipalities. French utilities buy power from small dam owners, and Danish law requires utilities to purchase power from small power producers. Even more far-reaching is the U.S. Public Utility Regulatory Policies Act (PURPA), which was passed in 1978 and which requires utilities to purchase electricity from small producers at a price equal to what it would cost the utility to produce electricity from new power plants. Already PURPA has transformed the economics of small-scale electricity generation in the United States, and an estimated 12,000 megawatts of new electricity production capacity will result by 1995. Dozens of small firms specializing in wind, geothermal, small-scale hydro, and cogeneration have sprung up to produce and sell power to utilities. Under attack by some utilities, PURPA is nonetheless critical to the development of new sources of electricity.[50]

Opening the grid to competition does not, of course, disqualify utilities from producing electricity from renewable sources. Many technologies—large wind machines, centralized photovoltaic systems, and solar thermal electric plants among them—enjoy economies of scale that make utility involvement particularly appropriate. In fact, despite the inertia and resistance of most utilities, a growing number have begun to look seriously at promoting renewable energy. In the United States 236 utilities spent $26 million on 943 projects in 1981, more than double the 1977 level. Under pressure from state regulators and faced with public opposition to coal and nuclear power plants, California utilities accounted for much of the spending. In Europe utilities have begun investigating wind turbines and solar thermal electric devices, while expanding reliance on small-scale dams, district heating, and municipal waste-to-energy plants.[51]

The fourth and most difficult utility reform needed is a reversal of the seemingly inexorable trend toward centralization that has prevailed since the 1920s. In the early days of electrification electric utilities were small. Typically, each served only one community, and many were owned or franchised by the people they served. The improved economics of large plants and the technology of long-distance transmission made centralized management of far-flung power networks possible, and utilities consolidated to exploit these possibilities. Before long, localized distribution systems had become mere appendages of larger systems. In France and the United Kingdom all electricity production and distribution is now performed by one government-owned company. In the Soviet Union and most Third World nations utilities have been centrally operated and state-owned since their creation.[52]

With the advent of new opportunities for conservation and small-scale renewable energy technologies, the optimum size of utilities is no longer "extra large." With key important investment decisions now revolving around demand management or small-scale power production, local networks will become ever more important and ever more difficult for centralized bureaucracies to manage efficiently. Then, too, as utilities begin buying power from locally owned sources and financing conservation-related building improvements, local governments will acquire a greater stake and ability to regulate them according to local goals.[53]

Since centralized organizations rarely break up of their own accord, governments will have to promote or force this diffusion of power. Once utilities become smaller, government's role will shift again. Then the key task will be maintaining minimum operating standards for all utilities and preserving long-distance transmission systems where they are economic.

Obviously this revitalization will be easiest where remnants of the early local systems still exist.

Among existing energy-supply institutions in the industrial world, the U.S. rural electric cooperatives (nonprofit corporations owned by the rural consumers of electric power) are well suited in terms of scale, structure, and supply opportunities to serve as prototypes of new utility systems for the renewable energy economy. The more than 1,000 cooperatives in forty-six states serve 75 percent of the land area of the United States and fill 15 percent of U.S. power needs. Since 90 percent of all coops have a capacity of no larger than 50 megawatts and 30 percent require less than 10 megawatts of power, a few small renewable energy projects can meet most or all of the average rural electric cooperative's power needs. The River Electric Cooperative in Gaffney, South Carolina, demonstrated this when it renovated an abandoned small hydro plant capable of meeting one-fifth of its total power demand.[54]

Another appealing attribute of the rural electric cooperatives is their diversity and flexibility. Coops are small enough to take advantage of whatever energy resources are available locally. Perhaps just as important, their history as industry pioneers makes coops natural agents of change.

Empowering People

Building the institutions needed to realize renewable energy's potential is a political task. The basic challenge is to empower people with the knowledge, resources, and freedom required to solve their own energy problems. Empowering people requires redirecting existing national energy programs, mobilizing local initiative, and unshackling individual effort. Powerful interests that benefit from the current state of affairs will resist change. Overcoming this entrenched opposition will be possible only if the beneficiaries of renewable energy—farmers, small businesses, environmentalists, consumers, homeowners, and the unemployed—band together and pursue coherent political objectives. During the last decade, institutions have been pioneered that foster the transformation of powerless energy consumers into self-reliant energy producers. The next politically more difficult step, is making these institutions the rule rather than the exception.

Today most energy decision making occurs at the national level, where the highly concentrated interests of centralized energy usually prevail. Their strength is reinforced by the widespread public perception that energy is a complex esoteric subject best left to energy experts and the energy industry. Therefore, it is no surprise that government spending the world over is skewed toward the large-scale centralized technologies favored by large corporations and large public agencies. Thus, Third World countries build large rather than small dams, the U.S. government spends R&D dollars on solar power towers instead of on passive solar design, and Brazil's alcohol program benefits nobody more than it does the powerful sugar barons.[55]

The key to counterbalancing the power of entrenched interests at the national level is formulating new political coalitions among the latent constituencies of renewable energy. In poll after poll the citizens of most countries show widespread preference for renewable energy and conservation over nuclear power and synthetic fuels. But the potential beneficiaries of renewable energy do not see themselves as bound by a common interest, so they pursue other political priorities.[56]

One lesson of the 1970s, however, is that declining rural incomes, housing stock deterioration, and many other issues that receive more attention than energy problems do are rooted partly in misguided energy policies.

With new fuel crops farmers could cut operating costs and capture new markets. An ambitious effort to add solar equipment to houses would help revive the housing industry, cut homeowners' bills, and put many unemployed people back to work. Recognizing the hidden links between energy and other problems could be a powerful catalyst for the formation of new political coalitions.

What objectives should these new coalitions pursue? Two tasks seem central. The first is to set up national goals and to redirect existing agencies to meet them. The second is to channel resources to individuals, community groups, corporations, and local governments while assuring that renewable energy is not unfairly excluded from the marketplace by the conventional fuel interests. Here national government's role is that of midwife. Whether in reforesting South Korea's hillsides or putting solar water heaters on Santa Monica's rooftops, local initiative and national government support have most often proved the winning combination.

If national leaders set ambitious but realistic goals, the effects can be far-reaching. Such goals serve as vital benchmarks against which various sectors' progress can be measured. They can pressure inert bureaucracies while imparting a clear sense of broader national purpose to the individuals, communities, and companies that must make change happen. In this respect, small countries probably have an advantage over larger ones. They can set goals with a clear, shared sense of what they will actually mean. In large countries regional differences intrude and national goals often sound like mere abstractions. Thus, for some countries national planning is most of what is needed; in others it is only the beginning.

Few countries have yet established firm goals for using renewable energy. In 1979 the United States established the goal of obtaining 20 percent of its energy from renewable sources in the year 2000—an increase from the current 6 percent—but by 1981 the Reagan administration had abandoned the target altogether. Official neglect notwithstanding, this goal has become an international benchmark and a measure that U.S. communities and states can use to chart their progress. Other countries have set ambitious but obtainable goals for particular renewable energy sources—Sweden in wood; Brazil in alcohol; and the Philippines in wood, hydropower, and geothermal energy.[57]

Achieving ambitious national goals requires more than establishing an energy agency or ministry with one branch assigned to conduct research and promote renewable energy. Amid bureaucratic wars over funding, publicity, and turf, diverse renewable energy sources (which have little in common beside their renewability) get treated to a blanket approach or lost in the shuffle. Oddly, the agencies best positioned to promote renewable energy are not energy agencies but, instead, those responsible for housing, agriculture, taxation, forestry, community development, and transportation. Housing agencies are better able to stimulate sales and acceptance of passive solar design and solar water heaters than are energy agencies with no ties to the housing construction industry. Similarly, agricultural agencies have research centers and extension networks of paramount importance in achieving biomass energy goals. Rather than consolidating all activities related to renewables within one agency, governments should form small high-level councils to coordinate, direct, and monitor progress. Organizationally, then, the challenge is to redirect existing national programs to new goals.

Within this framework governments should redirect resources to local institutions and private firms. This need for local institutions to help match local energy needs to local energy opportunities sets renewable energy apart from conventional energy technologies developed and regulated by national governments.

In an age of complex far-flung technological systems, few communities think of themselves as having energy choices. Fewer still have a clear idea of which renewable energy resources lie within their reach. Rooting out these misconceptions can be a powerful catalyst for change. The groundwork for empowering citizens to decide their community's energy future consists of a resource inventory and public discussion of all alternative energy paths. A good model for "change from within" is that of Franklin County in rural western Massachusetts.[58]

At first glance Franklin County does not appear to be a promising candidate for energy independence. With long severe winters and no local oil, gas, or coal, the area depends heavily on oil imported from around the world. But, looking closer, a team from the Future Studies Program at the University of Massachusetts found three foundations for local self-sufficiency: a building stock so leaky that cost-effective weatherization could cut energy use by 50 percent by the year 2000; some 157 developed and undeveloped hydroelectric sites that could make Franklin County a net exporter of electricity; and a poorly managed forest resource base with the potential to supply enough wood to run the local transportation system on methanol. Supplemented by passive solar design in new houses, cogeneration in industrial plants, and district heat in the denser towns, these resources could permit Franklin County to be completely self-sufficient in energy by the year 2000. Far from an economic drain, researchers found that such a "solar scenario" would substantially improve the local economy by creating jobs, strengthening the tax base, and eliminating the export of funds for fuel. Spurred by the study's conclusions, a broad coalition of citizens is today working to make use of these previously hidden energy resources. Involving people in a frank assessment of their community's energy future turned passive consumers of imported fuels into self-reliant energy activists.[59]

Local governments have a variety of powerful tools at their disposal. They control land use through zoning and building design through codes. In addition, they own, franchise, or regulate the collection and disposal of waste and the placement of water pipes, power lines, and district heating systems. Before large-scale energy systems became the norm, localities were largely self-reliant in energy. Some could be again.[60]

California, where 20 percent of all new housing in the U.S. is being built, now requires urban developers to take solar orientation into account in laying out streets and siting buildings. By planning subdivisions with an eye to solar access, communities make the later use of solar collectors easier. Building codes too can serve as powerful instruments: California's Title 24 building standards have, for example, helped reduce energy use in new residences by 50 percent since 1975. Going a step farther, San Diego County requires all new buildings to feature solar water heaters. The public supports these regulations because they are tailored to perceived needs.[61]

More often than not, municipal governments actually own key heat- and electricity-producing resources: the waste stream, utility corridors, and unused land. Neighborhood by neighborhood, local governments or local cooperatives can build source-separation programs that open the door to sophisticated recycling projects like those underway in West Germany. Or they can follow northern European and Soviet cities, which have planned and built elaborate district-heating systems. In the Indian state of Gujarat, committed officials turned roadsides and yards around government buildings into a fuelwood resource for the poor and a revenue source for the government. Dedicated to promoting renewable energy, such initiatives can turn neglected resources into the foundations of local energy systems.[62]

Market-responsive private firms are also pivotal actors in bringing renewable energy into widespread use. Like local governments, they can adapt their activities to diverse, site-specific opportunities. With profitability uppermost in mind, small firms vigorously seek innovative techniques and cost-cutting strategies, and they make sure things keep working after the ribbon-cutting ceremonies. Whereas governments too often throw good money after bad to escape political humiliation, businesspeople usually know how to cut their losses and learn from their mistakes.

The versatility of companies in taking advantage of profit-making opportunities enables them to create links between owners of resources, developers of technology, and energy users that would be hard for governments to match. Among the most dynamic of such linking institutions are the small firms that have sprung up to produce and sell electricity to American utilities. By bringing capital together with engineering know-how and regulatory acumen, these private companies catalyze much of the renewable energy activity occurring in the U.S. electricity sector today.[63]

Remarkably fragile despite their immense motivating power, market incentives can fail or backfire as a result of corporate or government actions. Typically, large corporations have excluded new sources of energy from the marketplace. For decades, the oil companies and the utilities have systematically used their power over distribution systems—in essence the marketplace for liquid fuels and electricity—to keep out ethanol, small dams, and cogeneration.

In 1977 mismanaged government aid dealt a crippling blow to the nascent solar water heater industry in the U.S. When President Carter announced a tax credit with great fanfare, consumer demand fell by almost one-half because would-be buyers waited for over a year for the government aid to become available. Within months many solar firms were forced into bankruptcy, and the industry's evolution was delayed by a year or two. Here the lesson is that while central governments are essential for setting goals, providing information, subsidizing research, and solving equity and environmental problems, they must leave much of the "doing" to those who do best—communities, individuals, and corporations.[64]

Those countries without either a market exchange system or a strong tradition of local problem solving will, of course, be at a decided disadvantage in promoting the widespread use of the smaller-scale renewable energy technologies. Centralized planning and capital allocation cannot effectively substitute for market incentives and local initiative when harnessing diverse, site-specific resources. In the Soviet Union, for example, central planners have been successful in building large dams and urban district-heating systems, but not at promoting industrial energy conservation. Clearly, bureaucratic commands and rewards are not enough. But where central planning is supplemented by strong local government initiatives as it is in China, chances for success are better.

Empowering individuals and providing those institutions nearest them with resources is a political challenge that will test the institutional flexibility and responsiveness of most societies. By acting to solve their own energy needs and banding together to force governments to support their efforts, individuals can break the momentum of existing institutions. The obstructions are many and powerful, but the first and most decisive step is an individual one—to recognize ourselves as the potential masters of our energy futures. Multiplied on a planetary scale, the self-empowered individual is the basic force for achieving a renewable energy future.

13

Shapes of a Renewable Society

Ever since people conceived of harnessing the power of the sun, waters, winds, and earth, they have also speculated on the shape of a renewable society. With a deeper grasp of the energy potential of the earth's self-renewing processes and new plans to harness them, we now ask what the implications of an energy transition are for society as a whole. Today's economies and societies have been shaped as much by the availability of inexpensive oil as by any other force. But this dominance is now slowly ending, and, whatever our future energy sources, changes are inevitable.

To rely heavily on coal and nuclear power is to narrow societies' future options. Environmental damage would limit the areas where people can live comfortably. The "security state" mentality would become pervasive as societies sought to protect thousands of central power plants from extremist groups. The power of a small elite over energy technologies would necessarily be strengthened, wrested away from ordinary people. Massive reliance on coal and nuclear power would mean that societies would increasingly have to sacrifice their other priorities to energy production.[1]

Renewable energy, on the other hand, can preserve options rather than close them. Harnessed by centralized or household technologies, renewable energy could boost employment, bring new life to declining rural areas, and enhance local and regional self-reliance.

New Landscapes

"Buy land," advised the American humorist Will Rogers, "they're not making any more of it." Behind the visible constraints of water, energy, and food, lies the more basic limit of land. Although there are still many empty deserts and uninhabited expanses of tundra, productive, livable land is already in short supply. Greater use of renewable energy technology can actually help moderate rising conflicts over land use. Energy production can be combined with existing land uses or concentrated on largely unused land. Indeed, intensifying land use to derive energy without obstructing other uses could give new primacy to the landscape arts and local planning.[2]

Because solar energy is so diffuse that large areas of collectors —whether plants, wind turbines, or solar panels—are necessary to capture significant quantities of energy, tomorrow's landscape will be far different from today's. However, the important question is not how much land a given energy system takes, but whether it uses land not otherwise useful or whether it can piggyback without displacing existing uses. Some renewable energy systems, most notably large fuel plantations that compete with food or fiber crops for prime farmland or hydropower projects that flood rich river bottom land, crowd out other valuable land uses. For this reason they will never approach their physical potential. But a greater number of renewable energy technologies will either make use of marginal land or intensify existing land use.

The ultimate in such dovetailing, of course, is the use of passive dwellings that essentially turn roofs, windows, and structural supports into heat collectors and regulators. With the building itself serving as a solar collector and solar water heaters and photovoltaic systems built into the roofs of houses, no additional land is needed for energy. Contrary to the image of "solar sprawl" stemming from the widespread use of solar collectors, most electricity and heat requirements of urban areas can be met using available roof space. Some idea of this potential is provided by a University of California study of several U.S. cities. Using detailed aerial photographs of Denver and Baltimore, the researchers found that the cities as a whole could come close to meeting their land requirements for energy with surpluses from the warehouse district making up for deficits in the central business district.[3]

Fast-growing trees planted along roads, streams, between houses, and under power lines will also collect solar energy. Besides providing energy on a cheap, maintenance-free basis, urban trees moderate temperatures, absorb noise, and clean the air. In a harbinger of this multi-purpose land use, Hagerstown, Maryland, is planting trees on 500 acres of marginal land. The trees will be fertilized with sludge from the town's sewage-treatment plant and later gasified to power the plant. In rural areas, trees planted along field borders to supply fuel for farm machinery will also provide wind breaks and moderate the local climate.[4]

The landscape of a solar society would be more natural than today's landscapes. Cities, suburbs, and rural areas would be blanketed with a canopy of trees pleasing to the eye and soothing to the spirit. Buildings are likely to be more climate-sensitive and more integrated into their surroundings. Landscape design will occupy a central place in energy production and consumption, and city dwellers will have the chance to reattune themselves to their natural surroundings. Overall, cities will have a gardenlike quality. Nature, in a controlled and limited fashion, will have reclaimed areas of human habitation and activity.[5]

This blurring of artifice and nature will also extend to lands today largely untouched by human hands. Solar ponds can turn otherwise unused desert salt flats or brine lakes into energy collectors. Taming rivers far from cities will prevent the flooding of farmland, though the primordial quality of many of the world's pristine waterways could be forever lost. Large windmills in mountain passes or along coastal areas can intrude on ocean and mountain vistas, so much care is needed to keep the earth's wild places wild. Harnessing the natural energies of the last wildernesses—Siberia, the great salt deserts of the American Southwest, the Amazon Basin—may make sense in the calculus of material need but not in the deeper logic of the human spirit. Our species' ancient dream of subjecting the wild spaces to human control will, without care, become the nightmare of a totally fabricated planet.[6]

Today's conflicts over the siting of oil, coal, or nuclear facilities in wild areas center around ecological and health threats, but controversies in the future could pit more benign renewable energy systems against aesthetic or spiritual values. Ultimately, only population control, improved energy efficiency, and restrained material appetites can insure that energy complexes do not literally cover the earth.

The intensification of land use to meet energy needs will slowly erase the line between energy production and other endeavors. The "energy sector" may lose its definition as farmers, homeowners, waste recyclers, and city governments become part of the energy system without devoting their full-time attention to it. Energy use and production will begin to merge with other activities. As individuals, communities, and countries pursue their own solutions, energy systems and energy policies could become increasingly sterile abstractions, largely irrelevant to practical living. Energy may not be an easily definable part of either the physical or the intellectual landscape in societies reliant on renewable sources.

Renewable Jobs

Too often neglected is the impact that different energy alternatives have on employment. A basic measure of economic and psychological well-being, employment is an essential standard against which renewable energy's viability should be measured and toward which renewable energy development should be directed.

The substitution of energy and capital to perform tasks once done with human hands has been a century-long trend that has helped boost the worldwide production of goods and services and freed millions from drudgery. Today, however, rising energy prices and rising unemployment call into question the simple formula that worked so well in the past. Already, high energy prices have contributed to job losses in the automobile, chemical, and steel industries. With 36 million people entering the global work force each year, technologies must be judged by their ability to create jobs as well as to supply energy.[7]

Studies conducted in the last several years show that developing one energy source creates very different numbers and types of jobs than developing another. Each approach results not only in direct jobs—say, in drilling oil wells, manufacturing wind machines, or installing insulation—but also in a certain number of indirect jobs in related industries and services. Moreover, some energy sources create mainly unskilled jobs, while others create ones that require specialized training. Some create jobs in urban communities, some in rural areas, and others in remote regions that become fast-fading boom towns.[8]

Evidence gathered so far suggests that renewable energy development will create more jobs than would the same amount of energy obtained from oil, natural gas, coal, or nuclear power. The most detailed case study, conducted by the Council on Economic Priorities, compared the job-creation potential of a solar/conservation strategy and two proposed nuclear power plants on Long Island. Solar energy alone was found to create nine times as many jobs per unit of energy produced as nuclear power, and overall the alternative strategy yielded nearly three times as many jobs and two times as much usable energy as would the nuclear plants. Studies sponsored by the state government in California indicate that active solar systems create more than twice as many jobs as either nuclear power or liquified natural gas. And according to the U.S. Office of Technology Assessment, deriving energy from forestry residues is 1.5 to 3 times as labor-intensive as using coal.[9]

The types of jobs created in renewable energy development run the gamut. Such technologies as solar collectors, wind generators, and wood gasifiers are likely to be manufactured on assembly lines that will be partly automated but still require numerous semiskilled workers. Installing most of these decentralized devices will also create jobs, some of them quite similar to existing jobs in construction, plumbing, and the installation of appliances. Renewable energy development also creates a demand for a wide array of more technical jobs in such fields as resource assessment, advanced research, and systems engineering.

Perhaps the most labor-intensive of renewable energy sources are the various forms of biomass. Fuelwood plantations, methanol plants, and biogas digesters all use less capital and more labor—most of it rural—than do conventional energy sources. Polycultures of mixed food and energy crops will require even more labor since they are less amenable to mechanization than are monocultures. The relatively small conversion facilities built to turn biomass energy sources into liquid and gaseous fuels will also offer new jobs. The widespread use of methanol in Canada, one study shows, would create 50,000 to 60,000 permanent jobs, three times as many as a comparable tar sands development. For developing countries wood gathering and charcoal production are already major sources of employment, and sustainable fuelwood projects will create many more jobs.[10]

An intriguing aspect of the employment-creating potential of "renewables" is the flexibility they allow. Each energy source requires a different amount of labor. Each can be developed in more than one way. The variables are local priorities and resource availability. Some industrial countries, Sweden and the United States among them, are seeking to mechanize the harvest of likely energy crops, while in the Philippines capital shortages and surplus inexpensive labor are bringing about a revival of traditional manual axes and pruning hooks.[11]

Renewable energy alone cannot resolve the world's immense employment problems. Yet it can take us beyond the boom-and-bust cycles that have for too long characterized energy development's impact on employment. Since renewable energy must be continuously harvested, jobs in manufacturing technologies and repairing them and in gathering fuels and processing them will always be available.

Rebalancing City and Country

For most of this century and especially since World War II, the relationship between cities and the countryside has been undergoing immense change. Throughout the world people have been moving away from farming and into urban industrial and service jobs. Only 5 percent of North Americans are farmers, but they produce enough food for the other 95 percent of the people, as well as grain exports for 100 other nations. In the developing countries the forces shaping the city-country balance are grimmer, the freedom of choice more constrained. Many people move to cities because rural areas cannot provide expanding populations with an adeuquate living.[12]

Renewable energy can begin to right the balance between cities and the countryside by helping to revitalize rural areas, particularly in the developing countries, where the high cost of conventional energy sources threatens to pauperize the majority. Small amounts of renewable energy could have a catalytic effect, raising agricultural productivity and stimulating the development of rural industries—giving rural areas the self-sustaining economic base they lack.

Biological sources of energy will probably be the most important in rejuvenating rural areas. But some social risk is involved. In Brazil and the United States alcohol fuels production has depended on large plantation-style farming, which can encourage further concentration of rich land among a few owners. Luckily, economics may succeed where land-tenure initiatives have not, since integrated food/energy farming is ideal for smaller farms. It reduces the need for fertilizers and pesticides, and makes mechanization less necessary to success.

By tapping renewable energy rural areas can develop a new economic support system and provide an important "export" to cities, as well as stabilize commodity markets and local economies. Agro-forestry systems may be particularly widespread, occupying marginal, now-underutilized land. Trees can be harvested for various uses besides energy or left standing if prices are depressed, thus reducing the small farmer's vulnerability to sudden changes in commodity prices. Forage and food-producing trees can also provide steady income.

The widespread use of renewable energy could, some say, lead to the dispersal of large, dense cities. Researchers at the Institute for Environmental Studies at the University of Wisconsin believe that switching to renewable energy will lower population densities, giving rise to communities of 35,000 or so surrounded by intensively cultivated bands of farm and forest lands. Another possibility is that large cities will undergo a renaissance, becoming more livable as competition for scarce resources stops escalating.[13]

The visionary architect Paulo Solari has sketched one new kind of city that is much denser than those of today. Most transportation would consist of walking, and heating and cooling needs would be met mainly by improved design. Solari envisions "green areas" around these cities, which would provide recreational opportunities for the residents as well as a means of growing food and harnessing energy. Like all such visions, Solari's is not likely to appeal to all people. Some would argue that cities should be less centralized and that homes should be privately owned and more architecturally diverse than Solari would allow. Yet Solari's ideas do confirm the need for a new and innovative balance between cities and the countryside.[14]

Besides their obvious commercial and cultural attractions, large cities have an enormous potential for energy efficiency. Even today, New York City uses half as much energy on a per capita basis as the United States as a whole, largely because its high density means that less energy is needed for heating, cooling, and transportation. All cities have opportunities to further improve energy efficiency, and these improvements can be supplemented by using renewable energy sparingly in the form of solar water heaters, passive solar design, and heating systems using municipal waste.[15]

Rising Self-Reliance

As renewable energy becomes more important in the world energy budget, patterns of international energy trade will shift, with far-reaching consequences for the global economy and the security of nations. No form of renewable energy (or, for that matter, nonpetroleum fossil fuel) is likely to ever replace petroleum as the driving force in global trade patterns. As the world shifts to renewable energy, this global energy trade will be gradually replaced by regional self-sufficiency and local self-reliance. For the next several decades, the most important energy supply opportunities will be within nations or between neighboring countries rather than between distant trading partners.

The years since World War II have seen an unprecedented growth in long-distance international trade. The engine of this globalization of economic exchange has been oil. Between 1950 and 1980 oil's share of total international trade ballooned from 1.5 to 19 percent. In 1980 the value of oil traded on the international market was twice the value of the second largest good, food. Even more sobering, one oil-importing country after another was forced during the 1970s to reshape—and often distort—its economy to produce for export. This chain reaction reaches deep into economies: America plows up marginal farmland to produce exportable food; Japan strives to sell high quality industrial goods; Tanzania plants fields with tobacco rather than with food crops.[16]

There are no Saudi Arabias of renewable energy. Most every place on earth has an abundance of either strong wind, intense sunlight, rich plant growth, heavy rainfall, or geothermal heat. Moreover, neither these energy sources nor the electricity or gas that will be generated from them can be cheaply carried across the oceans. While laying electric transmission cables under bodies of water the size of the English Channel or Lake Erie is feasible and could become more so as nations exploit regional energy trade opportunities, transmission of electricity under the oceans or for distances over 1,000 miles is unlikely both because of power losses and high costs. Building gas pipelines under large bodies of water is not feasible, and liquification and regasification is prohibitively expensive and dangerous.

Localities will be much more energy self-reliant in a world where renewable energy is dominant. David Morris of the

Institute for Local Self-Reliance observes that today "a dollar spent on energy is the worst expenditure in terms of its impact on the local economy—85 cents on a dollar leaves the economy." However, with renewable energy much of the money now leaving one nation for the purchase of fuel from another will be spent on locally gathered supplies. Heat for buildings in North America will come from the rooftops, not from the Middle East. Villagers in India will light their houses with electricity generated using the resources of the local environment instead of with kerosene from Indonesia's outer continental shelf. As the distance between users and suppliers of most energy supplies shrinks, self-reliant communities will be ever less subject to sudden price hikes, supply interruptions, or sabotage. Energy production will thus reinforce rather than undermine local economies and local autonomy.[17]

As energy becomes harder and harder to distinguish from real property such as farmland, buildings, or waste treatment plants, political attempts to redistribute it will not be as feasible as they are in a world where oil is almost as convertible to other goods as money itself. Taxation of energy flows by central governments will be more difficult as an increasing share of energy used never enters the marketplace or is provided in forms not readily comparable to energy used in other communities. And such attempts as the New International Economic Order to redistribute international wealth by redistributing resources will find energy less and less transferable.

Countries filled with communities that use small-scale dispersed energy technologies will be much less militarily vulnerable. As a recent U.S. government study points out, Japan—reliant on small hydro plants for 87 percent of its electricity—emerged from the bombing attacks of World War II with almost all its power-producing capacity intact. On the other hand, Germany's ability to wage war collapsed when allied bombers destroyed the relative handful of large coal and synthetic fuel plants that powered German industry. In an era when the threat of terrorist destruction is growing, countries heavily reliant on networks of small dams, wind turbines, and alcohol stills will be considerably more secure than those with large central power plants. Large dams, however, are societal jugular veins in wartime—inviting targets virtually impossible to protect from modern bomber or missile attacks.[18]

Localities may achieve increased self-reliance, but only larger regions will approach self-sufficiency. Energy in the form of methane, methanol, or electricity will be exchanged within regions to supplement locally harnessed energy. Regional trade may gradually supplant global trade. Thus, Germany will rely upon imported Danish wind power rather than on Persian Gulf oil, New England on hydropower from Quebec rather than on oil from Venezuela, and India on electricity from Nepal rather than on oil from Indonesia. This regional trade in energy will, of course, more readily register on the international trade ledgers in areas with many small nations than it will in multiregional countries like Brazil, China, or the United States.

The replacement of global by regional interdependency will be powerfully reinforced by basin-wide river development projects. The large dams rising on the Paranà, for example, will forge strong links between the economies of Paraguay, Brazil, and Argentina. By allowing navigation far upriver and providing power for new industries, the dams will draw these economies closer in the same way the St. Lawrence River projects did Canada and the United States. Economies in both Southeast Asia and the Indian subcontinent will become intertwined as the Mekong and the Ganges are harnessed. The need for all nations sharing a river basin to protect dams from sedimentation will give new impetus to regional reforestation and cropland protection. River valley inhabitants will take keen interest in the well-being and land-use practices of their highland neighbors. The easing of poverty in the uplands will be an act of regional self-interest rather than of global charity.[19]

The rise of regional interdependence will make peaceful

accommodation between neighboring countries more important than ever. Where deep cultural, religious, or political cleavages exist between regional neighbors, resources may remain undeveloped or rivalries may intensify as nations struggle for control over common resources. Increased reliance on large dams will affect regional politics by creating "mutual hostage" relations among neighboring countries. Yet large-scale hydroelectric development can bind together former enemies if their leaders cooperate to reach common goals. Brazil and Argentina overcame a tradition of hostility and suspicion to develop the rich energy resources of the Paranà River, which forms the boundary between the two countries. War between these neighbors would now be far more costly than before since a major part of the national wealth of each depends upon the continued operation of these expensive dams.

Shifting Power

Writing over a century ago, an early solar pioneer, John Ericsson, prophesied: "The time will come when Europe must stop her mills for want of coal. Upper Egypt, then, with her never ceasing sun power, will invite the European manufacturers to erect his [*sic*] mills . . . along the sides of the Nile." While Ericsson's vision has not materialized, the use of renewable energy has already shaped the location of industry and the relative power of nations. These patterns are most visible for the renewable source of energy thus far harnessed most extensively—hydropower.[20]

Throughout history, the availability of hydropower resources has been a key to the location of industry and cities and the relative power of nations. When waterwheels were the dominant hydropower technology, factories were small and dispersed throughout the countryside. Water's influence on urban location can best be seen in the eastern United States: Dozens of cities, from Springfield, Massachusetts, to Augusta, Georgia, cluster on the fall line where rivers drop from the Appalachian Piedmont to the coastal plain. Historians rank northern Europe's abundant hydropower resources as an important reason for the eclipse of the drier Mediterranean countries over the last few centuries.[21]

The explosive rise to influence and wealth of the petroleum-producing nations dominated world politics in the seventies. Within countries shifts of power have been just as dramatic—Australia, Canada, the Soviet Union, and the United States have all seen people and power move to their fossil fuel–producing regions. Yet these changes have drawn attention away from a slower—but more permanent—realignment of power and wealth to regions rich in water power. Just as factories and towns clustered around the mill sites of 100 years ago, so, too, new industries will locate near new dams in sparsely populated regions.

Quebec provides the most dramatic contemporary example of a region's rise to prominence due to its water resources. Launched ten years ago, La Grande Complex in northern Quebec will soon produce 11,400 megawatts, enough power to double the province's installed electrical capacity. Further additions on other rivers could bring the total to 27,500 megawatts. Using cheap power shipped south on giant 735-kilovolt transmission lines, Quebec hopes to revitalize its economy by attracting new industries. Electrification of oil-using industries and exports of power to the United States will reduce the area's dependence on costly imported oil and provide a permanent source of foreign exchange. In a pattern of dependence certain to grow, New York City by 1984 will receive 12 percent of its power from Quebec. This ambitious hydro program will give the French-speaking province new prominence within Canada and new leverage in its battle for greater autonomy and cultural independence.[22]

The development of hydropower in remote areas of the world will also have repercussions on the international economic system as important energy-intensive industries relocate near water resources. The most dramatic shift is occurring in the aluminum industry. Aluminum smelting requires prodigious amounts of electrical energy. In regions where hydropower is plentiful, aluminum smelters, which cluster around major dams on every continent, are often the major users of electricity.[23]

In the years ahead aluminum production will level off or even decline in Japan, the continental United States, and Western Europe as new aluminum smelters migrate to the world's peripheral regions, where major new hydroelectric complexes offer plentiful and cheap power. Aluminum companies are building major new plants in Brazil, Egypt, Sumatra, and Tanzania to take advantage of newly tapped hydro sources. Similarly, power shifts to the periphery can be seen within nations: Soviet aluminum production is moving deeper into Siberia, following new dams.[24]

The world's growing appetite for aluminum combined with the migration of smelters spells increased dependence between nations, and it could seriously affect many countries' balance of payments. New aluminum smelting sites could have a great impact on global employment if labor-intensive aluminum processing and finishing industries follow the producers of raw aluminum to major new dams. Just as nations that produce and export oil have sought to attract petrochemical and refinery activity, so too will primary aluminum producers try to entice related industries to the area. Already Venezuela, taking a cue from its success in attracting "downstream" oil industries, is seeking to force aluminum producers using its cheap hydropower to locate their highly profitable aluminum-processing and -fabricating facilities on Venezuelan soil. Should other hydro-rich nations follow suit, industrial nations could suffer job losses. These could be offset, however, by gains in countries that vigorously pursue recycling, which tends to be much more labor-intensive than either aluminum smelting or processing.[25]

Large wind machines, ocean thermal energy plants, and solar ponds could have similar effects on the balance of industrial and political power. Those remote regions of the world where strong winds blow or where the sun shines intensely may one day attract energy-intensive industry. The steady gale force winds howling off the Antarctic ice shelf are a far denser concentration of energy than the most sun-drenched equatorial desert. One day they may be a valuable enough lode of energy to lure humans to adapt to that forbidding climate. The salt-rich deserts of the world are also an early candidate for colonization by human energy systems.

It is, however, doubtful that large population centers will grow up alongside such large-scale energy development. Extremes of climate, lack of water, and forbidding transportation distances all stand in the way. More likely such areas will export energy in the form of electricity, ammonia, hydrogen, or smelted metals. Those nations containing such energy rich but inhospitable provinces will reap the benefits. The wastes of central Australia, southwest North America, and the vast desert stretching from Morocco to the Great Wall of China will change from blank spaces on the map to economic assets. Deserts that long served as buffer zones between nations may become objects of rivalry between them, adding a new dimension to international politics. Those energy resources that belong to no country—the winds of Antarctica or the thermal gradients of the ocean, for instance—could become subject to protracted Law of the Sea–type negotiations.

New Equalities

Societies relying on renewable energy will have more balance—between the rich and the poor, between nations, and between generations. The adequacy of energy supplies is determined not by the absolute amounts available, but by how they are distributed and whether individuals can afford the energy they need. Ours is a world of haves and have nots. Limousine owners in London have all the gasoline they need even during "shortages," while inadequate fuelwood or kerosene is a grim everyday reality in the mountain villages of Peru. Similarly, heating costs are hardly significant to apartment dwellers in Manhattan, while in the nearby South Bronx huddling around open ovens is the only way of keeping warm in the winter.[26]

As the equity question underscores, the major frontier today in most renewable energy technologies is mass production, which lowers the costs of solar collectors, biogas digesters, wind pumps, and even passive solar homes, bringing them within reach of low-income groups. If the initial capital costs are reduced and some subsidies are provided for the poorest, renewable energy could help narrow the gap in living standards. And once installed, renewable energy technologies are immune from fuel-price inflation, affording low-income people some protection from economic shocks and supply disruptions.

One successful example of renewable energy being harnessed by low income people is in the large but isolated San Luis Valley in Colorado. There several communities with a total population of 40,000—half of them with incomes below the poverty line—have mobilized to solve their energy problems. Since the late seventies 15 percent of the Valley's homes have been "solarized," and solar collectors, greenhouses, and crop dryers are now a common sight. Many of the systems are home-built at less than half the cost of comparable commercial equipment. Residents of the San Luis Valley are still not as wealthy as those in the Denver suburbs, but many of their energy needs are now met at an affordable price.[27]

Renewable energy can also help narrow the enormous inequalities between rich and poor nations. While some countries obviously have more abundant renewable energy sources than others, the gap between the best and least well endowed is not great compared to the inequalities of fossil fuel ownership. Indeed, some of the richest lodes of renewable energy are found in some of the poorest countries. For the more than fifty developing countries that are plagued both by extreme poverty and a complete lack of fossil fuels, renewable energy could represent the energy base needed to create wealth. Central America, for example, currently suffers from widespread poverty and staggering oil bills, but could fuel substantial economic growth by tapping abundant hydropower, geothermal, and biomass resources.

In energy terms all nations are "developing." To be sure some countries have advantages over others, but all the world's countries face a common need for a fundamental transition to new energy sources. Although lack of capital and scientific infrastructure will impede renewable energy development in the poorest countries, their lack of previous investment in large fossil fuel and nuclear systems could be an asset. Third World nations could enjoy some of the advantages that West Germany and Japan had after World War II since they will not be burdened with outmoded equipment. It could, for example, be easier for the Philippines to build cities and industries around geothermal sites than for France or Japan to rebuild its economy around new energy sources.[28]

After decades of developing greater dependency by importing technologies from the North, some countries of the South are—haltingly and in diverse ways—beginning to devise patterns of energy development appropriate to their own resources and circumstances. Using wood charcoal for steel-making, hydropower for cities and industries, and alcohol for automobiles, Brazil is building a modern industrial society that relies heavily on renewable energy—something no major country in the North is near achieving. Israel has reached the frontier in solar pond development and the Philippines in geothermal development, putting the industrial nations in the "less developed," learning situation. Serious obstacles notwithstanding, these countries and China are probably as far along in building a modern economy based on renewable resources as any industrialized nation.

Renewable energy can restore some balance between generations as well. Perhaps a quarter of all the petroleum that will ever be used on earth has been burned since World War II, and the rest is dwindling rapidly. In one generation a trove of energy accumulated in the earth over eons has been consumed, depriving future generations of their birthright. Fossil fuels—and particularly oil and gas—can continue to play a unique role in petrochemical production and some forms of transportation for a long time, and ours is the responsibility to ensure that many of our descendants have at least small amounts available. The more rapidly renewable energy is developed, the more likely that legacy. With nuclear power, however, the present generation goes beyond depletion, burdening the earth's inheritors with highly toxic wastes that will last millennia.

In contrast, the self-renewing energies of the sun and the earth can be harnessed by the living without diminishing the supply of energy available for the unborn. Properly executed renewable energy development bequeaths no legacy of environmental destruction and instead creates wealth for future generations. Some renewable energy technologies, like solar collectors, small wind machines, or wood stoves rest so lightly on the earth that the future may see no sign of their use. Other, more enduring changes, like dam building or forest planting, are assets bequeathed to the future by an otherwise profligate generation. We find it easy to forget that the early northern Europeans ran their factories with waterwheels or that southern Floridians heated their water using sunlight in the 1930s because the physical evidence—rotted or recycled—is gone from sight. But who will be able to ever forget that Chicago, Seoul, or Moscow relied on nuclear power for a brief moment in history?

Renewable energy development is surely the most conservative yet innovative energy course to follow. At our present crossroads we cannot maintain what we cherish unless we change the energy systems on which we rely. Only with renewable energy can our children enjoy at a reasonable cost the benefits we have enjoyed. Only with renewable energy development can we raise the living standards of most of the great majority. The power needed to make these choices is at hand.

Notes

Chapter 1. Introduction: The Power to Choose

1. International Institute for Applied Systems Analysis, *Energy in a Finite World,* (Laxenburg, Austria: 1981).
2. A good critique of the IIASA study is Florentin Krause, "IIASA's Fantasy Forecast," *Soft Energy Notes,* October/November 1981.
3. Amory B. Lovins, *Soft Energy Paths: Toward a Durable Peace* (Cambridge, Mass.: Ballinger, 1977).
4. This figure is based on the planet receiving sunlight at a rate of 1.38 kilowatts per square meter, which comes to approximately 1.55×10^{15} megawatt hours of energy for the entire planet each year (or 5.6 million exajoules). However, about 35 percent of this sunlight is reflected back into space, leaving 3.7 million exajoules of sunlight that are absorbed by the atmosphere, the oceans, the earth's surface and living plants. Total

human energy use, including noncommercial energy, comes to only about 350 exajoules. The sunlight calculations are included in Jack Eddy, *A New Sun: Solar Results from Skylab* (Washington, D.C.: National Aeronautics and Space Administration, 1979), and Vincent E. McKelvey, "Solar Energy in Earth Processes," *Technology Review*, March/April 1975.

5. A centralized, desert-based solar power system for the U.S. was advocated in Aden Baker Meinel and Marjorie Pettit Meinel, *Power for the People* (Tucson, Ariz.: privately published, 1970), and received considerable government attention. The "soft path" was originally described in Amory B. Lovins, *Soft Energy Paths*.

Chapter 2. Energy at the Crossroads

1. These numbers are developed more fully later in the chapter.
2. World historical figures are Worldwatch Institute estimates based on United Nations (U.N.) Department of International Economic and Social Affairs, *World Energy Supplies, 1950–74* (New York: 1976), and U.N. Department of International Economic and Social Affairs, *World Energy Supplies, 1973–1978* (New York: 1979). The 1980 figure is based on British Petroleum Company, *BP Statistical Review of World Energy 1981* (London: 1982). Non-energy uses of oil and gas (accounting for about 7 percent of the total) are included here. Japan used 241 million barrels of oil in 1960 and 1,850 million barrels in 1973 according to U.S. Department of Energy, *1980 Annual Report to Congress* (Washington, D.C.: U.S. Government Printing Office, 1981). Virtually all of the oil was imported.
3. Estimate based on U.S. Department of Energy, *1980 International Energy Annual* (Washington, D.C.: 1981). These numbers are developed more fully in Chapter 11.
4. Estimate based on United Nations, *World Energy Supplies*, and British Petroleum Company, *BP Statistical Review of World Energy 1981*.
5. World Bank, *Energy in the Developing Countries* (Washington, D.C.: 1980). Most energy statistics published include only commercial energy, thereby excluding the fuelwood and biomass that are so important in the Third World, providing over half the energy used in many African countries, for example. The energy trends of developing countries are also discussed in Thomas Hoffman and Brian Johnson, *The World Energy Triangle: A Strategy for Cooperation* (New York: Ballinger, 1981), and Joy Dunkerley and William Ramsey, "Energy and the Oil-Importing Developing Countries," *Science*, May 7, 1982.

6. Latest economic statistics are from Organisation for Economic Co-operation and Development, *OECD Economic Outlook* (Paris: July 1982), and Lawrence R. Klein, "World Economic Outlook: The Industrial Nations," *Business Week*, August 9, 1982.
7. International Energy Agency, *A Group Strategy for Energy Research, Development and Demonstration* (Paris: Organisation for Economic Co-operation and Development, 1980).
8. Alan Riding, "Costa Rica Finding the Fiesta is Over," *New York Times*, December 8, 1981. The costs of oil imports are from the World Bank, *World Development Report 1981* (New York: Oxford University Press, 1981).
9. Forest trends are from Adrian Summer, "Attempt at an Assessment of the World's Tropical Forests," *Unasylva*, Vol. 28, Nos. 112/113, 1976, and World Bank, *Forestry: Sector Policy Paper* (Washington, D.C.: 1978).
10. Oil analyst quoted is Sherman Clark in George Getschow, "More or Less Oil Will Go Up or Down or Maybe It Won't," *Wall Street Journal*, May 5, 1982.
11. Proven oil reserve figure is from the American Petroleum Institute, *Basic Petroleum Data Book* (Washington, D.C.: 1982). Future discovery figure is an estimate from Richard Nehring, "The Outlook for World Oil Resources," *Oil and Gas Journal*, October 27, 1980. One of the most detailed oil forecasts made in the late seventies was the Gulf Oil Company's "World Petroleum Outlook 1978–2000," unpublished, 1978. It projects rapid oil price increases by the late eighties and a peak in world oil production in the early nineties.
12. Oil use trends from U.S. Department of Energy, *Monthly Energy Review*, various issues. Developing country trends are from World Bank, *Energy in the Developing Countries*.
13. U.S. figure is from American Petroleum Institute, *Basic Petroleum Data Book*. Declining petroleum yield in the United States is described in U.S. Congress, Office of Technology Assessment, *World Petroleum Availability 1980–2000* (Washington, D.C.: 1980), and David H. Root and Lawrence J. Drew, "The Pattern of Petroleum Discovery Rates," *American Scientist*, November/December 1979.
14. Recent assessments of world oil prospects include U.S. Congress, Office of Technology Assessment, *World Petroleum Availability*, Richard Nehring, "The Outlook for World Oil Resources," and Exxon Corporation, *World Energy Outlook* (New York: December 1980).
15. Oil reserve statistics are from Larry Auldridge, "World Oil Flow Slumps, Reserves Up," *Oil and Gas Journal*, Worldwide Issue, December 29, 1980.

16. Estimate based on U.S. Department of Energy, *1980 International Energy Annual.*
17. Natural gas prices are discussed in "A Change in Philosophy in Natural Gas Pricing," *World Business Weekly,* February 11, 1980, Steve Mufson, "As Controls are Eased, Industrial Users Brace for Rises in Gas Prices," *Wall Street Journal,* February 11, 1982, and "Natural Gas," *Financial Times Energy Economist,* January 1982.
18. The prospects for gas from unconventional sources are discussed in J. Glenn Seay, "Gas Power: Its Promises & Problems," a research paper prepared by the Center for Industrial and Institutional Development, University of New Hampshire, February 1980.
19. Global gas reserve figures are from American Petroleum Institute, *Basic Petroleum Data Book.*
20. The dangers and costs of transporting large quantities of liquefied natural gas are discussed in Lee Niedringhaus Davis, "Gambling on 'Frozen Fire'," *New Scientist,* January 10, 1980. Gas export trends are discussed in International Energy Agency, *Natural Gas: Prospects to 2000* (Paris: 1982).
21. Authors' estimate based on U.S. Department of Energy, *1980 International Energy Annual.*
22. Reserve figures are from the World Coal Study, *Coal—Bridge to the Future* (Cambridge, Mass.: Ballinger, 1980). Use figures are estimates based on U.S. Department of Energy, *1980 International Energy Annual.*
23. The World Coal Study, *Coal—Bridge to the Future.*
24. Ibid.
25. International coal mining accident statistics that exclude China and the Soviet Union are found in International Labor Organization, *Yearbook of Labor Statistics 1980* (Washington, D.C.: 1981). The worldwide estimates used in this book are from Curtis Seltzer, Institute for Policy Studies, private communication, September 21, 1982. Soviet coal mining accidents are discussed in U.S. Congress, Office of Technology Assessment, *Technology and Soviet Energy Availability* (Washington, D.C.: 1981). Chinese accidents are discussed in Vaclav Smil, *China's Energy* (New York: Praeger Publishers, 1976). Detailed U.S. figures are found in U.S. Department of Commerce Mine Safety and Health Administration, *Mine Injuries and Work Time* (Washington, D.C.: 1981).
26. Coal pollution fatality statistics are at best imprecise since often pollution interacts with other factors or exacerbates other health problems in causing death. It strikes hardest at the very young and the elderly. The Ohio Valley statistics are from U.S. Environmental Protection Agency,

Ohio River Basin Energy Study (Washington, D.C.: U.S. Government Printing Office, 1980). Third World statistics are particularly difficult to find, so the figure of a half million deaths worldwide is an order of magnitude estimate by Curtis Seltzer, Institute for Policy Studies, private communication, September 21, 1982. The health effects of coal burning in China are described in Vaclav Smil, *China's Environment* (New York: Praeger Publishers, 1982).

27. A detailed study of the effects of acid rain is Ross Howard and Michael Perley, *Acid Rain* (New York: McGraw-Hill, Inc., 1982).
28. Good overviews of the carbon dioxide problem include W.S. Broecker et al., "Fate of Fossil Fuel Carbon Dioxide and the Global Carbon Budget," *Science*, October 26, 1979, Council on Environmental Quality, *Global Energy Futures and the Carbon Dioxide Problem* (Washington, D.C.: U.S. Government Printing Office, January 1981), and William W. Kellogg and Robert Schware, "Society, Science, and Climate Change," *Foreign Affairs*, Summer 1982.
29. The World Coal Study, *Coal—Bridge to the Future.*
30. The prospects for synthetic fuels are discussed in E.J. Hoffman, *Synfuels: The Problems and the Promise* (Laramie, Wyo.: The Energon Co., 1982), and William Houseman, "Synfuels: No Barrel of Fun," *Audubon*, September 1980.
31. The early "bandwagon market" for nuclear reactors is described in Irvin C. Bupp and Jean-Claude Derian, *The Failed Promise of Nuclear Power: The Story of Light Water* (New York: Basic Books, 1978).
32. "Nuclear Power: World Status," *Financial Times Energy Economist*, November 1981.
33. The nuclear waste problem is discussed in Todd R. LaPorte, "Nuclear Waste: Increasing Scale and Sociopolitical Impacts," *Science*, July 7, 1978, U.S. Congress, Office of Technology Assessment, *Managing Commercial High-Level Radioactive Waste* (Washington, D.C.: 1982), Ronnie D. Lipschutz, *Radioactive Waste: Politics, Technology and Risk* (Cambridge, Mass.: Ballinger, 1980), and Fred C. Shapiro, *Radwaste: A Reporter's Investigation of a Growing Nuclear Menace* (New York: Random House, 1981).
34. The proliferation of nuclear materials and weapons and efforts to slow it are described in U.S. Congress, Office of Technology Assessment, *Nuclear Proliferation and Safeguards* (Washington, D.C.: Praeger Publishers, 1977), Albert Wohlstetter, *Moving Towards Life in a Nuclear Armed Crowd?* (Chicago: University of Chicago Press, 1979), Joseph A. Yager, ed., *Nonproliferation and U.S. Foreign Policy* (Washington, D.C.: The Brookings Institution, 1980). The efforts of the Interna-

tional Atomic Energy Agency are discussed in Judith Miller, "Trying Harder to Block the Bomb," *New York Times Magazine*, September 12, 1982.

35. Irvin C. Bupp, "The Nuclear Stalemate," in *Energy Future: Report of the Energy Project of the Harvard Business School* (New York: Random House, 1979).
36. Charles Komanoff, *Power Plant Cost Escalation: Nuclear and Coal Capital Costs, Regulation, and Economics* (New York: Komanoff Energy Associates, 1981). Similar conclusions have been reached about the cost of British nuclear plants in J.W. Jeffrey, "The Real Cost of Nuclear Electricity in the UK," *Energy Policy*, June 1982.
37. "France: Nuclear Over-Capacity Even Before 1985," *European Energy Report*, January 7, 1980.
38. U.S. plant orders and cancellations are from Mary Ellen Warren, Atomic Industrial Forum, private communication, August 5, 1982. Projections are authors' estimates based on plants that are now in the planning and construction phases.
39. The nuclear power situation in various European countries is described in Peter Bunyard, "Nuclear Power—The Grand Illusion," *The Ecologist*, April/May 1980, "The Nuclear Option that Won't Go Away," *World Business Weekly*, February 2, 1981, "Sweden Proposes Phasing Out Its Nuclear Power Plants," *World Business Weekly*, April 27, 1981, and "France: A Commitment to Nuclear Power," *World Business Weekly*, January 12, 1981.
40. P. Feuz, "Nuclear Power in Eastern Europe: Part I," *Energy in Countries with Centrally Planned Economies*, March 1980; Jim Harding, "Soviet Nuclear Setbacks," *Society*, July/August 1981.
41. "Nuclear Power: World Status," *Financial Times Energy Economist*, Jane House, "The Third World Goes Nuclear," *South*, December 1980.
42. The 1981 figures are from "Nuclear Power: World Status," *Financial Times Energy Economist.* The 1990 and 2000 estimates are the authors' and assume that most of the nuclear power plants now planned and under construction will be completed, but that no additional ones will be started. It is more likely that the actual figures will be lower rather than higher than these.
43. The economic feasibility of breeder reactors is discussed in U.S. Congress, General Accounting Office, *The Liquid Metal Fast Breeder Reactor—Options for Deciding Future Pace and Direction* (Washington, D.C.: 1982), and "Cheap Uranium Dampens Breeder Interest," *Energy Economist*, December 1981. The proliferation dangers posed by breeder reactors are discussed in Amory B. Lovins and L. Hunter Lovins, *Brittle*

Power: Energy Strategy for National Security (Andover, Mass.: Brick House, 1982).

44. John P. Holdren, "Fusion Energy in Context: Its Fitness for the Long Term," *Science*, April 4, 1978; John F. Clarke, "The Next Step in Fusion: What It Is and How It Is Being Taken," *Science*, November 28, 1980.
45. The 19 percent efficiency improvement estimate is based on a 19 percent increase in gross national product during the period and no growth in energy use as shown in British Petroleum Company, *BP Statistical Review of World Energy 1981*. Oil use figures come from the same source.
46. Nairobi hotel example from Lee Schipper, Lawrence Berkeley Laboratory, private communication, January 5, 1982; Japanese appliance efficiencies from "Progress and Tradition in Energy Conservation," *Chikyu no Koe* (published by Friends of the Earth, Japan), November 1981; U.S. air travel efficiency data from Eric Hirst et al., "Energy Use from 1973 to 1980: The Role of Improved Energy Efficiency," Oak Ridge National Laboratory, December 1981.
47. Solar Energy Research Institute, *A New Prosperity: Building a Renewable Future* (Andover, Mass.: Brick House, 1981).
48. "Energy Conservation: Spawning a Billion-Dollar a Year Business," *Business Week*, April 6, 1981.
49. Amory B. Lovins, *Soft Energy Paths: Toward a Durable Peace* (Cambridge, Mass.: Ballinger, 1977).
50. World Bank, *Energy in the Developing Countries.*

Chapter 3. Building with the Sun

1. The 75 to 90 percent figure is a rough estimate based on information supplied by numerous architects and builders. In all but the most severe climates it is possible virtually to eliminate the need for supplemental heating and cooling.
2. Architect quoted is Belinda Reeder in a presentation to the U.S. Department of Energy, Passive and Hybrid Solar Energy Program Update Meeting, Washington, D.C., September 21–24, 1980 (referred to in following notes as DOE Passive Solar Update Meeting). The 60,000 passive solar houses figure is an estimate by the U.S. Passive Solar Industries Council. See note 32 for details.
3. Socrates quote from Ken Butti and John Perlin, *A Golden Thread: 2000 Years of Solar Architecture and Technology* (New York: Van Nostrand Reinhold Co., 1980).

4. The historical material in this section draws heavily on Butti and Perlin, *A Golden Thread,* Amos Rapaport, *House Form and Culture* (Englewood Cliffs, N.J.: Prentice-Hall, 1969), and Victor Olgyay, *Design with Climate: A Bioclimatic Approach to Architectural Regionalism* (Princeton, N.J.: Princeton University Press, 1963).
5. Information on China was provided by Sarah Balcomb after a trip there, private communication, December 3, 1980. Spanish galerías are described in Ricardo Piñon and Fernando Bores, "Las Casas de Galerías," *Sunworld,* Vol. 4, No. 5, 1980.
6. Information on thatch is from Malcolm Lillywhite, Domestic Technology Institute, private communication, July 8, 1980. Information on cooling is from Mehedi N. Bahadori, "Passive Cooling Systems in Iranian Architecture," *Scientific American,* February 1978.
7. Richard G. Stein, *Architecture and Energy* (Garden City, N.Y.: Anchor Press/Doubleday, 1978).
8. European data are based on Building Research Establishment, "Energy Conservation: A Study of Energy Consumption in Buildings and Possible Means of Saving Energy in Housing," A Building Research Establishment Working Party Report, Watford, England, undated. Patrick J. Minogue, *Energy Conservation Potential in Buildings* (Dublin: An Foras Forbartha, 1976), and Stig Hammarsten, "A Survey of Swedish Buildings from the Energy Aspect," *Energy and Buildings,* April 1979. U.S. data from U.S. Congress, Office of Technology Assessment, *Residential Energy Conservation* (Washington, D.C.: 1979).
9. Energy use figures are authors' estimates based on data from Organisation for Economic Co-operation and Development (OECD), *Energy Balances of OECD Countries* (Paris: 1978), Joy Dunkerley, ed., *International Comparisons of Energy Consumption* (Washington, D.C.: Resources for the Future, 1978), U.S. Department of Energy, *Annual Report to Congress 1980* (Washington, D.C.: 1980), and U.S. Congress, Office of Technology Assessment, *Residential Energy Conservation.*
10. Information on traditional architecture from Hassan Fathy, *Architecture for the Poor* (Chicago: University of Chicago Press, 1973). Information on energy problems of Third World buildings from Alan Jacobs, Energy Office, U.S. Agency for International Development, private communication, July 9, 1980.
11. U.S. Department of Energy, *Monthly Energy Review,* July 1982; West German figure is from William W. Hogan, "Dimensions of Energy Demand," in Hans H. Landsberg, ed., *Selected Studies on Energy, Background Paper for Energy: The Next Twenty Years* (Cambridge, Mass.: Ballinger, 1980). Swedish figure is from "Energy Conservation:

Results and Prospects," *OECD Observer,* November 1979.

12. Joanne Omang, "TVA's Electrifying Change: Cheap Power Yields to 'Quality Growth'," *Washington Post,* June 1, 1980.
13. "The Energy Saving Look in Buildings," *U.S. News and World Report,* 1982 (precise date unknown).
14. The fundamentals of solar architecture are discussed in detail in Bruce Anderson, *The Solar Home Book: Heating, Cooling, and Designing with the Sun* (Andover, Mass.: Brick House, 1976), Bruce Anderson, *Solar Energy: Fundamentals in Building Design* (New York: McGraw-Hill Inc., 1977), Edward Mazria, *The Passive Solar Energy Book* (Emmaus, Pa.: Rodale Press, 1979), and Los Alamos Scientific Laboratory, *Passive Solar Buildings* (Springfield, Va.: National Technical Information Service, July 1979).
15. The Chicago house is described in Anderson, *Solar Home Book. Business Week* quote is from Butti and Perlin, *A Golden Thread.*
16. Wade Green, "A Conversation with the New Alchemists," *Environment,* December 1978.
17. The Trombe wall is discussed in Anderson, *Solar Home Book,* J.D. Walton, Jr., "Space Heating with Solar Energy at the CNRS Laboratory, Odeillo, France," in *Proceedings of the Solar Heating and Cooling for Buildings Workshop,* U.S. National Science Foundation, Washington, D.C., 1973, and Ian Hogan, "Solar Building in the Pyrenees," *Architectural Design,* January 1975. Information on Ladakh is from Helena Norberg-Hodge, private communication, July 26, 1982.
18. Underground houses are described in detail in the Underground Space Center, University of Minnesota, *Earth Sheltered Housing* (New York: Van Nostrand Reinhold Co., 1979).
19. Harold R. Hay, "Roof Mass and Comfort" and "Skytherm Natural Air Conditioning for a Texas Factory," in *Proceedings of the 2nd National Passive Solar Conference,* American Section of the International Solar Energy Society, Philadelphia, March 16–18, 1978; "Roof Ponds Can Work Anywhere," *Solar Energy Intelligence Report,* June 9, 1980; William A. Shurcliff, *Superinsulated and Double-Envelope Houses* (Cambridge, Mass.: privately published, 1980).
20. Shurcliff, *Superinsulated and Double-Envelope Houses;* Karl E. Munther, "Three Experimental Energy Houses in Ostersund," in *Swedish Building Research Summaries* (Stockholm: Swedish Council for Building Research, 1978); Per Madsen and Kathy Goss, "Low-Energy Houses in Denmark," *Solar Age,* February 1980; Robert W. Besant, Robert S. Dumont and Greg Schoenau, "Saskatchewan House: 100 Percent Solar in a Severe Climate," *Solar Age,* May 1979; "Operational

Saskatchewan Solar-Conservation House Yields Further Data on Energy Efficient Building Designs," *Soft Energy Notes,* May 1979; William A. Shurcliff, "Air-to-Air Heat Exchangers for Houses," *Solar Age,* March 1982.

21. Passive cooling research efforts are described in Darian Diachok and Dianne Shanks, *International Passive Architectural Survey,* Solar Energy Research Institute, Golden, Colo., unpublished, September 1980. General discussion of solar cooling including the Jeffrey Cook quote is from Joe Kohler, "A Fresh Look at Solar Cooling," *Solar Age,* July 1982.
22. Darian Diachok, Solar Energy Research Institute, private communication, September 18, 1980.
23. General information on architectural situation in developing countries is from Alan Jacobs, Energy Office, U.S. Agency for International Development, private communication, July 9, 1980, and Malcolm Lillywhite, Domestic Technology Institute, private communication, July 8, 1980.
24. Darian Diachok, Solar Energy Research Institute, private communication, September 18, 1980.
25. For a full discussion of this issue, see "Round Table: A Realistic Look at 'The Passive Approach'—Using Natural Means to Conserve Energy," *Architectural Record,* August 1980.
26. J. Douglas Balcomb, "Energy Conservation and Passive Solar: Working Together," Los Alamos Scientific Laboratory, Los Alamos, New Mex., unpublished, 1978.
27. The line between active and passive solar systems is a fuzzy one. Here we treat systems as passive if they do not rely primarily on fans and pumps to gather solar energy. Data developed in the last few years indicate that for space heating of buildings, passive systems (sometimes employing fans and collectors) are more economical than truly active systems based on water-filled solar collectors. See Larry Sherwood, "Passive Solar Systems . . . The Economic Advantages," in *Proceedings of the Solar Energy Symposia,* American Section of the International Solar Energy Society, Denver, Colo., August 1978, and Harrison Fraker, Jr., and William L. Glennie, "A Computer Simulated Performance and Capital, Cost Comparison of 'Active vs. Passive' Solar Heating Systems," in *Proceedings of the Passive Solar Heating and Cooling Conference,* American Section of the International Solar Energy Society, Albuquerque, New Mex., May 18–19, 1976.
28. Listing of solar designers is from National Solar Heating and Cooling Information Center, "Passive Solar Building and Design Professionals," Rockville, Md., unpublished, 1980. Among the leading schools in teaching solar architecture are the Massachusetts Institute of Technology and

Arizona State University in the U.S., the Architectural Association in England, the University of Alberta in Canada, the University of Auckland in New Zealand and the Technical University in Denmark. See Harrison Fraker, Jr. and Donald Prowler, "Evaluation of Energy Conscious and Climate Responsive Design Curricula in Professional Schools of Architecture," Princeton Energy Group, unpublished, March 1979.

29. These problems are discussed in Daniel W. Talbott and Ralph J. Johnson, "What the Builder Needs Before He Will Use Passive Solar Techniques," in *Proceedings of the 4th National Passive Solar Conference*, American Section of the International Solar Energy Society, Kansas City, Mo., October 3–5, 1979, and Rick Schwolsky, "Energy Conscious Construction," *Solar Age*, February 1980.
30. Studies of the economics of alternative building designs include Rosalie T. Ruegg et al., *Life-Cycle Costing: A Guide for Selecting Energy Conservation Projects for Public Buildings* (Washington, D.C.: U.S. Government Printing Office, 1978), James W. Taul, Jr., Carol F. Moncrief and Marcia L. Bohannon, "The Economic Feasibility of Passive Solar Space Heating Systems" and Mark A. Thayer and Scott A. Noll, "Solar Economic Analysis: An Alternative Approach," in *Proceedings of the 3rd National Passive Solar Conference*, American Section of the International Solar Energy Society, San Jose, Calif., January 11–13, 1979, and Peter F. Chapman, "The Economics of UK Solar Energy Schemes," *Energy Policy*, December 1977. Fuel costs for various buildings are described in A.H. Rosenfeld et al., "Building Energy Use Compilation and Analysis (BECA) and an International Compilation and Critical Review," Lawrence Berkeley Laboratories, Berkeley, Calif., unpublished, July 1979.
31. This is both an approximate and a conservative figure. With a good design in a relatively mild climate it is possible to eliminate artificial heating and cooling entirely at virtually no additional cost.
32. Figure on proportion of U.S. builders doing passive solar work is based on a survey of approximately 500 builders done for the Passive Solar Industries Council (PSIC) in 1982. The estimate of 60,000 to 80,000 passive solar houses in the U.S. is a conservative composite figure developed by PSIC in 1982. If all buildings with solar orientation or attached greenhouses were included, the number would be at least twice as high. The figure on proportion of new starts in the U.S. that incorporate passive solar features is from a survey by the National Association of Home Builders Research Foundation in 1982. All of the above were described by David Johnston, Passive Solar Industries Council, private communication, July 22, 1982. Information on West Germany is from

Ken Butti, private communication, June 4, 1982. Information on Scandinavia from Per Madsen and Kathy Goss, "Low-Energy Houses in Denmark." Information on France is from French architect Pierre Diaz-Pedregal, private communication, September 3, 1981.

33. Energy problems of large buildings are discussed extensively in Stein, *Architecture and Energy.*
34. Douglas Bulleit and George E. Way quotes are included in extensive discussion of climate-sensitive design for large buildings in "Round Table: A Realistic Look at 'The Passive Approach'." For a good discussion of daylighting of commercial buildings, see "An Interview with William Lam," *Solar Age,* August 1980. The IBM building is described in "IBM Uses Multiple Options Approach for Massive Energy Savings Program," *Solar Engineering,* September 1981.
35. Authors' estimate is based on U.S. data in U.S. Congress, Office of Technology Assessment, *Residential Energy Conservation,* European data in "Energy Conservation: Results and Prospects," *OECD Observer,* Irish data in Minogue, *Energy Conservation Potential in Buildings,* and Swedish data in U. Thunberg, "Improved Thermal Performance of Existing Buildings," prepared for United Nations Seminar on the Impact of Energy Considerations on the Planning and Development of Human Settlements, Ottawa, October 3–14, 1977.
36. Ken Bossong, "Homeowner's Guide to Passive Solar Retrofit," Citizens Energy Project, Washington, D.C., unpublished, 1978. Financial data are from promotional material, Green Mountain Homes, 1980.
37. Retrofit possibilities are described in Scott F. Keller, Arthur V. Sedrick and William C. Johnson, "Solar Experiments with Passive Retrofit," *ASHRAE Journal,* November 1978, Larry Sherwood, "Commercial Building Retrofits," *Sunpaper,* May 1980, Jeanne W. Powell, *An Economic Model for Passive Solar Design in Commercial Environments* (Washington, D.C.: U.S. Government Printing Office, 1980). The systems used in the San Luis Valley were described by Bob Dunsmore, San Luis Valley Solar Energy Association, press briefing, July 19, 1982.
38. Information on the effects of the Nandy design competition is from French architect Pierre Diaz-Pedregal, private communication, September 3, 1981. The Solar Energy Research Institute builders program began in 1980 and had a brief life, eliminated by the Reagan administration in 1981.
39. Activities of the National Association of Home Builders and the Home Improvement Council were described by David Johnston, Passive Solar Industries Council, private communication, July 22, 1982.

40. A strong case for requiring energy labeling of buildings is made by Arthur Rosenfeld of the University of California at Berkeley. His proposal is described in "House Energy Labels Needed, Researcher Says," *Energy Conservation Digest,* April 27, 1981.
41. "Interest Reduced on Solar Loans," *Solar Law Reporter,* May/June 1980; "Hanover Insurance Cuts Rates 10% for Owners of Solar Homes," *Energy Conservation Digest,* September 14, 1981.
42. Kathleen Vadnals, "Light of Financial Breaks Shines on Passive Solar," *Earth Shelter Digest and Energy Report,* January/February 1980; "Incentives for Passive Solar Lag Behind," *Energy in New England Forum,* Spring 1980.
43. U.S. Department of Energy goal is from talk by Frederick H. Morse, Director, Office of Solar Applications for Buildings, at DOE Passive Solar Update Meeting. The National Association of Home Builders (NAHB) goal is from Michael Bell of NAHB, private communication, March 2, 1981.
44. According to U.S. Congress, Office of Technology Assessment, *Residential Energy Conservation,* 146 gigajoules of primary energy is needed to heat and cool a typical gas-heated, electrically-cooled residence in Baltimore. A passive solar design could reduce energy requirements by at least half, or 73 gigajoules. Ten million such designs would save 0.7 exajoules. Fifty million would save 3.7 exajoules. One hundred million would save 7.3 exajoules.
45. Designer quoted is Fred Dubin in "Round Table: A Realistic Look at the Passive Approach."

Chapter 4. Solar Collection

1. Ken Butti and John Perlin, *A Golden Thread: 2000 Years of Solar Architecture and Technology* (New York: Van Nostrand Reinhold Co., 1980).
2. Ibid.
3. Ibid.
4. R. Melicher et al., *Solar Water Heating in Florida* (Washington, D.C.: National Science Foundation, 1974); Ethan B. Kapstein, "The Transition to Solar Energy: An Historical Approach," in Lewis J. Perelman, August W. Gielbelhaus and Michael D. Yokell, eds., *Energy Transitions* (Washington, D.C.: American Academy for the Advancement of Science, 1981).
5. For history of solar collectors in Israel, see "Israel's Place in the Sun,"

Nature, October 19, 1978. Israel's export markets are described in "Exports of Solar Ponds Gather Steam," *World Business Weekly,* September 22, 1980.

6. For an overview of the variety of solar collectors, see U.S. Congress, Office of Technology Assessment, *Application of Solar Technology to Today's Energy Needs, Vol. 2* (Washington, D.C.: 1978).
7. On storage of solar heat, see William D. Metz, "Energy Storage and Solar Power: An Exaggerated Problem," *Science,* June 30, 1978.
8. Bruce J. Walker, "Market: Challenges for Solar Products," *Business,* March/April 1979; "Quality of Solar Installers Held Higher Than Experience/Training Gained Elsewhere," *Solar Energy Intelligence Report,* December 14, 1981; R. Roy, "Comparison of Commercial and Do-It-Yourself Solar Collectors," *Sun at Work in Britain*
9. The economics of solar heating are discussed in Eric Hirst, "Is Solar Really the Best Way?," *ASHRAE Journal,* January 1980; Avraham Shama, "The Solar High Potential/Low Adoption Paradox," *Solar Engineering,* December 1981 and Office of Technology Assessment, *Application of Solar Energy to Today's Energy Needs.*
10. Roger H. Bezdek, Alan S. Hirshberg and William H. Babcock, "Economic Feasibility of Solar Water and Space Heating," *Science,* March 23, 1979.
11. Overall U.S. energy use patterns are described in Solar Energy Research Institute, *A New Prosperity: Building a Renewable Future* (Andover, Mass.: Brick House, 1981). Energy use patterns in developing countries are described in V. Smil and W.E. Knowland, eds., *Energy in the Developing World* (New York: Oxford University Press, 1980).
12. Kevin Bell, "Heat Pump Water Heaters: Goodbye to Active Solar?," *RAIN,* April 1981.
13. Jim Harding, "Staying Out of Hot Water," *Soft Energy Notes,* October/November 1980; Joe Carter and Robert G. Flower, "The Micro-Load," *Solar Age,* September 1980; and Raymond W. Bliss, "Why Not Just Build the House Right in the First Place?" in Robert H. Williams, ed., *Toward a Solar Civilization* (Cambridge, Mass.: MIT Press 1978).
14. Abraham Rabinovich, "More Than One in Ten Israelis Heat with Solar Energy," *World Environment Report,* March 24, 1980.
15. "Production of Solar Heaters in Japan," *World Environment Report,* August 11, 1980; "Japanese Lead World in Solar Hot Water Use," *Solar Engineering,* February 1981; "MITI Bares 16-Year Program to Develop Alternative Energy," *Japan Times,* September 19–23, 1979; Sara Terry, "Sun Power from the Land of the Rising Sun," *Christian Science Monitor,* August 12, 1981.

16. "Alternative Energy Use Growth Since 1975 equals 22% of Total U.S. Energy Growth: TRM," *Solar Energy Intelligence Report,* June 22, 1981; "Solar Heating and Cooling Equipment Sales," *Solar Energy Intelligence Report,* October 13, 1980; U.S. Department of Energy, *Solar Collector Manufacturing Activity, January Through June 1981* (Washington, D.C.: Energy Information Administration 1981) and SolarWork Institute, *Status Report on California's Solar Collector Industry* (Sacramento, Calif.: Office of Appropriate Technology 1982).
17. "In Europe, Solar Power Poised for Great Leap Forward," *The Energy Daily,* January 5, 1982; Peter E. Firth, "Growth of the U.K. Solar Energy Industry," *Sunworld,* Vol. 5, No. 3, 1981; "Solar Power, Used by Ancient Greeks, Spreads Like Fire Today," *Christian Science Monitor,* October 15, 1981 and J. B. Kirkwood, "Solar Energy Alternatives," in *Liquid Fuels: What Can Australia Do?* (Canberra: Australian Academy of Science, 1981).
18. Trevor Drieberg, "India Outlines Plans for Solar Energy Use," *Journal of Commerce,* September 25, 1980; "Installation of the Month: Korean Resort to Get 1-Billion BTU/Year from the Sun," *Solar Energy Intelligence Report,* March 15, 1982; Arnaldo Vieira de Carvalho, Jr. et al., "Solar Energy for Steam Generation in Brazil," *Interciencia,* May/June 1979.
19. M. M. Hoda, "Solar Cookers," in International Solar Energy Society, *Sun: Mankind's Future Source of Energy* (New York: Pergamon Press, 1978); M. M. Hoda, "Solar Cooker," *Appropriate Technology,* August 1977.
20. T. A. Lawand, "The Potential of Solar Cooking in Developing Areas," United Nations Industrial Development Organization Conference, Vienna, Austria, February 14–18, 1977; "Upper Volta Women Testing Dutch Sun Oven," *The Washington Star,* April 22, 1978.
21. B. A. Stout, *Energy for World Agriculture* (Rome: Food and Agriculture Organization, 1979) and Arjun Makhijani and Alan D. Poole, *Energy and Agriculture in the Third World* (Cambridge, Mass.: Ballinger, 1975).
22. J. D. Walton, Jr., "Solar Energy for Rural Development in the Asia and Pacific Region," Georgia Institute of Technology Engineering Experiment Station, Atlanta, October 15, 1980; T. A. Lawand and B. Saulnier, "The Potential of Solar Agricultural Dryers in Developing Areas," presented to the United Nations Industrial Development Organization Conference, Vienna, Austria, February 14–18, 1977; "Drying Grain with Solar Saves Money, But Few Farmers Are Aware of Potential," *Solar Energy Intelligence Report,* January 25, 1982; "Experimental Solar

Drying Leaves Good Quality Corn," *Journal of Commerce*, December 7, 1976. For an excellent overview of the current state of solar technology for use in agriculture see Walter G. Heid, Jr. and Warren Trotter, *Progress of Solar Technology and Potential Farm Uses* (Washington, D.C.: U.S. Department of Agriculture, September 1982).

23. Brace Research Institute, *Technical Report T99: A Survey of Solar Agricultural Dryers* (Ste. Anne de Bellevue, Canada: McGill University, December 1975); C. Stuart Clark, "A Solar Food Dryer for Bangladesh," *Appropriate Technology*, March 1982; "Sri Lanka Tries Solar Tea Drier," *China Daily*, November 21, 1981.
24. "Chapter 5: Distillation Methods," from K. S. Spiegler, *Salt Water Purification*, 2nd ed. (New York: Plenum Press, 1977).
25. J. R. Willianis, *Solar Energy: Technology and Applications* (Ann Arbor: Science Publishers, 1974); "Table 1. Existing Large Solar Stills, 1969," in *Solar Distillation as a Means of Meeting Small-Scale Water Demands* (New York: United Nations, 1970); Rolf Grunbaum, "Alternative Energy Sources in the USSR," *Ambio*, Vol. 7, No. 2.
26. S. D. Gomkale and R. L. Datta, "Solar Distillation in India," *Annals of Arid Zone*, Vol. 15(3), No. 208, 1976; S. D. Gomkale and H. D. Goghari, Central Salt and Marine Chemicals Research Institute, "Solar Distillation in India," Bhavnagar, India, 1979; R. Alward and T. A. Lawand, "Solar Distillation: How One Village Got Involved," *CERES*, November/December 1980.
27. R. K. Saksena and J. V. S. Mani, "Solar Collectors for Rural Use," *Sunworld*, Vol. 4, No. 5, 1980; "Installation of the month: Papua New Guinea Hotel Gets Solar DHW System," *Solar Energy Intelligence Report*, July 19, 1982; "Hong Kong Planning to Offer Less Expensive Solar Devices," *Journal of Commerce*, October 11, 1978.
28. India pumping figure is from Douglas Smith, "Rural Electrification or Village Energization," *Interciencia*, March/April 1980. Water pumping energy requirements in California are described in Robert Ranzel, *A River No More* (New York: Random House, 1981).
29. J. D. Walton, Jr., A. H. Roy, and S. H. Bomar, Jr., "A State of the Art Survey of Solar Powered Irrigation Pumps, Solar Cookers, and Wood Burning Stoves for Use in Sub-Sahara Africa," Georgia Institute of Technology Engineering Experiment Station, Atlanta, 1978.
30. Max G. Clemot, "Contribution of Solar Energy to the Development of Arid Zones: Solar Water Pumps," Societe Francaise d'Etudes Thermiques et d'Energie Solaire, n.p., September 1978.
31. An optimistic assessment of solar pump potential is "Solar Pumps *Can* Compete," *World Solar Markets*, August 1981.

32. The directions of solar collection innovation are described in William A. Shurcliff, "Active-Type Solar Heating Systems for Houses: A Technology in Ferment," in Robert E. Williams, *Toward a Solar Civilization* (Cambridge, Mass.: MIT Press, 1978).
33. Per Madsen and Kathy Goss, "Report on Non-Metallic Solar Collectors," *Solar Age*, January 1981; "Plastic Films and Laminates in Collectors Seen Slashing Installed Costs of HC Systems," *Solar Energy Intelligence Report*, April 5, 1982; John W. Andrews and William G. Wilhelm, "Thin-Film Flat-Plate Solar Collectors for Low-Cost Manufacture and Installation," Brookhaven National Laboratory, Upton, New York, 1980.
34. Elyse Axell, "The Solar Sandwich: Cheap Film Laminates Replace Copper and Glass," *Soft Energy Notes*, October/November 1980.
35. Barry Butler and Rob Livingston, "Fusion-Drawn Glass: Super Glazing for Solar Applications," *The SERI Journal*, Spring 1981.
36. David Godolphin, "Rising Hopes for Vacuum Tube Collectors," *Solar Age*, June 1982; Wendy Peters, "Evacuated Tube Collectors 'Logical Progression,'" *Canadian Renewable Energy News*, March 1981.
37. Charles Drucker, "Some Like it Hotter," *Soft Energy Notes*, October/November 1980; Ken Brown, "The Use of Solar Energy to Produce Process Heat for Industry," Solar Energy Research Institute, Golden, Colo., 1980.
38. Everett D. Howe, "Solar Thermal Power: Overview," *Sunworld*, Vol. 5, No. 3, 1981; Allen L. Hammond and William D. Metz, "Capturing Sunlight: A Revolution in Collector Design," *Science*, July 7, 1978; Frank Kreith, "A Technical and Economic Assessment of Three Solar Conversion Technologies," Solar Energy Research Institute, Golden, Colo., July 1979.
39. Alan T. Marriott, "Parabolic Dish Systems at Work: Applying The Concepts," *Sunworld*, Vol. 5, No. 3, 1981; Vincent C. Truscello, "Parabolic Dish Collectors: A Solar Option," *Sunworld*, Vol. 5, No. 3, 1981; Elyse Axell and Katy Slichter, "Solar Pilots Take Off," *Soft Energy Notes*, October/November 1980; E. Kenneth May, "The Potential for Supplying Solar Thermal Energy to Industrial Unit Operations," Solar Energy Research Institute, Golden, Colo., April 1980.
40. For progress in Fresnel lens development, see Charles Drucker, "Roll Out The Lenses," *Soft Energy Notes*, October/November, 1980.
41. A good overview of the solar central receiver technology is Richard S. Caputo, "Solar Power Plants: Dark Horse in the Energy Stable," in Robert H. Willianis, ed., *Toward a Solar Civilization* (Cambridge, Mass.: MIT Press, 1978). Smelter project is discussed in George Stern

and Paul Allen Curto, "Central Receiver Applications for Utilities and Industry," Gibbs and Hill, Inc., New York, N.Y. Martin Marietta Denver Aerospace, "Advanced Conceptual Design for Solar Repowering of the Seguaro Power Plant," Denver, Colo., October 1981; R. Raghavan, R. T. Nehar, and J. Corcoran, "Central Receiver Solar Retrofit System Proposed for Refinery," *Modern Power Systems,* November 1981; E. K. May, "Solar Energy and the Oil Refining Industry," Solar Energy Research Institute, Golden, Colo., March 1980.

42. For California solar thermal electric projects, see Burt Solomon, "Power Tower Ushers In Commercial Solar Age," *The Energy Daily,* April 19, 1982 and "SoCal Edison Seeking Proposals for 100-MEW Solar Thermal Power Plant," *Solar Energy Intelligence Report,* May 10, 1982. For a representative critique of the "power towers," see Brian Gallagher, "Solar Thermal Power Towers," Report Series No. 107, Citizens Energy Project, Washington, D.C., 1981.
43. *Solar Industrial Process Heat,* Conference Proceedings, Houston, Texas, December 16–19, 1980, sponsored by the Solar Energy Research Institute, Golden Colo. For specific projects, see "Food Industry Uses Solar Power for Processing and Saves Energy Dollars," *Chilton's Oil and Gas Energy,* September 1980; Sara Terry, " 'Cooking' Frozen Orange Juice: Solar Comes to Food Industry," *Christian Science Monitor,* April 2, 1980; Susan C. Frey, "On West Coast, More Complex Solar Preheats Water to Wash Old Cans," *Christian Science Monitor,* August 13, 1980.
44. E. J. Rattin and P. K. Munjal, "Solar Enhanced Oil Recovery," *Sunworld,* Vol. 5, No. 3, 1981; Sara Terry, "Solar Energy May Help Pry Hard-to-Recover 'Heavy Oil' Out of Old Fields," *Christian Science Monitor,* April 23, 1980.
45. "Prospects for Solar Industrial Process Heat," *Solar Engineering,* March 1980; Kenneth Brown, "Cost and Performance Vary Widely," *Solar Engineering,* June 1981.
46. For an overview of the technology of solar air conditioning, see David Venhuizen, "Will Active Cooling End Up High and Dry?," *Solar Age,* September 1982 and "Solar Powered Air Conditioning," *Heating-/Piping/Air Conditioning,* January 1980. For specific commercial benchmarks and prospects, see Joe Szostak, "Canadian Firm and U.S.-Japan Team Market Zeolite," *Canadian Renewable Energy News,* October 1981. "Dessicant Solar Cooling Could be Competitive in Late 1980's, ET-9 Told," *Solar Energy Intelligence Report,* February 22, 1982; Emilie Tavel Livezey, "Something New Under the Sun," *Christian Science Monitor,* October 27, 1981.

47. U.S. air conditiong figures from Solar Energy Research Institute, *Building a Sustainable Future* (Andover, Mass.: Brick House, 1981).
48. Christine Sutton, "Solar Energy—The Salty Solution," *New Scientist,* September 17, 1981; Thomas H. Maugh II, "Solar With A Grain of Salt," *Science,* June 11, 1982; B. Nimmo, A. Dabbagh, and S. Said, "Salt Gradient Solar Ponds," *Sunworld,* Vol. 5, No. 4, 1981.
49. For specific solar pond projects, see "Dead Sea Project to Supply Multi Megawatts of Power," *Solar Engineering,* April 1980; "Salton Sea Study to Determine Electrical Generation Potential," *Solar Engineering,* April 1980; "Australia's First Solar Pond Starts Up," *World Solar Markets,* October 1981; ". . . And Portugal Opens First Solar Pond," *World Solar Markets,* December 1981; "Solar Perspectives: Solar Pond Power, The Israel-California Connection," *Sunworld,* Vol. 4, No. 5, 1980.
50. Allan S. Krass and Roger LaViale, III, "Community Solar Ponds," *Environment,* July/August 1980; Layton J. Wittenberg and Marc J. Harris, "City of Miamisburg Heats Pool with Salt Gradient Solar Pond," *Solar Engineering,* April 1980.
51. The environmental and land use impacts of solar ponds are discussed in T. S. Jayadev and M. Edesess, "Solar Ponds and Their Applications," Solar Energy Research Institute, Golden, Colo., March 1980. Jet Propulsion Laboratory, *Salton Sea Solar Pond Project,* (Pasadena, Calif.: California Institute of Technology, 1981).
52. For an overview of ocean thermal energy, see Gerald L. Wich and Walter R. Schmitt, eds., *Harvesting Ocean Energy* (Paris: The UNESCO Press, 1981) and Tom Johnson, "Electricity from the Sea," *Popular Science,* May 1981.
53. Lyle E. Dunbar, "Market Potential for OTEC in Developing Nations," (LaJolla, Calif.: Science Applications, Inc., 1981).
54. Problems with OTEC technology are discussed in Clarence Zener "Solar Sea Power," Robert H. Williams, ed., *Toward A Solar Civilization* (Cambridge, Mass.: MIT Press, 1978); U.S. Congress, Office of Technology Assessment, *Ocean Thermal Energy Conversion* (Washington, D.C.: 1979).
55. "Report of the Technical Panel On Ocean Energy On Its Second Session," prepared for the United Nations Conference on New and Renewable Sources of Energy, Nairobi, Kenya, August 10–21, 1981. U.S. OTEC program is discussed in Beverly Karplus Hartline, "Tapping Sun-Warmed Ocean Water for Power," *Science,* August 15, 1980. Japan OTEC program is discussed in Glyn Ford and Luke Georghiou, "Japan Stakes its Industrial Future in the Sea," *New Scientist,* June 3, 1980 and Tony Marjoram, "Energy Pipe Dreams in the Pacific," *New*

Scientist, August 12, 1982. Nauru project is discussed in "Japan turns the dream into reality," *New Scientist,* August 12, 1982. The impact of OTEC in Hawaiian energy prospects is described in John W. Shupe, "Energy Self-Sufficiency for Hawaii," *Science,* June 11, 1982.

56. The environmental effects of OTEC facilities are discussed in "Ocean Energy," special issue of *Oceanus,* Winter 1979/80.

57. For an overview of solar energy R&D programs in various countries, see George F. W. Telfer, "France Bares Solar Power Muscle," *Journal of Commerce,* March 12, 1979, Robert Richards, "Solar Prototype Developments in Spain Show Great Promise," *Modern Power Systems,* April 1982, Robert Gibbs, "Potable Water from Briny Seas and Brackish Lakes," *R & D in Mexico,* December 1981/January 1982. For specific numbers see International Energy Agency, *Energy Research, Development and Demonstration in the IEA Countries, 1981 Review of National Programmes* (Paris: Organisation for Economic Co-operation and Development, 1982).

58. For an analysis of the kinds of questions facing solar energy research programs, see U.S. Congress, Office of Technology Assessment, *Conservation and Solar Energy Programs of the Department of Energy: A Critique* (Washington, D.C.: U.S. Government Printing Office, June 1980). An interesting case study of how political machinations undercut technological progress is Byron Harris, "No Profit in Politics," *Atlantic,* May 1982.

59. An excellent study of the impact of government R&D strategies on small innovative private companies in the United States is Committee on Small Business, House of Representatives, *Role of Government Funding and Its Impact on Small Business in the Solar Energy Industry* (Washington, D.C.: U.S. Government Printing Office 1980).

60. An analysis of long-term vs. short-term research goals is found in Solar Lobby, *Blueprint for a Solar America* (Washington, D.C.: 1979).

61. Irene F. Olson, "State Tax Incentives for Solar Energy," *Journal of Energy and Development,* Spring 1981; Alan Chen, "Spotlight on Solar Subsidies," *Soft Energy Notes,* August/September 1981; Jonathan Lloyd-Owen, "Japan," *Canadian Renewable Energy News,* August 1981.

62. For examples of effective local programs to promote solar collector use, see "San Diego County Basking in Benefits of Solar Energy Legislation," *New York Times,* February 10, 1981 and Allan Mazur, "Solar Heaters in Israel," *Bulletin of the Atomic Scientists,* February 1981; Sara Terry, "Solar Energy Gets A New 'Lease' in Oceanside, Calif." *Christian Science Monitor,* December 30, 1981. "California Town Plays Middle-

man in $20 Million Solar Leasing Plan," *The Energy Daily,* January 5, 1982.

63. For a discussion of industrial investment motivations, see G. N. Hatsopoulos et al., "Capital Investment to Save Energy," *Harvard Business Review,* March/April 1978.
64. New leasing strategies are discussed in Sara Terry, "Solar Energy: Leasing Cuts High Cost of Providing 'Harness' for the Sun," *Christian Science Monitor,* July 20, 1981. Barnaby J. Feder, "On Marketing the Unknown," *New York Times,* September 3, 1981; Aryeh Wolman, "Israeli Solar Firm Sells Steam to U.S. Textile Plants," *Canadian Renewable Energy News,* October 1981; Christopher Pope, " 'Solar Utility' Leases Collectors," *Renewable Energy News,* February 1982.
65. Various economic disincentives to use solar energy systems are described in Solar Lobby, *Blueprint for a Solar America* (Washington, D.C.: 1979).
66. J. B. Kirkwood, "Solar Energy Alternatives," *Liquid Fuels: What Can Australia Do?,* (Canberra: Australian Academy of Science, 1981); Randall J. Feurestein, "Utility Rates and Solar Commercialization," *Solar Law Reporter,* July/August 1979; D. Spencer, "Solar Energy: A View From an Electric Utility Standpoint," *American Power Conference,* April 21–23, 1975; Joseph G. Asbury and Ronald O. Mueller, "Solar Energy and Electric Utilities: Should They Be Interfaced?" *Science,* February 4, 1977 and "Colorado PUC Orders New Electric Backup Rates for Solar Energy," *Solar Law Reporter,* September/October 1979.
67. Modesto A. Madique and Benson Woo, "Solar Heating and Electric Utilities," *Technology Review,* May 1980; Norman L. Dean and Alan S. Miller, "Plugging Solar Power Into the Utility Grid," *Environmental Law Reporter,* July, 1977; Stephen Feldman and Bruce Anderson, "Financial Incentives for the Adoption of Solar Energy Design: Peak Load Pricing of Back-Up Systems," *Solar Energy,* April 1975.
68. TVA's solar program is described in S. David Freeman, "After the Joyride," *The Futurist,* December 1980. The impact of greater use of renewable energy on the TVA system is analyzed in U.S. Congress, General Accounting Office, "Electric Energy Options Hold Great Promise for the Tennessee Valley Authority," November 1978.
69. "Utility Makes Gain in Renewables Plan," *Solar Law Reporter,* January/February 1982; Larry Goldberg, "Bringing Solar Down To Earth," *Critical Mass Energy Journal,* June 15, 1982; "SoCal Edison: Up the Soft Path," *The Energy Daily,* March 19, 1981.
70. A review of 1980 utility activity in the solar field is found in Margaret Laliberte, "Solar Update," *EPRI Journal,* June 1981.

71. The case against utility and oil company involvement in solar energy is made in Scott Denman and Ken Bossong, "Big Oil Makes Its Move," *Amicus Journal*, Spring 1981, and Ken Bossong, "Opposing Utility Involvement in Solar Commercialization," Citizen's Energy Project, Washington, D.C.: July 1979.
72. Examples of recent sales of solar subsidiaries are described in Peter Steinhart, "Standard Oil Sells DHW Technology," *Solar Times*, November 1981. "American Solar King Acquires Daystar for $2.2-Million From Exxon Enterprises," *Solar Energy Intelligence Report*, February 2, 1981; "Olin Gets Out of Solar Energy Business; Will No Longer Make Collectors, Absorbers," *Solar Energy Intelligence Report*, June 29, 1981; Ron Scherer, "Solar's Bright Promise: Corporate Second Thoughts," *Christian Science Monitor*, February 12, 1981.
73. Studies of solar markets in Europe and Japan are summarized in "Best European Solar Markets Seen in France, Greece, Spain, Italy, Germany," *Solar Energy Intelligence Report*, January 11, 1982, and "1981: Solar In a State of Flux," *World Solar Markets*, January 1982.
74. InterTechnology Corporation, *Analysis of the Economic Potential of Solar Thermal Energy to Provide Industrial Process Heat* (Warrenton, Va.: InterTechnology Corp. 1977) and *A Response Memorandum to the President, Domestic Policy Review of Solar Energy* (U.S. Department of Energy: Washington, D.C., February 1979). Other studies of solar market potential are summarized in "Active Cooling to Snare 20% Market Share," *Solar Energy Intelligence Report*, November 30, 1981, Peter Steinhart, "Local Firms and Low-Cost Collectors Will Lead to $20-Billion Solar Industry," *Solar Times*, October 1981, and "Fast Growth to Huge Industry Forecast for Alternate Energy," *Solar Engineering*, December 1981.
75. *Regional Applicability and Potential of Salt-Gradient Solar Ponds in the United States*, Vol. I: Executive summary (Pasadena, Calif.: Jet Propulsion Laboratory, California Institute of Technology, 1981). Dr. Tabor's survey of 14 countries is described in Sandra Winsberg "Solar Pond Systems," *Sunworld*, Vol. 5, No. 4, 1981. "Large-Scale Solar Ponds Proposed for Australia by University of Sydney," *Solar Energy Intelligence Report*, March 9, 1981.

Chapter 5. Sunlight to Electricity: The New Alchemy

1. Estimate of 10,000 solar-powered houses is from Paul Maycock, a U.S. photovoltaics consultant, private communication, March 8, 1982.
2. The photovoltaics industry worldwide manufactured approximately 500

kilowatts of solar cells in 1977 and 8,000 kilowatts in 1982. Solar cell cost figures fell from $19 per peak watt in 1977 to $10 per peak watt in 1982. See William J. Murray, "PV in 1980: Growth Accelerates," *Solar Engineering,* January 1981, and "Photovoltaics: Ready Kilowatts from Today's Electric Alternative," *Solar Engineering & Contracting,* June 1982.

3. Detailed descriptions of photovoltaics technology and the history of its development are included in Paul D. Maycock and Edward N. Stirewalt, *Photovoltaics: Sunlight to Electricity in One Step* (Andover, Mass.: Brick House, 1981), Bruce Chalmers, "The Photovoltaic Generation of Electricity," *Scientific American,* October 1976, "The Promise of Photovoltaics," *Solar Energy Research Institute Journal,* Spring 1981, and Ed Roberton, ed., *The Solarex Guide to Solar Electricity* (Rockville, Md.: The Solarex Corporation, 1981).
4. "Solar Cell Is Ready for Commercial Jobs," *Business Week*, July 20, 1957.
5. For a more detailed discussion of silicon solar cell technology, see Yvonne Howell and David Adler, "How Silicon Solar Cells Work," *Sunworld,* Vol. 4, No. 1, 1980, and Jeffrey L. Smith, "Photovoltaics," *Science,* June 26, 1981.
6. The costs cited in this chapter are for solar cell "modules," unless otherwise noted. Photovoltaic "systems" costs are nearly twice as high because they include the cost of assembly and support materials as well as the cells themselves. Cost and market figures here are the authors' estimates based on discussions with industry representatives and market analysts. The 1500 houses estimate is the authors' and is based on average household electricity use of 800 kilowatt hours per month and the photovoltaics operating on average at 20 percent of their rated peak capacity, which is their output in full sunlight.
7. This is the authors' estimate based on the photovoltaics operating on average at 20 percent of capacity and having a 20-year life. Approximately half the kilowatt-hour cost is for the "balance of system," including battery storage.
8. Investment estimate is the authors' based on government budgets of major countries and discussions with industry analysts.
9. U.S. Government funding levels are from Center for Renewable Resources, *The Solar Agenda: Progress and Prospects* (Washington, D.C.: 1982). The research program is described in Henry Kelly, "Photovoltaic Power Systems: A Tour Through the Alternatives," *Science,* February 10, 1978, Paul Maycock, "Overview of the U.S. Photovoltaic Program," in *Fifteenth IEEE Photovoltaic Specialists Conference—1981 Proceed-*

ings, Kissimmee, Florida, May 12–15, 1981, and U.S. Department of Energy, *Photovoltaic Energy Systems Program Summary* (Springfield, Va.: National Technical Information Service, 1982).

10. Information on European and Japanese photovoltaics programs is from R. R. Ferber, U.S. Jet Propulsion Laboratory, "The Status of Foreign Photovoltaics R&D," in U.S. House of Representatives, Subcommittees on Energy Development and Applications, and Investigations and Oversight, Committee on Science and Technology, Joint Hearings, June 3, 1982.
11. Smith, "Photovoltaics"; H.L. Durand, "Photovoltaics: Present Status and Future Prospects," *Sunworld,* Vol. 4, No. 1, 1980; Charles F. Gay, "Solar Cell Technology: An Assessment of the State of the Art," *Solar Engineering,* March 1980.
12. Advanced silicon manufacturing processes are described in Joseph L. Loferski, "Photovoltaics I: Solar Cell Arrays," *IEEE Spectrum,* February 1980.
13. Amorphous silicon technology is described in J. Richard Burke, "Photovoltaics: Down to Earth at Last," *Solar Energy Research Institute Journal,* Spring 1981.
14. "Sanyo Electric, Fuji Electric Boast Amorphous Si PV Efficiencies of 6.91%, 6.47%," *Solar Energy Intelligence Report,* April 27, 1981; "Japan Dominates PV Calculator Market," *World Solar Markets,* September 1981; Paul Danish, "Japanese Produce Amorphous Cells with Over 7.5% Conversion Efficiency," *Solar Times,* October 1981. The Japanese produced 700 kilowatts of amorphous silicon cells in 1982 and lead the world in production, product testing, and marketing according to Paul Maycock, "My Fact Finding Tour of Japan," *Solar Age,* September 1982.
15. Paul Blythe, Jr., "Thin Film Solar Cell Research Progresses," *Solar Engineering,* April 1981; "Boeing Achieves Highest Efficiency Ever for True Thin-Film Photovoltaic Cell," *Solar Energy Intelligence Report,* August 4, 1980; Solar Energy Research Institute, *Environmental Health, Safety, and Regulatory Review of Selected Photovoltaic Options* (Springfield, Va.: National Technical Information Service, 1982).
16. Concentrators are described in E. C. Boes, B. D. Shafer and D. G. Schueler, "Economic Motivation for Photovoltaic Concentrator Technology," Sandia National Laboratory Report, unpublished, 1981. A strong case for concentrator technology's future competitiveness is made in Harbinger Research Corporation, *Photovoltaic Power Systems Patents: A Technical and Economic Analysis* (White Plains, N.Y.: Madsen Russell Associates, Ltd., 1982).

17. U.S. Department of Energy, *Multi-Year Program Plan, National Photovoltaics Program* (Washington, D.C.: U.S. Government Printing Office, 1980).
18. These figures are the authors' estimates based on assessments of various analysts. See, for example, "Maycock Predicts PV Future," *World Solar Markets*, August 1982, and "IEEE Conference Report," *Photovoltaic Insider's Report*, November 1982.
19. Figures on industry size are authors' estimates based on discussion with industry analysts.
20. The photovoltaics industry is well-described in U.S. Department of Commerce, "Photovoltaics Industry Profile," unpublished, 1981, Barrett Stambler and Lyndon Stambler, *Competition in the Photovoltaics Industry: A Question of Balance* (Washington, D.C.: Center for Renewable Resources, 1982), and Science Applications, Inc., *Characterization and Assessment of Potential European and Japanese Competition in Photovoltaics* (Springfield, Va.: National Technical Information Service, 1979).
21. This figure is an estimate by Paul Maycock in remarks to the American Section of the International Solar Energy Society, Annual Conference, Philadelphia, Pa., May 26, 1981.
22. Critical views of oil companies' involvement in photovoltaics are included in Ray Reece, *The Sun Betrayed: A Report on the Corporate Seizure of U.S. Solar Energy Development* (Boston: South End Press, 1979), Stambler and Stambler, *Competition in the Photovoltaics Industry*, and Ralph Flood, "Big Oil Reaches for the Sun," *New Scientist*, November 12, 1981. Morris Adelman is quoted in Ralph Flood, "Big Oil Reaches for the Sun."
23. The Japanese success in amorphous silicon is described in Douglas L. Finch, "The Japanese Photovoltaic Threat," *Solar Age*, February 1981, R. Ferber and K. Shimada, "Japanese Photovoltaic R&D," U.S. Jet Propulsion Laboratory, Pasadena, Calif., unpublished, 1982, and Maycock, "My Fact Finding Tour." European firms' progress is described in William J. Murray, "The Europeans are Coming," *Solar Engineering*, September 1981. A particularly exciting joint venture in amorphous silicon technology was announced by the Sharp Corporation in Japan and Energy Conversion Devices in the United States in 1982, described in Burt Solomon, "Sharp, ECD to Produce Silicon Cells in Japan," *Energy Daily*, June 22, 1982.
24. A good analysis of the evolving international market is Science Applications, Inc., *Characterization and Assessment of Potential European and Japanese Competition*.

25. David Morris, *Self-Reliant Cities* (San Francisco: Sierra Club Books, 1982).
26. Louis Rosenblum, "Status of Photovoltaic Systems for Applications in Developing Countries," National Aeronautics and Space Administration, unpublished, 1981.
27. The Papago Indian Reservation system is described in Bill D'Alessandro, "Villagers Light the Way: Solar Cell Power in Gunsight, Arizona," *Solar Age,* May 1979. The potential of photovoltaics in developing countries is discussed in Charles Drucker, "Third World Briefing: Photovoltaics Debated," *Soft Energy Notes,* May/June 1982 and Dennis Elwell, "Solar Electricity Generation in Developing Countries," *Mazingira,* Vol. 5/3, 1981.
28. "PV Equipment Manufacturers See Growing Market in Third World," *Solar Energy Intelligence Report,* February 22, 1980; Rebecca Kauffman, "India Promotes Local PV for Space and Pumping," *Renewable Energy News,* April 1982; "Pakistan Gets First of 14 PV Generators," *World Solar Markets,* October 1981.
29. Miles C. Russell, "An Apprentice's Guide to Photovoltaics," *Solar Age,* July 1981.
30. Photovoltaic residences are described in Miles C. Russell, "Residential Photovoltaic System Designs," *Solar Engineering,* November 1981, Charles H. Cox, III, "Power Producing Homes: Making the Utility Connection," *Solar Age,* December 1981, Gorden F. Tully, J. Stewart Roberts and Thomas A. Downer, "The Design Tradeoff for the Mid to Late 1980's: Photovoltaics versus Passive," in *Proceedings of the 5th National Passive Solar Conference,* and Burt E. Nichols and Steven J. Strong, "The Carlisle House: Solar Electric Residence is Energy Self-Sufficient," *Solar Engineering,* November 1981.
31. The various large solar power projects are described in "Major Solar Projects Round the World," *World Solar Markets,* December 1981, "Fresnel Lens PV Systems Starting Up at Saudi Villages," *Solar Energy Intelligence Report,* May 17, 1982, "ARCO to Build 1-MWe PV Array, Sell Power to SoCal Ed," *Solar Energy Intelligence Report,* April 5, 1982, and "100 MW—Phase One," *Renewable Energy News,* February 1982.
32. The original proposal for an orbiting solar electric system is P.E. Glaser, "Power from the Sun: Its Future," *Science,* November 22, 1968. Detailed critical assessments of the concept are National Science Foundation, *Electric Power from Orbit: A Critique of a Satellite Power System* (Washington, D.C.: 1981), and U.S. Congress, Office of Technology Assessment, *Solar Power Satellites* (Washington, D.C.: 1981).

33. The U.S. photovoltaics goals are set forth in U.S. Congress, "Solar Photovoltaic Energy Research, Development, and Demonstration Act of 1978," Washington, D.C., 1978 and U.S. Department of Energy, *Multi-Year Program Plan*. A "maximum practical" figure of 1.0 quads for the year 2000 was used by the landmark study, *Domestic Policy Review of Solar Energy* (Washington, D.C.: U.S. Department of Energy, 1979). A good 1982 review of these goals and the meager chances for attaining them is U.S. Congress, General Accounting Office, *Probable Impacts of Budget Reductions.* The Japanese goals are described in Maycock, "My Fact Finding Tour." It is assumed here that photovoltaics operate on average at 20 percent of their rated capacity so that 1.0 megawatts of cells generates 1752 megawatt hours per year. Moreover it is assumed that 10.7 exajoules of primary energy (that is 10.1 quads or 366 million tons of coal) is required to generate a billion megawatt hours of electricity.
34. These estimates are the authors' and use the same energy conversion factors noted in note 33.

Chapter 6. Wood Crisis, Wood Renaissance

1. For role of wood fuel and wastes in the Third World, see E.M. Mnzava, "Fuelwood: The Private Energy Crisis of the Poor," *Ceres,* July/August 1981. For a description of charcoal consumption in Third World, see J.E.M. Arnold and Jules Jongma, "Fuelwood and Charcoal in Developing Countries," *Unasylva,* Vol. 29, No. 118, 1978.
2. In areas with "acute scarcity" people suffer hardship from lack of fuelwood. And in "deficit" areas the rate of fuelwood burning exceeds the rate of regeneration. Figures are from United Nations (UN), Food and Agriculture Organization, "Report of the Technical Panel on Fuelwood and Charcoal to the U.N. Conference on New and Renewable Sources of Energy," Nairobi, Kenya, August 1981. Information on electrification in the Third World is from Douglas V. Smith, "Rural Electrification or Village Energization?," *Interciencia,* March/April 1980.
3. For an examination of the roots of wood fuel scarcity and deforestation, see Erik Eckholm, *Losing Ground: Environmental Stress and World Food Prospects* (New York: W.W. Norton & Co., 1976).
4. Philip Wardle and Massimo Palmieri, "What Does Fuelwood Really Cost?," *Unasylva,* Vol. 33, No. 131, 1981.
5. E.M. Mnzava, "Village Industries vs. Savanna Forest," *Unasylva,* Vol. 33, No. 131, 1981.
6. K.S. Salariyna, "Fuel Conservation in Domestic Consumption," pre-

sented to the First Conference of the Asia-Pacific Confederation of Chemical Engineers (APCCHE), Jakarta, November 21–23, 1978; Gyan Sagar, "A Fuel-Efficient, Smokeless Stove for Rural India," *Appropriate Technology*, September 1980; I.Evans and D. Wharton, "The Lorena Mudstove: A Wood-Conserving Cookstove," *Appropriate Technology*, August 1977.

7. "Smoke in the Kitchen," *Cookstove News*, May 1981; Don Shakow, Cara Seiderman, and Philip O'Keefe, "Kenya's Lesson in Conservation," *Soft Energy Notes*, January/February 1982.
8. Bina Agarwal, *The Wood Fuel Problem and the Diffusion of Rural Innovations*, preliminary draft (Sussex: University of Sussex, Science Policy Research Unit, October 1980; Dales V. Shaller, "Socio-Cultural Assessment of the Lorena Stove and Its Diffusion in Highland Guatemala," Volunteers in Technical Assistance, unpublished, 1979.
9. H.E. Booth, "Realities of Making Charcoal," *Unasylva*, Vol. 33, No. 131, 1981; D.E. Earl, *Forestry, Energy and Economic Development* (Oxford: Clarendon Press, 1975); E. Uhart, "The Wood Charcoal Industry in Africa," memorandum, African Forestry Commission, Fourth Session, Bangui, Central African Republic, March 22–27, 1976.
10. For Brazil example, see *World Environment Report*, September 28, 1981; for Sri Lanka example, see *New Scientist*, August 13, 1981.
11. John U. Nef, "An Early Energy Crisis and Its Consequences," *Scientific American*, November 1977; W.G. Youngquist and H.O. Fleischer, *Wood in American Life: 1776–2076* (Madison, Wisc.: Forest Products Research Society, 1977).
12. Energy Information Administration, *Estimates of U.S. Wood Energy Consumption from 1949 to 1981* (Washington, D.C.: U.S. Department of Energy, August 1982); Colin High, "New England Returns to Wood," *Natural History*, February 1980; Mark R. Bailey and Paul R. Wheeling, *Wood and Energy in Vermont* (Washington, D.C.: U.S. Department of Agriculture, Economic Research Service, April 1982); H. Swain et al., "Canadian Renewable Energy Prospects," in *Solar Energy* (Oxford: Pergamon Press, 1979).
13. Kit Prins, "Energy derived from Wood in Europe, the USSR, and North America," *Unasylva*, Vol. 31, No. 123, 1979. An estimated 25 percent of all homes in the Soviet Union are heated with wood; see Leslie Dienes and Theodore Shabad, *The Soviet Energy System: Resource Use and Policies* (Washington, D.C.: V.H. Winston & Sons, 1979).
14. William R. Day and Kurt A. Schloth, "Development in Woodburning Technology," *RAIN*, May 1981; Fred Stiebeigh, "Hottest Stoves in America," *Quest*, December 1980; and Peter Tonge, "This Could be

'Ultimate' Wood Stove," *Christian Science Monitor,* February 22, 1989.

15. For a survey of the economics of wood energy use in the United States by region, see William T. Glidden, Jr., "Wood Energy Logs a Comeback," *Soft Energy Notes,* September/October 1982, and David A. Tillman, *Wood as an Energy Resource* (New York: Academic Press, 1978).
16. Edwin McDowell, "A Glow on Wood Furnaces," *New York Times,* November 11, 1979.
17. M. Allaby and J. Lovelock, "Wood Stoves: The Trendy Pollutant," *New Scientist,* November 13, 1980; Philip Shabecoff, "Wood Fires Arouse Fear of Pollution," *New York Times,* December 13, 1981; Stephen Budiansky, "Bioenergy: The Lesson of Wood Burning," *Environmental Science and Technology,* July 1980; and "Stoves Pollute Worse than Industry," *Canadian Renewable Energy News,* July 1981.
18. Elissa Krzeminski, "The Catalytic Combuster," *New Roots,* January 1982; Peter Tonge, "Wood Savings: Burn the Smoke, Too," *Christian Science Monitor,* January 28, 1981; Jon Vera "Woodburning: The Catalytic Combuster Comes of Age," *Country Journal,* November 1982.
19. Jeff Nadherny, "A Survey of Industrial Wood-Fired Boilers in New England: Analysis of Responses," Thayer School of Engineering, Dartmouth College, Hanover, N.H., February 1979.
20. Stephen Grover, " 'Jaws' Invades Forest-Products Industry as Use of Wood Waste Displaces Fuel Oil," *Wall Street Journal,* August 12, 1981; The Swedish Institute, "Forestry and Forest Industry in Sweden," Stockholm, November 1980. For a description of FIRE program, see *World Solar Markets,* June 1981.
21. Jourdan Houston, "Industry (Re)Discovers Wood," *Wood N' Energy,* April 1981.
22. John Zerbe, "The Many Forms of Wood as Fuel," *American Forests,* October 1978; R. L. Berry, "An Ancient Fuel Provides Energy for Modern Times," *Chemical Engineering,* April 21, 1980.
23. Jourdan Houston, "Pelletized Fuels: The Emerging Industry, *Wood N' Energy,* January 1981.
24. Tom Reed and Tom Milne, "Biomass Gasification: New Approach to Old Technology, *The SERI Journal,* Spring 1981; "Wood Gasifiers Offer Better Payback Than Any Other Solar System, Backers Say," *Solar Energy Intelligence Report,* January 18, 1982; National Research Council, *Producer Gas: Another Fuel for Motor Transport* (Washington, D.C.: National Academy Press, 1981); R. Datta and G. Dutt, "Producer Gas Engines in Villages of Less-Developed Countries," *Science,* August 14, 1981.

25. Robert Burgess, "Potential of Forest Fuels for Producing Electrical Energy," *Journal of Forestry*, March 1978; Burlington Electric Department, *Wood Fired Electric Power Generation: A New England Alternative* (Burlington, Vt.: 1979); and Michael Harris, "Can Burlington Turn on With Trees?," *Yankee Magazine*, January 1979.
26. Ministry of Energy, *Ten-Year Energy Program, 1980–1989* (Manila, Republic of the Philippines: 1980); summarized in "Curious Ipil-Ipil the Key to Wood Fueled Power Plants," *Christian Science Monitor*, September 18, 1980.
27. R. Powell and A. Hokanson, "Methanol from Wood: A Critical Assessment," in Sarkanen and Tillman, ed., *Progress in Biomass Conversion* (New York: Academic Press, 1979); and D. L. Hagen, "Methanol: Its Synthesis, Use as a Fuel, Economics and Hazards," U.S. Department of Commerce, Washington, D.C., December 1976.
28. Description of methanol technology is from E.G. Baker, D.J. Stevens, and D.A. Easkin, "Assessment of the Technology for Producing Mobility Fuels from Wood," Battelle Laboratories, Columbus, Ohio, February 1980, and A. Hokanson and R. Powell, "Methanol from Wood Waste: A Technical and Economic Study," Forest Service, U.S. Department of Agriculture, Washington, D.C., 1977.
29. U.S. Congress, Office of Technology Assessment, *Energy from Biological Processes*, Vol. II (Washington, D.C.: September 1980).
30. "Brazil Out to Show Methanol Grows on Trees," *Chemical Week*, March 14, 1979; "Direct Wood Gasification Method Seen Leading to Methanol Production in Brazil," *Solar Energy Intelligence Report*, October 13, 1980; "Biomass Gasifier Developed at SERI Converts 60% Wood Feed to Methanol," *Solar Times*, October 1981.
31. Ken R. Stamper, "50,000 Mile Methanol/Gasoline Blend Fleet Study —A Progress Report" (Springfield, Va.: National Technical Information Service, 1979); J. Finegold et al., "Demonstration of Dissociated Methanol as an Automotive Fuel: System Performance," Solar Energy Research Institute, Golden, Colo., April 1981.
32. "Methanol Fueled Car," *World Solar Markets*, October 1981; Joseph G. Finegold, J. Thomas McKinnon, and Michael E. Karpuk, "Decomposing Methanol As Consumable Hydride for Automobiles and Gas Turbines," Solar Energy Research Institute, Golden, Colo., March 1982.
33. The Canadian study is summarized in Peter Love and Ralph Overend, *Tree Power: An Assessment of the Energy Potential of Forest Biomass in Canada* (Ottawa: Ministry of Energy, Mines and Resources, 1978), and "Major Study Finds Enormous Potential in Canadian Biomass," *Soft Energy Notes*, October 1978.

34. The optimal scaling of plants is discussed in U.S. Congress, Office of Technology Assessment, *Energy from Biological Processes.* International Harvester's plans are described in David F. Salisbury, "Methanol: Ready to Muscle Its Way Into U.S. Gas Tanks?," *Christian Science Monitor,* March 10, 1982.
35. Charles E. Hewett and Colin J. High, *Construction and Operation of Small, Dispersed, Wood-fired Power Plants,* (Hanover, N.H.: Thayer School of Engineering, Resource Policy Center, September 1978).
36. Maureen Robb, "Producer-User Pacts Held Vital to Methanol Fuel Development," *Journal of Commerce,* March 31, 1982.
37. For overall world forestry trends, see Council on Environmental Quality and U.S. Department of State, *The Global 2000 Report to the President* (Washington, D.C.: U.S. Government Printing Office, 1980); R. Nigh and J. Nations, "Tropical Rainforests," *Bulletin of the Atomic Scientists,* March 1980; Norman Myers, "The Present Status and Future Prospects of Tropical Moist Forests," *Environmental Conservation,* Summer 1980; Smith, *Wood: An Ancient Fuel with* a Bright Future. For an example of emerging plans to harvest energy from tropical rain forests, see "Brazil's CNE Considers Deforestation Plan," *Latin American Energy Report,* April 1981.
38. For an assessment of the prospects for forestry in northern regions see John Hanrahan and Peter Gruenstein, *Lost Frontier: The Marketing of Alaska* (New York: W.W. Norton & Co., 1977)
39. U.S. Congress, Office of Technology Assessment, *Energy from Biological Processes.*
40. For discussion of the ecological implications of energy plantations and intensive silviculture, see A. Carlilse and I. Methven, "The Environmental Consequences of Intensive Forestry and the Removal of Whole Trees," in Stephen G. Boyce, ed., *Biological and Sociological Basis for a Rational Use of Forest Resources for Energy and Organics,* proceedings of an International Workshop, Michigan State University, East Lansing, Mich., May 6–11, 1979 (Asheville, N.C.: Forest Service, U.S. Department of Agriculture, 1979); Carl F. Jordan, "The Environmental Consequences of Intensive Forestry and the Removal of Whole Trees from Forests: The Situation in Latin America," in Boyce, ed., *Biological and Sociological Basis for a Rational Use of Forest Resources;* G. E. Likens et al., "Recovery of a Deforested Ecosystem," *Science,* February 3, 1978; David Pimentel et al., "Biomass Energy from Crop and Forest Residues," *Science,* June 5, 1981.
41. *Ibid.*
42. *Ibid.*

43. Steven Price, "You Don't Need Trees to Make Paper," *New York Times*, September 13, 1981; Dan Morgan, "The Forests of Tomorrow," *Washington Post*, November 21, 1977.
44. Sven Bjork and Wilhelm Graneli, "Energy Needs and the Environment," *Ambio*, Vol, 7, No. 4, 1978; J. Henry, "The Silvicultural Energy Farm in Perspective," in Sarkanen and Tillman, eds., *Progress in Biomass Conversion.*
45. For Eucalyptus Forestry, see B. Knowland and C. Ulinski, *Traditional Fuels: Present Data, Past Experience and Possible Strategies* (Washington, D.C.: Agency for International Development, 1979); and National Academy of Sciences, *Firewood Crops: Shrub and Tree Species for Energy Production* (Washington, D.C.: National Academy Press, 1980).
46. National Academy of Sciences, *Leucaena: Promising Forage and Tree Crops for the Tropics* (Washington, D.C.: National Academy Press, 1977).
47. For an excellent review of the literature on the impacts of intensive forestry practice, see Joseph F. Coates, Henry H. Hitchcock and Lisa Heinz, *Environmental Consequences of Wood and Other Biomass Sources of Energy*, EPA-600/8-82-017 (Washington, D.C.: U.S. Environmental Protection Agency, April 1982).
48. Helmut Sick, "Aves Brasileiras Ameaçadas de Extinção e Noções Gerais de Conservação no Brazil," in Simpósio Sobre Conservação de Natureza e *Restauração do Ambiente Natural do Homem*, Rio de Janeiro, 1979; David Pimentel et al., "Energy from Forests: Environmental and Wildlife Implications," *Interciencia*, September/October 1981.
49. For an overview of social forestry, see Erik Eckholm, *Planting for the Future: Forestry for Human Needs*, Worldwatch Paper 26 (Washington, D.C.: Worldwatch Institute, February 1979); B. Knowland and C. Ulinski, *Traditional Fuels: Present Data, Past Experience and Possible Strategies* (Washington, D.C.: Agency for International Development, 1979). For a country-by-country survey of social forestry, see Forest Products Laboratory, U.S. Forest Service, *Forestry Activities and Deforestation Problems in Developing Countries* (Washington, D.C.: U.S. Government Printing Office, 1980).
50. B. Ben Salem and Tran van Nao, "Fuelwood Production in Traditional Farming Systems," *Unasylva*, Vol. 33, No. 131, 1981; David Spurgeon, "Agroforestry: A New Life for Farmland," *Christian Science Monitor*, February 13, 1980.
51. James W. Howe and Frances A. Gulick, "Fuelwood and Other Renewable Energies in Africa: A Progress Report on the Problem and the Response," Overseas Development Council, Washington, D.C., 1980.

52. Descriptions of China's community forestry efforts are provided in Food and Agriculture Organization, Forestry Department, *Forestry for Local Community Development* (Rome: Food and Agriculture Organization, 1980); see also Jack C. Westoby, "Making Trees Serve People," *Commonwealth Forestry Review,* Vol. 54, Nos. 3/4, 1975. Estimates of new forest area from Reidar Persson, "Need for a Continuous Assessment," and "Summary Report," FAO/Nepal Study Tour on Multiple Use Mountain Forestry to the People's Republic of China, November 26-December 11, 1978, unpublished, Rome, undated; "The Green Wall of China," *Development Forum,* July/August 1981. A more pessimistic assessment is contained in reports by Vaclav Smil to the World Bank summarized in Libby Bassett, "Special Report: China's Ecosystems are Deteriorating Rapidly," *World Environment Report*, May 30, 1982, and Bayard Webster, "China's Progress Hurting Land," *New York Times*, October 3, 1982.
53. For South Korean social forestry, see Eckholm, *Planting for the Future.*
54. John Madeley, "The Trees of Life," *Development Forum,* July/August 1981; Modhumita Mojumdar, "India's Lost Woodlands," *Christian Science Monitor,* September 9, 1981; "Indian Social Forestry Study Finds Poor are Not Helped," *World Environment Report,* November 14, 1981; "Forests for Fuels," *Science, Today,* October 1981; Gunnar Poulsen, *Man and Trees in Tropical Africa,* Publication No. 101c (Ottawa: International Development Researcher Center, 1978); "Tanzania's Tree Planting Campaign Called a Success," *World Environment Report,* November 15, 1981.
55. An excellent discussion of the social and cultural factors in community forestry can be found in Raymond Noronha, "Why Is It So Difficult to Grow Fuelwood," *Unasylva,* Vol. 33, No. 131, 1981.
56. Agency for International Development, *The Socio-Economic Context of Fuelwood Use in Small Rural Communities,* Evaluation Special Study No. 1 (Washington, D.C.: August 1980); Eckholm, *Planting for the Future.*
57. India's new social forestry goals are described in "Faced with Huge Fuelwood Loss, India Pushes Social Forestry," *World Environment Report,* July 20, 1981. Some of the limitations on social forestry in India are described in Vandana Shiva, H.C. Sharatchandra, and J. Bandyopadhyay, "Social Forestry-No Solution within the Market," *Ecologist*, August 1982.
58. "Review of World Bank Financed Forestry Activity, FY 80," World Bank, Washington, D.C., June 1980; John Spears, Forestry Advisor, World Bank, "Overcoming Constraints to Increased Investment in For-

estry," presented to the 11th Commonwealth Forestry Conference, Trinidad, September 1980.

59. Maurice Strong and Mahbub ul Haq, *The Castel Gandolfo Report on Renewable Energy: Policies and Options,* presented to the North-South Roundtable Seminar at the United Nations Conference on New and Renewable Sources of Energy, Nairobi, Kenya, August 10–21, 1981.
60. "Deforestation in Himalayan Region Cause of India's Worst Flood," *World Environment Report,* November 6, 1978; James Sterba, "Chinese Say Deforestation Caused Flood Damage in Sichuan," *New York Times,* August 22, 1981; Richard Pascoe, "Flooding Blamed on Deforestation," *Washington Post,* September 2, 1981.
61. Solar Energy Research Institute, *A New Prosperity: Building a Renewable Future* (Andover, Mass.: Brick House, 1981); Marion Clawson, *The Economics of U.S. Nonindustrial Private Forests* (Washington, D.C.: Resources for the Future, 1979).
62. Forestry cooperatives are described in Robert Kilborn, Jr., "Giant Forest Companies Help Small Landowners Harvest Their Trees," *Christian Science Monitor,* November 24, 1981; Chris Wood, "Nova Scotia Entices Private Woodlot Owners to Sell Trees for Fuel," *Canadian Renewable Energy News,* December 1980; Clayton Jones, "Timberman Sees Land of Dixie as Future 'Wood Basket' of the World," *Christian Science Monitor,* May 11, 1981.
63. Christopher H. Holmes, *An Analysis of the New England Pilot Fuelwood Project,* (Washington, D.C.: U.S. Department of Agriculture, Forest Service, October 1980); John Forbes, "Helping the Small Woodlot Owner," *Country Journal,* January 1982.
64. These figures are the authors' estimates.
65. William W. Kellogg and Robert Schware, *Climate Change and Society* (Boulder, Colo.: Westview Press, 1981); Don Scroggin and Robert Harris, "Reduction at the Source," *Technology Review,* November/December 1981; Freeman J. Dyson, "Can We Control the Carbon Dioxide in the Atmosphere?," *Energy,* Vol. 2 (New York: Pergamon Press, 1977); Freeman Dyson and Gregg Moorland, "Technical Fixes for the Climatic Effects of CO_2," in *Workshop on the Global Effects of Carbon Dioxide from Fossil Fuels,* March 7–11, 1977 (Washington, D.C.: U.S. Department of Energy, 1979).
66. Amulya K.N. Reddy, "Alternative Energy Policies for Developing Countries: A Case Study of India," in Robert A. Bohm, Lillian A. Clinard, and Mary R. English, eds., *World Energy Production and Productivity,* Proceedings of the International Energy Symposium I, October 14, 1980 (Cambridge, Mass.: Ballinger, 1981); summarized in Amulya

Reddy, "India—A Good Life Without Oil," *New Scientist,* July 9, 1981.

67. Subsidies for nuclear power are described in Battelle Memorial Institute, *An Analysis of Federal Incentives Used to Stimulate Energy Production* (Richland, Wash.: 1978).

Chapter 7. Growing Fuels: Energy from Crops and Waste

1. Material on overview of photosynthetic energy sources is found in Alan D. Poole and Robert H. Williams, "Flower Power: Prospects for Photosynthetic Energy," *Bulletin of the Atomic Scientists,* May 1976.
2. A survey of current alcohol fuels activities is Bill Kovarik, "Third World Fuel Alcohol Push Shows Mixed Results," *Renewable Energy News,* September 1982.
3. For historical use of alcohol fuels, see Charles A. Stokes and Gale D. Waterland, "Alcohols: The Old New Fuels," *Technology Review,* July 1981; Hall Bernton, William Kovarits, Scott Sklar, *The Forbidden Fuel: Power Alcohol in the Twentieth Century* (New York: Boyd Griffin, 1982), and August W. Giebelhaus, "Resistance to Long-Term Energy Transition: The Case of Power Alcohol in the 1930's," in Lewis J. Perelman, August W. Giebelhaus and Michael D. Yokell, eds., *Energy Transitions* (Washington, D.C.: American Academy for the Advancement of Science, 1981).
4. For a description of alcohol technology, see U.S. Congress, Office of Technology Assessment, *Gasohol, A Technical Memorandum* (Washington, D. C.: September 1979).
5. The net energy balance of alcohol is discussed in U.S. Department of Energy, *Report of the Alcohol Fuels Policy Review* (Washington, D.C.: June 1979); U.S. National Alcohol Fuels Commission, *Fuel Alcohol: An Energy Alternative for the 1980's* (Washington, D.C.: 1981); and J. G. DaSilva et al., "Energy Balance for Ethyl Alcohol from Crops," *Science,* September 8, 1978.
6. Frederick F. Hartline, "Lowering the Cost of Alcohol," *Science,* October 5, 1979; M. Ladisck and K. Dyck, "Dehydration of Ethanol: New Approach Gives Positive Energy Balance," *Science,* August 31, 1979; Meg Cox, "Researchers Accelerate Search for Way to Use Energy in Making Gasohol," *Wall Street Journal,* January 31, 1980; and Stewart Wallace, "Tropical Bacterium May Halve Fermentation Costs," *Canadian Renewable Energy News,* October 1981.
7. "New Process Provides Key to Low-Cost Fuel Alcohol," *Journal of Commerce,* March 24, 1980; M.R. Ladisch et al., "Cellulose to Sugars: New Path Gives Quantitative Yield," *Science,* September 21, 1978.

8. Cissy Wallace, "Brazilian National Alcohol Program," *Soft Energy Notes,* July 1979; Leon G. Mears, "The Brazilian Experiment," *Environment,* December 1978; Leon G. Mears, "Brazil's Agricultural Program Moving Ahead," *Foreign Agriculture,* July 17, 1978; Warren Hoge, "Brazil's Shift to Alcohol as Fuel," *New York Times,* October 13, 1980.
9. World Bank, *Alcohol Production from Biomass in Developing Countries,* September 1980. World Bank, *World Development Report 1979* (Washington, D.C.: 1979).
10. World Bank, *Brazil: Human Resources Special Report* (Washington, D.C.: 1979); Kenneth Freed, "Brazil's Dream of Cheap Fuel is Running Out of Gas," *Los Angeles Times,* December 20, 1981; Lester R. Brown, *Food or Fuel: New Competition for the World's Cropland* (Washington, D.C.: Worldwatch Institute, March 1980).
11. "Gasohol Plans Face Problems," *World Business Weekly,* January 19, 1981; "Alcohol Fuels Hit a Rocky Road," *World Business Weekly,* July 6, 1981.
12. John Madeley, "Ethanol Production Runs into Environmental Difficulties," *Mazingira,* Vol. 5/3, 1981.
13. Madeley, "Ethanol Production"; Kenneth Freed, "Brazil's Dream of Cheap Fuel is Running Out of Gas."
14. "Carter Lays Out Multi-Billion Dollar Plan for Gasohol," *Solar Energy Intelligence Report,* January 14, 1980; "Biomass and Alcohol Fuels in the Energy Security Act," *Solar Energy Law Reporter,* November/December 1980; G. Alan Petzet, "Alcohol Blends Claim Small but Rising Share of U.S. Motor Fuel Market," *Oil and Gas Journal,* September 6, 1981.
15. Earl Butz, "U.S. Farmers Mount Gasohol Bandwagon," *Journal of Commerce,* November 14, 1979; U.S. Congress, Office of Technology Assessment, *Energy from Biological Processes* (Washington, D.C.: 1980).
16. U.S. Department of Agriculture, *Small-Scale Fuel Alcohol Production* (Washington, D.C.: March 1980); "Big Awards Go to Big Business," *Biofuels Report,* June 8, 1981.
17. Fred Sanderson, "The High Cost of Gasohol," *Resources* (Washington, D.C.: Resources for the Future, July 1981).
18. "The Gasohol Boom Dries Up," *Business Week,* March 30, 1981; Douglas Martin, "Budget Cuts, Weak Market Hurt Gasohol," *New York Times,* December 14, 1981.
19. T. J. Goering, "Root Crops: Potential for Food and Energy," *Finance and Development,* June 1980; "Pilot Alcohol Distillery Using Manioc

Nearing Break-Even Point," *Solar Energy Intelligence Report,* May 11, 1981; Richard A. Nathan, *Fuels from Sugar Crops* (Washington, D.C.: Battelle Institute for U.S. Department of Energy, 1978).

20. G. Stewart, W. Rawlins, and R. Quick, "Oilseeds as a Renewable Source of Diesel Fuel," *Search,* May 1981; D. D. Hall, G.W. Barnard, and P.A. Moss, *Biomass for Energy in the Developing Countries* (Oxford: Pergamon Press, 1982).
21. Joseph Bonney, "Substitutes for Diesel Sought in Brazil," *Journal of Commerce,* November 30, 1979; George Hawrylyshyn, "Diesel Next Target of Brazil Liquid Fuels Program," *Canadian Renewable Energy News,* April 1981.
22. Caryle Murphy, "South Africa's Quest for Energy Leads to Sunflowers," *Washington Post,* August 29, 1979; "Sunflower Oil as a Fuel Alternative," North Dakota State University, undated; David Hall, "Put a Sunflower in Your Tank," *New Scientist,* February 26, 1981.
23. Andy McCue, "Filipino's Cocodiesel is Fueling Vehicles and Buoying Price of Coconut Oil Crop," *Wall Street Journal,* May 28, 1981.
24. U.S. Congress, Office of Technology Assessment, *Energy from Biological Processes.*
25. Jack Johnson quoted in Bill Paul, "On the Arid Plains of the Southwest, Gopherweed Grows as Energy Hope," *Wall Street Journal,* August 27, 1980; see also Melvin Calvin, "Petroleum Plantations," in Hautala, ed., *Solar Energy: Chemical Conversion and Storage* (Clifton, N.J.: The Humana Press, 1979); John Noble Wilford, "Agriculture Meets the Desert on its Own Terms," *New York Times,* January 15, 1980; Jack Johnson and C. Wiley Hinnian, "Oils and Rubber from Arid Land Plants," *Science,* May 2, 1980.
26. The classic work in temperate-zone agro-forestry was originally published in 1929 but is still relevant today. See Russel J. Smith, *Tree Crops —A Permanent Agriculture* (New York: Harper & Row, 1978).
27. B. Wolverton and R.C. McDonald, "The Water Hyacinth: From Prolific Pest to Potential Provider," *Ambio,* Vol. 8, No. 1, 1979; "Aquatic Plants Clean Wastewater Lagoons," *World Water,* September 1980. B. Wolverton and R.C. McDonald, "Taking Advantage of the Water Hyacinth," *Water Power and Dam Construction,* July 1979; B. Wolverton and R.C. McDonald, "Don't Waste Waterweeds," *New Scientist,* August 12, 1976; W. J. North, "Giant Kelp: Sequoias of the Sea," *National Geographic,* No. 142, 1972; "Seaweed is Studied as Source of Natural Gas," *New York Times,* September 20, 1981; U.S. Congress, Office of Technology Assessment, *Energy from Open Ocean Kelp Farm* (Washington, D.C.: September 1980).

28. Norman Myers, *The Sinking Ark* (Oxford: Pergamon Press, 1979); National Academy of Sciences, *Firewood Crops: Shrub and Tree Species for Energy Production* (Washington, D.C.: 1980).
29. Ward Sinclair, " 'Merlins' of Corn Improve Yields Dramatically," *Washington Post*, June 9, 1981.
30. U.S. Congress, Office of Technology Assessment, *Impacts of Applied Genetics on Micro-Organisms, Plants and Animals* (Washington, D.C.: 1981); Harold M. Schmeck, Jr., "Gene-Splicing is Said to be Key to Future Agricultural Advances," *New York Times*, May 20, 1981.
31. "Report of the Ad Hoc Group on Rural Energy, Including the Utilization of Energy in Agriculture," prepared for the United Nations Conference on New and Renewable Sources of Energy, Nairobi, Kenya, August 10–21, 1981.
32. Joy Dunkerly, "Patterns of Energy Consumption by Rural and Urban Poor in Developing Countries," *Natural Resources Forum*, November 4, 1979.
33. Andrew Barnett, Leo Pyle, and S.K. Subramanian, *Biogas Technology in the Third World: A Multidisciplinary Review* (Ottawa: International Development Research Centre, 1978); National Academy of Sciences, *Methane Generation for Human, Animal and Agricultural Wastes*, report of an Ad Hoc Panel of the Advisory Committee on Technology Innovation (Washington, D.C.: 1977).
34. V.V. Bhatt, "The Development Problem, Strategy, and Technology Choice: Sarvodaya and Socialist Approaches in India," in Long, Oleson, eds., *Appropriate Technology and Social Values: A Critical Appraisal* (Cambridge, Mass.: Ballinger, 1980).
35. S.K. Subramanian, "Biogas Systems and Sanitation in Developing Countries," presented to a conference on Sanitation in Developing Countries Today, sponsored by Oxfam, Pembroke College, Oxford, July 1977.
36. Edgar J. Da Silva, "Biogas: Fuel of the Future?," *Ambio*, No. 1, 1980.
37. Peter Hayes and Charles Drucher, "Community Biogas in India," *Soft Energy Notes*, April 1980; Jyoti K. Parikh and Kirit Parikh, "Mobilization and Impacts of Bio-gas Technologies," *Energy*, Vol. 2, 1977.
38. Vaclav Smil, "Energy Solution in China," *Environment*, October 1977; Food and Agriculture Organization, "China: Recycling of Organic Wastes in Agriculture," *FAO Soils Bulletin* 40, 1977. Smil's recent doubts are expressed in Vaclav Smil, "Chinese Biogas Program Sputters," Soft Energy Notes, July/August 1982.
39. Ibid.
40. H.R. Srinivasan, "Biogas Development in India," *Mazingira*, August 1981.

41. Ariane van Buren, "Biogas Beyond China," *Ambio,* 1980; "India Plans to Build Millions of Village Biogas Plants," *World Environment Report,* September 14, 1981; "Biogas Powered Piggery in the Philippines," *Soft Energy Notes,* August 1978; "Brazil Plans Biogas Generator Installation Throughout Country," *Latin American Energy Report,* May 21, 1981.
42. "When the Chips are Down . . . ," *Soft Energy Notes,* December/January 1981.
43. A.G. Hashimoto, Y.R. Chen, and R.L. Prior, "Methane and Protein from Animal Feedlot Wastes," *Journal of Soil and Water Conservation,* January/February 1979.
44. U.S. Department of Energy, *Report of the Alcohol Fuels Policy Review* (Washington, D.C.: June 1979).
45. Wendy Peters, "High-Growth Eucalyptus Offers Island Energy," *Canadian Renewable Energy News,* May 1981; "Costa Rica: New Plans to Cut the Oil Bill," *World Business Weekly,* January 19, 1981; Jawaharhall Baguant, "Electricity from the Sugar Cane Industry in Nicaragua," *Interciencia,* March/April 1981; "High Potential for Ethanol from Whey Seen for Several Areas of New York State," *Solar Energy Intelligence Report,* May 4, 1981.
46. U.S. Congress, Office of Technology Assessment, *Materials and Energy from Municipal Waste* (Washington, D.C.: 1979).
47. James Sterba, "Garbage Into Energy: Now a Seller's Market," *New York Times,* May 16, 1978.
48. Neil Seldman and Jon Huls, "Beyond the Throwaway Ethic," *Environment,* November 1981.
49. "Worldwide Inventory of Waste-to-Energy Systems," in *Refuse-Fired Energy Systems in Europe: An Evaluation of Design Practices* (Washington, D.C.: U.S. Environmental Protection Agency, November 1979); "Recycling," *World Environment Report,* November 17, 1980; Gary Yerkey, "Denmark Saves on Energy with Waste Energy," *Christian Science Monitor,* June 15, 1981.
50. U.S. Environmental Protection Agency, *Refuse-Fired Energy Systems in Europe: An Evaluation of Design Practices* (Washington, D.C.: November 1979; Susan Traill, "Municipal Waste: An Energy Source for Europe," *Energy International,* November/December 1980.
51. Council on Environmental Quality, "Municipal Solid Waste," in *Environmental Quality, the Tenth Annual Report of the Council on Environmental Quality* (Washington, D.C.: 1979); Jerry Knight, "Saugus, Mass.: Not Enough Refuse for Plant to Operate at Capacity," *Washington Post,* August 19, 1979; Robert Kaiser, "New England Plant Turns Trash to Energy," *New York Times,* October 27, 1979.

52. Frances Cerra, "Garbage-to-Fuel Recycling in U.S. Moves Slowly, Mixed in Problems," *New York Times,* August 19, 1980; Congressional Research Service, *Status Report on Resource Recovery Facilities* (Washington, D.C.: Committee on Science and Technology, U.S. House of Representatives, 1979).
53. Bill Peterson, "Baltimore's Trash Plant is Costly Failure," *Washington Post,* March 20, 1977.
54. Jane Rochman, "Trash Power: A Worthy Notion that Doesn't Yet Pay," *New York Times,* November 11, 1979; Council on Environmental Quality, *Environmental Quality, the Tenth Annual Report.*
55. U.S. Congress, Office of Technology Assessment, "Resource Recovery from Municipal Solid Waste," in *An Assessment of Technology for Local Development* (Washington, D.C.: January 1981); "Recyclers Charge Energy Department with Acting at Cross Purposes," *Journal of Commerce,* March 25, 1981.
56. Christopher Dickey, "Burning Refuse Dumps Chokes Mexico City, Arouses Ragpickers' Ire," *Washington Post,* April 12, 1981; John Vogler, *Work from Waste: Recycling Wastes to Create Employment* (London: Intermediate Technology Publications and Oxfam, 1981); S.V. Sethuraman, ed., *The Urban Informal Sector in Developing Countries: Employment, Poverty and Environment,* World Employment Programme Research Study (Geneva: International Labour Organization, 1981); and Daphne Miller, "Making Waste Less Wasteful," *Development Forum,* July/August 1982.
57. James Barron, "Garbage is Garbage, Burning it for Energy is Difficult," *New York Times,* August 2, 1981.
58. "Getty Bullish on Refuse," *Biofuels Report,* October 5, 1981; "Getty Says U.S. Ruins 'Garbage' Gas Business," *Wall Street Journal,* February 23, 1981.
59. "Delhi Gas Project," *World Environment Report,"* October 17, 1981. William Dietrich, "Sewage Plants Clean Up," *Soft Energy Notes,* February/March 1981.
60. Vaclav Smil, *China's Energy: Achievements, Problems, Prospects* (New York: Praeger, 1976). The projections are the authors' estimates.
61. David Pimentel and Marcia Pimentel, *Food, Energy and Society* (New York: John Wiley & Sons, 1979); D.O. Hall, G.W. Barnard, and P.A. Moss, *Biomass for Energy in the Developing Countries* (Oxford: Pergamon Press, 1982); David Pimentel et al., "Biomass Energy from Crop and Forest Residues," *Science,* June 5, 1981.
62. Art Candell and Libby Bassett, "Washington, D.C., Sewage Sludge May Green Haitian Plantation," *World Environment Report,* Septem-

ber 14, 1981; Joanne Omang, "U.S. Faces the Problem of Disposing the Undisposable," *Washington Post,* March 14, 1980; Joanne Omang, "Maryland Trying to Squeeze Usable Energy from Waste," *Washington Post,* November 11, 1977.

Chapter 8. Rivers of Energy

1. For contribution of various energy sources on a country-by-country basis, see United Nations (UN), Department of International Economic and Social Affairs, Statistical Office, *1979 Yearbook of World Energy Statistics* (New York: 1981).
2. For history of hydropower, see Norman A. F. Smith, "Water Power," *History Today,* March 1980, and Norman A. F. Smith, *Man and Water: A History of Hydro-Technology* (London: Peter Davies, 1975). L. Sprague DeCamp, *The Ancient Engineers* (New York: Doubleday & Co., 1963); Antipater quoted in *Encyclopedia Britannica,* 15th ed., s.v. "Hydro-Power."
3. For discussion of social conflict in medieval France, see Jean Gimpel, *The Medieval Machine: The Industrial Revolution of the Middle Ages* (New York: Holt, Rinehart & Winston, 1976); Kathleen Earley, "When America Ran on Water," *The Sciences,* December 1975.
4. Norman Smith, "The Origins of the Water Turbine," *Scientific American,* January 1980; Harold I. Sharlin, *The Making of the Electrical Age* (New York: Abelard-Schuman, 1963).
5. For trends in hydropower development, see Louis H. Klotz, "Water Power, Its Promises and Problems," in Louis H. Klotz, ed., *Energy Sources: The Promises and Problems* (Durham, N.H.: Center for Industrial and Institutional Development, University of New Hampshire, 1980).
6. R. S. Kirby et al., *Engineering in History* (New York: McGraw-Hill, Inc., 1956). Recent attempts at revival are discussed in "The Water Wheel Makes a Return for Power Generation," *Chilton's Energy,* August 1980; George S. Erskine, "A Future for Hydropower," *Environment,* March 1978; Yvonne Howell, "New Straight-Flow Turbine," *Sunworld,* February 1977.
7. World Energy Conference, *Survey of Energy Resources, 1980,* (Munich: 1980).
8. *Ibid.*
9. Lars Kristoferson, "Waterpower—A Short Overview," *Ambio,* Vol. 6, No. 1, 1977. For a survey of world water resources by region, see World Energy Conference, *Survey of Energy Resources, 1974* (Munich: 1975).

10. "Tens of Thousands of MWs of New Hydro-Generating Capacity Planned for Siberia," *Energy in Countries with Planned Economies,* January 1981; "Africa Could Install 15,000-MW Hydro Capacity in 1980s," *World Water,* August 1980; Ronald Antonio, "Western Venezuela Focus on Major Hydro Development," *Modern Power Systems,* September 1981; "Harnessing Energy in the Amazon," *Engineering News Record,* July 10, 1980; Kenneth Adelman "Energy in Zaire," *Africa Today,* October/December 1976.
11. "Report of the Technical Panel on Hydropower," prepared for the United Nations Conference on New and Renewable Sources of Energy, Nairobi, Kenya, August 10–21, 1981; Denis Hayes, *Rays of Hope: The Transition to a Post-Petroleum World* (New York: W. W. Norton & Co., 1977).
12. Richard Lawrence, "Energy Self-Sufficiency Seen for Latins," *Journal of Commerce,* September 15, 1980; United Nations, *1979 Yearbook.* Ghana, Norway and Zambia receive 99 percent of their electricity from hydropower. Portugal, New Zealand, Nepal and Switzerland, three-fourths; Austria and Canada, two-thirds.
13. E. Fels and R. Keller, "World Register of Man-Made Lakes," in William Ackermann et al., eds., *Man-Made Lakes: Their Problems and Environmental Effects* (Washington, D.C.: American Geophysical Union, 1973); T. W. Mermel, "Major Dams of the World," *Water Power and Dam Construction,* November 1979.
14. Information on China is from "A Hydroelectric Bonanza," *Energy Daily,* December 12, 1980. "Egypt Planning Vast Desert Hydroelectric Project," *New York Times,* March 28, 1981; Philip P. Micklin, "Soviet Plans to Reverse the Flow of Rivers: The Koma-Vychegda-Pechora Project," *Canadian Geographer,* Vol. 18, No. 3, 1969; R. Partl, *Power from Glaciers: The Hydropower Potential of Greenland's Glacial Waters* (Laxenburg, Austria: International Institute for Applied Systems Analysis, November 1977).
15. Information on the establishment and history of the TVA is from Philip Selznick, *TVA and the Grass Roots* (New York: Harper Torchbooks, 1966), "Hydroelectric and Electric Power," in Robert Engler, ed., *America's Energy: Reports from the Nation on 100 Years of Struggles for the Democratic Control of Our Resources* (New York: Pantheon Books, 1980), and Read A. Elliott, "The TVA Experience: 1933–1971," in Ackermann, *Man-Made Lakes.*
16. For ongoing hydro projects in Third World countries, see G. V. Eckenfelder, "Hydro Power Plants in Nigeria," *Energy International,* July 1977; Jonathan Kandell, "Itaipu Dam: Brazil's Giant Step," *New York*

Times, September 6, 1976; Clayton Jones, "Harnessing Sri Lanka's Nile," *Christian Science Monitor,* May 6, 1982.

17. Istvan Kovács and László Dávid, "Joint Use of International Water Resources," *Ambio,* Vol. 6, No. 1, 1977; United Nations, *Register of International Rivers* (New York: Pergamon Press, 1978).
18. "Newfoundland Hydroelectric Plans Stalled," *Journal of Commerce,* December 18, 1980; Newfoundland situation is also discussed in Henry Giniger, "A Dispute on Power Escalates," *New York Times,* November 22, 1980. John Bardack, "Some Ecological Implications of Mekong River Development Plans," in M. T. Farvar and J. P. Milton, eds., *The Carless Technology: Ecology and International Development* (Garden City, N.Y.: Natural History Press, 1972); Ann Crittenden, "Aid Bank Weighs Disputed Guyana Project," *New York Times,* October 30, 1980; Peter Crabb, "There Is More to Canada's Constitutional Problems than Albertan Oil," *Energy Policy,* December 1981; Aryeh Wolman, "Israel, Jordan in Hydro Conflict," *Renewable Energy News,* October 1981.
19. Story of Nile's development from John Waterbury, *Hydropolitics of the Nile Valley* (Syracuse, N.Y.: Syracuse University Press, 1980).
20. A. A. Abul-Ata, "After Aswan," *Mazingira,* Vol. 7, No. 11, 1979; J. N. Goodsell, "Power Grid Generates a Million Jobs in a Former Brazilian Wasteland," *Christian Science Monitor,* March 24, 1981; "Energy Profile of Brazil," *Energy International,* September 1976; Eric Jeffs, "Hydro to be Main Source for Venezuela," *Energy International,* June 1980; Ministry of Energy, *Ten-Year Energy Program 1980–1989* (Metro Manila, Philippines: 1980).
21. Henry Kamm, "Dam Project Brings Little Gain for Sumatra's People," *New York Times,* October 2, 1980; for Zaire, see Jay Ross, "A Tale of Two Projects: One Winner, One Loser," *Washington Post,* April 25, 1982.
22. For a general discussion of environmental changes caused by large dams, see Ackermann, *Man-Made Lakes.*
23. "Washing China's Wealth Out to Sea," *Technology Review,* October 1980; Waterbury, *Hydropolitics of the Nile Valley;* Fabian Acker, "Saving Nepal's Dwindling Forests," *New Scientist,* April 9, 1981.
24. Waterbury, *Hydropolitics of the Nile;* Susan Waton, "U.S.-Egypt Nile Project Studies High Dam's Effects," *Bioscience,* January 1981; "Irrigation and Water Development, A Discussion" in Farvar and Milton, *The Careless Technology.*
25. Nigel J. H. Smith, *Man, Fishes and the Amazon* (New York: Columbia University Press, 1981). Information on Columbia River is from Colin Nash, "Fisheries Should be Important Part of Water Resource Plan-

ning," *World Water*, October 1980. Letitia E. Obeng, "Environmental Impacts of Four African Impoundments," in *Environmental Impacts of International Civil Engineering Projects* (New York: American Society of Civil Engineers, 1977); Carl J. George, "The Role of the Aswan High Dam in Changing the Fisheries of the Southeastern Mediterranean," in Farvar and Milton, *Careless Technology.*

26. For a general discussion of schistosomiasis, see Erik Eckholm, *The Picture of Health: Environmental Sources of Disease* (New York: W. W. Norton & Co., 1977). For the role of dams in the disease's spread, see A.W.A. Brown and J. O. Deom, "Summary: Health Aspects of Man-Made Lakes," and B.B. Waddy, "Health Problems of Man-Made Lakes: Anticipation and Realization, Kainji, Nigeria, and Kossou, Ivory Coast," in Ackermann, *Man-Made Lakes;* Diana Gibson, "The Blue Nile Project," *World Health,* August/September 1980; Nigel Pollard, "The Gezira Scheme—A Study in Failure," *The Ecologist,* January/February 1981.
27. Thayer Scudder, "Summary: Resettlement" and Charles Takes, "Resettlement of People from Dam Reservoir Areas," in Ackermann, *Man-Made Lakes;* Eugene Balon, "Kariba: The Dubious Benefits of Large Dams," *Ambio,* Vol. 7, No. 2, 1978; Letitia E. Obeng, "Should Dams be Built? The Volta Lake Example," *Ambio,* Vol. 6, No. 1, 1977; Nancie Gonzalez, "The Sociology of a Dam," *Human Organization,* Vol. 31, No. 4, 1972. For a discussion of China's Three Gorges project, see "China Hydro Agreement Gives U.S. Edge for Jobs," *Engineering News Report,* September 6, 1979.
28. Sheldon Davis, *Victims of the Miracle: Development and the Indians of Brazil* (Cambridge: Cambridge University Press, 1977); Paul Aspelin, "Electric Colonialism," *Soft Energy Notes,* July/August 1982; "Filipino Hill Tribes Fight to Save Homes From Dam," *New York Times,* August 12, 1980. For a discussion of the Inuit conflict, see Clayton Jones, "Quebec Turns Water Into Gold," *Christian Science Monitor,* July 30, 1980. For information on Survival International, see Warren Hoge, "Brazil's Indians Have Seen Too Much Progress," *New York Times,* February 15, 1981.
29. For a discussion of the economic value of endangered species, see Norman Meyers, *The Sinking Ark* (New York: Pergamon Press, 1979) and Erik Eckholm, *Disappearing Species: The Social Challenge,* Worldwatch Paper 22 (Washington, D.C.: Worldwatch Institute, July 1978). For cases of dams flooding endangered species habitats, see Philip Shabecoff, "Behold the Tiny Snail-Darter: An Ominous Legal Symbol?," *New*

York Times, October 7, 1979. Jane Naczynski-Phillips, "Tasmania Saves a Wilderness by Modifying a Power Scheme," *World Environment Report*, October 6, 1980; W. E. Scott, "Tasmania Posed for Hydropower Expansion," *Energy International*, October 1979. For environmental planning in Quebec's James Bay, see R. Paehlke "James Bay Project: environmental assessment in the planning of resource development" in O. P. Dwivedi, ed., *Resources and the Environment: Policy Perspectives for Canada* (Toronto: McClelland & Stewart Ltd., 1980).

30. Jean-Marc Fleury, "Aswan-on-Senegal?," *IDRC Report*, February 1981; F. M. G. Budweg, "Environmental Engineering for Dams and Reservoirs in Brazil," *Water Power and Dam Construction*, October 1980; Alan Grainger, "Will the Death Knell Sound in Silent Valley?", *The Ecologist*, August 1982. For a discussion of the role of international lending groups in the environmental planning for large dams, see Robert E. Stein and Brian Johnson, *Banking on the Biosphere?* (Lexington, Mass.: Lexington Books, 1979).
31. Thayer Scudder, "Ecological Bottlenecks and the Development of the Kariba Lake Basin," in Farvar and Milton, *Careless Technology*.
32. D. E. Abramowitz, "The Effect of Inflation on the Choice Between Hydro and Thermal Power," *Water Power & Dam Construction*, February 1977.
33. Cost estimates are from Waterbury, *Hydropolitics of the Nile*. J. R. Cotrim et al., "The Bi-National Itaipu Hydropower Project," *Water Power and Dam Construction*, October 1977; Central Intelligence Agency, *Electric Power for China's Modernization: The Hydroelectric Option* (Washington, D.C.: May 1980); World Bank, *Energy in Developing Countries* (Washington, D.C.: August 1980).
34. T. W. Berrie, "The Role of International Lending Agencies," *Water Power and Dam Construction*, August 1979; "Brazil Builds Up Her Industrial Capability," *Energy International*, September 1976; U.S. Central Intelligence Agency, *Electric Power for China's Modernization;* Trevor Drieberg, "India Seeks Collaboration on Hydroelectric Projects," *Journal of Commerce*, May 26, 1982.
35. For links between mineral and aluminum projects and hydropower, see "The Amazon Dilemma," *Energy International*, September 1976. W.E. Scott, "Hydro Power Leads Energy Development Plans for Papua New Guinea," *Energy International*, January 1977; Leslie Dienes and Theodore Shabad, *The Soviet Energy System* (Washington, D.C.: V. H. Winston & Sons, 1979); Monica M. Cole, "The Rhodesian Economy in Transition and the Role of Kariba," *Geography*, Vol. 47, No. 2, 1962;

Peter Crabb "Hydro Power on the Periphery: A Comparison of Newfoundland, Tasmania and the South Island," *Alternatives*, Summer 1982.

36. Dan Morgan, "Pacific Northwest Faces Kilowatt Crisis," *Washington Post*, October 28, 1979; Roger Gale, "Kaiser Squabbles With Ghana Over Cheap Power," *Energy Daily*, October 29, 1980.
37. For cost figures on various fuels, see U.S. Energy Information Administration, *Annual Report to Congress 1981* (Washington, D.C.: Department of Energy, 1981) and U.S. Congress, General Accounting Office, "Region of the Crossroads—The Pacific Northwest Searches for New Sources of Electric Energy," Washington, D.C. August 1978. For nuclear cost overruns in the U.S. Pacific Northwest, see Leslie Wayne, "Utility Setbacks on the Coast," *New York Times*, November 3, 1981.
38. "Projected De-Controlled Hydroelectric Revenues for 56 Developing Countries Based on 1979 Output," Center for Development Policy, Washington, D.C., unpublished, 1981. In 1981 the World Bank loaned $8.8 billion.
39. Roger Gale, "Kaiser Squabbles With Ghana Over Cheap Power," *Energy Daily*, October 29, 1980; Nicholas Burnett, "Kaiser Shortcircuits Ghanian Development," *Multinational Monitor*, February 1980; "Imperialism and the Volta Dam," *West Africa*, March 24, 1980; Cheryl Payer, *The World Bank: A Critical Analysis* (New York and London: Monthly Review Press, 1982); David Hart, *The Volta River Project* (Edinburgh: University Press, 1980).
40. Kai Lee and Donna Lee Klemka, *Electric Power and the Future of the Pacific Northwest* (Pullman, Wash.: State of Washington Water Resource Center, March 1980). For a discussion of recent pricing disputes, see "Environmentalists' Roles in Northwest Power Talks May Put the Bite on Industry," *Energy Daily*, May 1, 1981. For some of the adjustment problems posed by higher prices without revenue recycling, see Victor F. Zonama, "Power Prices Upset Business in Northwest," *Wall Street Journal*, September 3, 1982.
41. For a discussion of similar proposals, see Ralph Cavanaugh et al., *Choosing an Electrical Energy Future for the Pacific Northwest: An Alternative Scenario* (San Francisco: Natural Resources Defense Council, August 1980) and Kojo Arthur, "Electric Power Rates in Ghana: The Case for More Revenue," Africa Research and Publications Project, November 27, 1980.
42. World Bank, *Energy in Developing Countries.*
43. Cost estimates are from Allen Inversin, "A Pelton Micro-Hydro Proto-

type Design," Appropriate Technology Development Unit, Papua New Guinea, June 1980, and Ueli Meier, "Development of Equipment for Harnessing Hydro power on a Small Scale," presented to the Workshop in Mini/Micro Hydroelectric Plants, Kathmandu, Nepal, September 10–14, 1979.

44. For a general picture of China's centralized energy system, see U.S. Central Intelligence Agency, *Electric Power for China's Modernization;* and William Clark, "China's Electric Power Industry," in *Chinese Economy Post Mao* (Washington, D.C.: Joint Economic Committee, U.S. Congress, 1978).
45. For overview of China's small-scale hydro program, see Robert P. Taylor, *Rural Energy Development in China* (Washington, D.C.: Resources for the Future, 1981).
46. Taylor, *Rural Energy Development in China;* Mao Wen Jing and Deng Bing Li, "An Introduction of Small Hydro-Power Generation in China," United Nations Industrial Organization, New York, May 1980.
47. Mao and Deng, "Introduction to Small Hydro-Power in China"; Alexander Ansheng Tseng et al., "The Role of Small Hydro-Electric Power Generation in the Energy Mix Development for the People's Republic of China," Oriental Engineering and Supply Co., Palo Alto, Calif., October 31, 1979.
48. Vaclav Smil, "Intermediate Energy Technology in China," *Bulletin of the Atomic Scientists,* February 1977.
49. The Papua New Guinea project is discussed in Allen Inversin, "Technical Notes on the Baindoang Micro Hydro and Water Supply Scheme," PNG University of Technology, Lae, Papua New Guinea, unpublished, 1981. Ed Arata, "Micro-Hydroelectric Projects for Rural Development in Papua New Guinea," in Donald Evans and Laurie Nogg Adler, eds., *Appropriate Technology for Development: A Discussion and Case Histories* (Boulder, Colo.: Westview Press, 1979).
50. Fabian Archer, "Saving Nepal's Dwindling Forests," *New Scientist,* April 9, 1981. For activities in Peru, see memorandum on "Small Decentralized Hydropower Program," International Programs Division, National Rural Electric Cooperative Association, Washington, D.C., November 13, 1980. For a country-by-country review of small-scale hydro efforts, see United Nations Industrial Development Organization, "Draft Report of the Seminar-Workshop on the Exchange of Experiences and Technology Transfer on Mini Hydro Electric Generation Units," Kathmandu, Nepal, September 10–14, 1979. For a worldwide breakdown of energy aid programs, see Thomas Hoffmann and Brian

Johnson, *The World Energy Triangle: A Strategy for Cooperation* (Cambridge, Mass.: Ballinger, 1981). World Bank figures are from Edward Moore, World Bank, private communication, April 30, 1981.

51. For a discussion of new small-scale hydro assistance approaches, see Ken Grover, "Small Decentralized Hydropower for Developing Countries," GSA International, Katonah, N.Y., unpublished, undated, and Rupert Armstrong-Evans, "Micro-Hydro as an Appropriate Technology in Developing Countries," presented to the International Conference on Small-Scale Hydropower, Washington, D.C., October 1–3, 1979. For the hydropower development corporation idea, see S. David Freeman, "Hydropower for Development and Energy," presented to the North-South Roundtable at the United Nations Conference on New and Renewable Sources of Energy, Nairobi, Kenya, August 10–21, 1981.
52. For examples of conflicts over hydro development, see E. W. Kenworthy, "Kleppe Moves to Block Dam in Carolina," *New York Times,* March 13, 1976. J. B. Kirkpatrick, *Hydro-Electric Development and Wilderness in Tasmania* (Department of the Environment, Hobart, Tasmania, 1979); B. Connolly, *The Fight for the Franklin: The Story of Australia's Last Wild River* (Sydney: Cassell Australia Ltd., 1981). For ecological merits of preserving rivers, see Committee on Environmental Effects of the United States Committee on Large Dams, *Environmental Effects of Large Dams* (New York: American Society of Civil Engineers, 1978). Mark M. Brinson, "Riparian and Floodplain Ecosystems: Functions, Values, and Management," Fish and Wildlife Service, U.S. Department of the Interior, April 1980. For the moral case for bequeathing some rivers to future generations, see John McPhee, *Encounters with the Archdruid* (New York: Doubleday, 1975).
53. S. Angelin and H. Boström, "Hydro Development in Sweden," *Water Power and Dam Construction,* June 1980. Information on U.S. scenic rivers is from Warren Viessman, Jr., "Hydropower," in Congressional Research Service for U.S. House of Representatives, Committee on Interstate and Foreign Commerce, and U.S. Senate, Committee on Energy and Natural Resources and Committee on Commerce, Science, and Transportation, *Project Independence: U.S. and World Energy Outlook Through 1990,* Committee Print, November 1977.
54. For a critique of the misuse of cost-benefit analysis in water projects, see Brent Blackwelder, "In Lieu of Dams," *Water Spectrum,* Fall 1977. For an overview of the dominant water development mentality in the United States, see Richard L. Berkman and W. Kip Viscuse, *Damming the West* (New York: Grossman Publishers, 1973).

55. There is no standard definition of "mini," "micro" and "small" hydro. Generally, however, "micro" refers to a plant under 50 kilowatts, "mini" under 10 megawatts and "small" anything under 100 megawatts. Information on France is from J. Cotillon, "Micropower: An Old Idea For a New Problem," *Water Power and Dam Construction,* January 1979. D. G. Birkett, "Review of Potential Hydroelectric Development in the Scottish Highlands," *Electronics and Power,* May 1979; Jim Harding, "Soft Paths for Difficult Nations: The Problem of Japan," *Soft Energy Notes,* June/July 1980. Information on Wales is from *World Environment Report,* February 16, 1981. "Rumania's Maxi Strategy on Mini Hydro Plants," *Energy in Countries with Planned Economies,* April 1981; E. G. Greunert, "Crash Program for Mini Hydro in Spain," *Modern Power Systems,* February 1981; "Mini Hydropower for Sweden," *Water Power and Dam Construction,* May 1978.
56. Comparison on energy from Rhone and Ohio rivers is from U.S. Congress, General Accounting Office, "Hydropower—An Energy Source Whose Time Has Come Again," Washington, D.C., January 1980. Estimates of potential at small U.S. dams are from Institute of Water Resources, *Preliminary Inventory of Hydropower Resources,* Vol. 1 (Fort Belvoir, Va.: U.S. Army Corps of Engineers, July 1979), and R. J. McDonald, "Estimates of the National Hydroelectric Power Potential at Existing Dams," U.S. Army Corps of Engineers, Institute of Water Resources, Fort Belvoir, Va., July 1977.
57. New England River Basins Commission, "Potential for Hydropower Development at Existing Dams in the Northeast," *Physical and Economic Findings and Methods,* Vol. I (Boston: January 1980).
58. For a discussion of the limited power producers section of the Public Utility Regulatory Policy Act of 1978, see Reinier H. J. H. Lock, "Encouraging Decentralized Generation of Electricity: Implementation of the New Statutory Scheme," *Solar Law Reporter,* November/December 1980. For a discussion of the institutional barriers to small-scale hydro renovation, see Peter Brown, "Federal Legal Obstacles and Incentives to the Development of the Small Scale Hydroelectric Potential of the Nineteen Northeastern United States," Energy Law Institute, Concord, N.H., 1979. For a summary of federal government small-scale hydro power programs, see U.S. Congress, General Accounting Office, "Hydropower—Energy Source Whose Time Has Come Again," January 11, 1980.
59. Solar Energy Research Institute, "Hydroelectric Power," Appendix C of Solar Energy Research Institute, *A New Prosperity: Building a Sustain-*

able Future (Andover, Mass.: Brick House, 1981); U.S. Congress, General Accounting Office, "Non-Federal Development of Hydroelectric Resources at Federal Dams—Need to Establish a Clear Federal Policy," Washington, D.C., September 1980.

60. Todd Crowell, "Grand Coulee Dam: Growing, Changing Hydropower Giant," *Christian Science Monitor,* August 27, 1980; G. Weber et al., "Uprating Switzerland's Hydro Plants," *Water Power and Dam Construction,* February 1978; Gladwin Hill, "U.S. May Add 200 Feet to a California Dam," *New York Times,* January 18, 1979.
61. For information on use of hydro plants for peaking, see Gordon Thompson, "Hydroelectric Power in the USA: Evolving to Meet New Needs," Center for Energy and Environmental Studies, Princeton University, 1981. Global pumped storage figure is from "Report of the Technical Panel on Hydropower," prepared for the United Nations Conference on New and Renewable Sources of Energy, Nairobi, Kenya, August 10–21. For a discussion of the role of pumped storage in highly developed electricity systems, see A. M. Angelini, "The Role of Pumped-Storage in Western Europe," *Water Power and Dam Construction,* June 1980 and Orval Burton, "Hydro-Power and Pumped Storage in the Northwest," *Energy and Water Resources* (Washington, D.C.: Water Resources Research Institute, January 1977).
62. "Canadians Utilize Hydroelectric Power," *Journal of Commerce,* February 25, 1981; Costa Rican information from "Power on Tap," *Euromoney,* August 1980.
63. K. Goldsmith, "The Role of Swiss Hydro in Europe," *Water Power and Dam Construction,* June 1980; Richard Critchfield, "Nepal: Majestic Mountains, Snowcapped Riches," *Christian Science Monitor,* November 1, 1978. For a discussion of the parallels between Nepal and Switzerland, see Robert Rhoades, "Cultural Echoes Across the Mountains," *Natural History,* January 1979.
64. Hydropower provides 90 percent of Brazil's electricity. Output increased at an average annual rate of 11 percent over the last decade. Assy Wallace, "Brazil Moves Toward Energy Self-Sufficiency," *Soft Energy Notes,* April 1980; Eric Jeffs, "Hydro to be Main Power Source for Venezuela," *Energy International,* June 1980; Conrad Manly, "Mexico Turns to Water Power," *Journal of Commerce,* September 17, 1975; P. M. Belliappa, "Planning of Hydropower and Future Prospects of Hydropower Development in India," *Indian Journal of Power and River Valley Development,* July/August 1981.
65. World Energy Conference, *Survey of Energy Resources.* For a discussion of foreign exchange considerations in Chinese hydro development, see

U.S. Central Intelligence Agency, *Electric Power for China's Modernization.*

Chapter 9. Wind Power: A Turning Point

1. Wind patterns worldwide are discussed in Carl Aspliden, "Technical Note on Wind Energy" (Preliminary), World Meteorological Organization, Geneva, July 1981, and Nicholas P. Cheremisinoff, *Fundamentals of Wind Energy* (Ann Arbor, Mich.: Ann Arbor Science Publishers, 1978).
2. M. R. Gustavson, "Limits to Wind Power Utilization," *Science,* April 6, 1979.
3. Wind power's early history is discussed in Walter Minchinton, "Wind Power," *History Today,* March 1980, E. W. Golding, *The Generation of Electricity by Wind Power* (London: E. & F.N. Spon, Ltd., 1955), and Volta Torrey, *Wind-Catchers: American Windmills of Yesterday and Tomorrow* (Brattleboro, Vt.: Stephen Greene Press, 1976).
4. Ibid.
5. The six million wind pumps estimate is from Frank R. Eldridge, *Wind Machines* (New York: Van Nostrand Reinhold Co., 1980). The Fraenkel quote is from Peter L. Fraenkel, "The Use of Wind Power for Pumping Water," a contribution for *The British Wind Energy Association Position Paper on Wind Power,* 1980.
6. Cheremisinoff, *Fundamentals of Wind Energy;* Eldridge, *Wind Machines;* V. Daniel Hunt, *Windpower: A Handbook on Wind Energy Conversion Systems* (New York: Van Nostrand Reinhold Co., 1981).
7. The NASA Lewis Research Center, *Wind Energy Developments in the Twentieth Century* (Cleveland, Oh.: National Aeronautics and Space Administration, 1979).
8. Efficiency figures are from Hunt, *Windpower.* The five-year energy payback period is a conservative estimate; in most cases it will be even lower. See Lockheed California Company, *Wind Energy Mission Analysis* (Burbank, Calif.: October 1976), and Institute for Energy Analysis, "Net Energy Analysis of Five Energy Systems," Oak Ridge Associated Universities, Oak Ridge, Tenn., unpublished, September 1977.
9. Cheremisinoff, *Fundamentals of Wind Energy;* Hunt, *Windpower.*
10. A broad view of worldwide wind availability is included in Pacific Northwest Laboratory, "World-Wide Wind Energy Resource Distribution Estimates," a map prepared for the World Meteorological Organization, 1981.
11. The estimate of approximately one million wind pumps is widely used

(see, for example, Fraenkel, "The Use of Wind Power for Pumping Water") but is not based on an actual survey. The energy-capacity estimates are based on data from Kenneth Darrow, Volunteers in Asia, private communication, May 28, 1981.

12. A good overview of wind-pump technology and related issues is found in Steve Blake, "Wind Driven Water Pumps—Economics, Technology, Current Activities," Sunflower Power Company, Oskaloosa, Kansas, prepared for the World Bank, unpublished, December 1978.
13. Information on the wind-pump industry is from Alan Wyatt, Volunteers in Technical Assistance, and Peter Fraenkel, Intermediate Technology Development Group (ITDG), private communications, April 28 and May 13, 1981.
14. These points are discussed in Marshal F. Merriam, "Windmills for Less Developed Countries," *Technos,* April/June 1972, and in H. J. M. Beurskens, "Feasibility Study of Windmills for Water Supply in Mara Region, Tanzania," Steering Committee on Wind-Energy for Developing Countries, Amersfoort, Netherlands, March 1978. Zambia example is from Alan Wyatt, Volunteers in Technical Assistance, private communication, April 28, 1980.
15. Peter L. Fraenkel, "The Relative Economics of Windpumps Compared with Engine-Driven Pumps," ITDG, London, unpublished, 1981. Studies in India are described in "Report of the Technical Panel on Wind Energy," prepared for the United Nations Conference on New and Renewable Sources of Energy, Nairobi, Kenya, August 10–21, 1981.
16. Various simple wind pump designs are described in Blake, "Wind Driven Water Pumps," and Ken Darrow, "Locally Built Windmills as an Appropriate Technology for Irrigation of Small Holdings in Developing Countries," Volunteers in Asia, Stanford, Calif., unpublished, February 26, 1979.
17. Victor Englebert, "The Wizard of Las Gaviotas," *Quest,* May 1981. Information on the ITDG design from Peter Fraenkel, private communication, May 13, 1981.
18. See for instance Darrow, "Locally Built Windmills."
19. The most detailed study of the feasibility of using sails on modern commercial ships is Lloyd Bergeson, *Wind Propulsion for Ships of the American Merchant Marine* (Washington, D.C.: U.S. Maritime Agency, 1981).
20. The Phoenix and the Mini Lace are described in Christopher Pope, "Saving on Sail," *Renewable Energy News,* June 1982. The Japanese tanker is described in Wesley Marx, "Seafarers Rethink Traditional

Ways of Harnessing the Wind for Commerce," *Smithsonian,* December 1981.

21. Philippines and Sri Lanka efforts are described in "Fishers Take Forward Step Back to Sailing," *Christian Science Monitor,* April 19, 1982. Lloyd Bergeson is quoted in Pope, "Saving on Sail."
22. The estimate of 20,000 wind turbines is from Carl Aspliden, U.S. Department of Energy, private communication, June 4, 1981. Cost estimate of over 20¢ per kilowatt hour is from Theodore R. Kornreich and Daryl M. Tompkins, "An Analysis of the Economics of Current Small Wind Energy Systems," presented to the U.S. Department of Energy's Third Wind Energy Workshop, Washington, D.C. May 1978. A figure of 50¢ to $1 per kilowatt-hour is used in "Report of the Technical Panel on Wind Energy."
23. Small wind turbines are discussed in detail in Dermot McGuigan, *Harnessing the Wind for Home Energy* (Charlotte, Vt.: Garden Way Publishing, 1978), and in Jack Park and Dick Schwind, *Wind Power for Farms, Homes, and Small Industry* (Springfield, Va.: National Technical Information Service, 1978). Information on Denmark from B. Maribo Pedersen, "Windpower in Denmark," presented to the California Energy Commission Wind Energy Conference, Palm Springs, Calif., April 6–7, 1981 (referred to in following notes as Palm Springs Wind Energy Conference).
24. U.S. small wind systems on the market today are described in Rockwell International, "Commercially Available Small Wind Systems and Equipment," Golden, Colo., unpublished, March 31, 1981. The 2,400 wind machines figure is from Tom Grey of the American Wind Energy Association, private communication, September 21, 1982.
25. W. S. Bollmeier et al., *Small Wind Systems Technology Assessment: State of the Art and Near Term Goals* (Springfield, Va.: National Technical Information Service, 1980).
26. Ned Coffin is quoted in Frank Farwell, "New Energy: A Burgeoning Business in Windmills," *New York Times,* April 27, 1980. Potential for assembly line manufacturing is discussed in Wind Energy Program, "Commercialization Strategy Report for Small Wind Systems," U.S. Department of Energy, Washington, D.C., unpublished, undated, and Bollmeier et al., *Small Wind Systems Technology Assessment.*
27. Cost-of-electricity figures are at best approximate, and several manufacturers claim their machines can generate electricity for less than 15¢ per kilowatt-hour. The actual figures, however, are usually higher. See Bollmeier et al., *Small Wind Systems Technology Assessment,* and "Report of the Technical Panel on Wind Energy."

28. Eldridge, *Wind Machines;* Kurt H. Hohenemser, Andrew H. P. Swift, and David A. Peters, *A Definitive Generic Study for Sailwing Wind Energy Systems* (Springfield, Va.: National Technical Information Service, 1979); Irwin E. Vas, *A Review of the Current Status of the Wind Energy Innovative Systems Projects* (Springfield, Va.: National Technical Information Service, 1980).
29. D. E. Cromack, *Investigation of the Feasibility of Using Wind Power for Space Heating in Colder Climates* (Springfield, Va.: National Technical Information Service, 1979).
30. The U.S. Department of Energy wind power program is based on this premise, which is discussed in Lockheed California Company, *Wind Energy Mission Analysis,* and in Grant Miller, "Assessment of Large Scale Windmill Technology and Prospects for Commercial Application," working paper submitted to National Science Foundation, Washington, D.C., unpublished, September 8, 1980. Some wind power experts disagree with these claims for the economic superiority of large wind machines, noting that theoretical cost savings may not be realized because of the complex engineering that must go into large turbines.
31. The 1000 homes figure is based on the assumption that an average house uses 800 kilowatt-hours of electricity per month and that a 4-megawatt wind machine operates at a capacity factor of 30 percent. Assuming that a coal or nuclear plant operates at a capacity factor of 60 percent, it takes a 2,000-megawatt wind farm (500, 4-megawatt machines) to generate as much power as a large 1,000-megawatt coal or nuclear plant.
32. Large-scale turbine technology is discussed in NASA Lewis Research Center, "Wind Energy Developments," in Eldrige, *Wind Machines,* and in "Going with the Wind," *EPRI Journal,* March 1980.
33. The NASA program is discussed in NASA Lewis Research Center, "Wind Energy Developments," and in Miller, "Assessment of Large Scale Windmill Technology." See also Ray Vicker, "PG&E Hopes to be Buying Electricity from Biggest U.S. 'Wind Farm' by 1985," *Wall Street Journal*, November 11, 1982.
34. Bendix program is described by David Taylor, Bendix Corporation, private communication, March 23, 1981. Hamilton Standard program is described by Mr. Wolf, Hamilton Standard Company, private communication, March 23, 1981.
35. B. Maribo Pedersen, "The Danish Large Wind Turbine Program," presented to the Wind Energy Workshop on Large Wind Turbines, NASA Lewis Research Center, Cleveland, Ohio, April 1979; "Great Britain to Build 3 MW, 250 KW WECS," *Wind Energy Report,* January 1981.

36. National Research Council of Canada, "Canadian Wind Energy Research and Development," Ottawa, unpublished, undated; Paul N. Vosburgh and John E. Primus, "Wind Energy for Industrial Applications," *IE Conference Proceedings*, conference site unknown, Fall 1980.
37. The involvement of U.S. utility companies in wind-power projects is described in Elizabeth Baccelli and Karen Gordon, *Electric Utility Solar Energy Activities: 1981 Survey* (Palo Alto, Calif.: Electric Power Research Institute, 1982). British utility programs are described in R. H. Taylor and D. T. Swifthook, "Windpower Studies in the C.E.G.B.," presented to the U.S. Department of Energy's Fourth Biennial Conference and Workshop on Wind Energy, Washington, D.C., October 29–31, 1979. The Dutch program is described in "Dutch to Build 10 MW Windfarm," *World Solar Markets*, March 1982.
38. U.S. Windpower's program is discussed in Ellen Perley Frank, "Breaking the Energy Impasse," *New Roots*, (month unknown) 1979, and in "World's First SWECS Windfarm Built on Mountain Ridge in New Hampshire," *Wind Energy Report*, January 1981. The Hawaii project is described in Jane Wholey, "Hawaii: New Dynasty in Renewable Energy," *Solar Age*, May 1981. See also Christopher Pope, "Windfarming Gains Ground in U.S. Market," *Renewable Energy News*, February 1982.
39. For more information on California's wind farms see, California Energy Commission, *Wind Energy Program Progress Report* (Sacramento, Calif.: 1982) and "Development Fast and Furious in Four of California's Wind-Swept Mountain Passes," *Renewable Energy News*, May 1982. The 1982 California wind farm statistics and goals are from Kathleen Gray, California Energy Commission, private communication, September 27, 1982.
40. Cost estimates for Denmark are from Pedersen, "Windpower in Denmark." British estimate is from David Lindley of Taylor Woodrow Construction Ltd. at the Palm Springs Wind Energy Conference. U.S. estimate is from Robert Lowe, "Expected Electricity Costs for the U.S. Mod-2 Windmill," *Energy Policy*, December 1980, and Miller, "Assessment of Large Scale Windmill Technology."
41. Roderick Nash, "Problems in Paradise: Land Use and the American Dream," *Environment*, July/August 1979. Land use figures are authors' estimates based on wind resource surveys in California. See California Energy Commission, "Wind Energy Assessment of California," Sacramento, Calif., unpublished, March 1981, and Michael Dubey and Ugo Coty, *Impact of Large Wind Energy Systems in California* (Sacramento, Calif.: California Energy Commission, 1981).

42. Dubey and Coty, *Impact of Large Wind Energy Systems.* The wind turbines cannot be packed too densely onto a site since blade-caused wind turbulence causes them to interfere with one another. This leaves ample room for livestock grazing between the wind machines. One area where land use conflicts have arisen is the arid San Gorgonio Pass in California, one of the state's richest wind power sites. The U.S. Bureau of Land Management has recommended that wind farm development there be limited so that scenic values are maintained. See U.S. Department of the Interior, Bureau of Land Management, *Final Environmental Impact Statement on the San Gorgonio Pass Wind Energy Project* (Washington, D.C.: 1982). Another land use conflict is described in Lincoln Sheperdson, "Vermont Citizens Move to Protect National Forest from Wind Turbine," *Canadian Renewable* Energy News, December 1980. The 10 to 25 percent figure is the authors' estimate based on data described above.
43. Noise was a particular problem for one of the U.S. Department of Energy's early experimental wind machines erected at Boone, North Carolina, in 1979. Called a Mod-1, the machine has several flaws and has operated only intermittently. Recently developed large turbines such as the Mod-2 have not had this problem. Wind turbine regulations for Lincoln, Nebraska, and Boulder County, Colorado, are described in David Morris, *Self-Reliant Cities* (San Francisco: Sierra Club Books, 1982).
44. D. L. Sengupta and T. E. A. Senior, "Electromagnetic Interference to TV Reception Caused by Windmills," presented to the U.S. Department of Energy's Third Wind Energy Workshop, Washington, D.C., May 1978.
45. Favorable results of surveys carried out at experimental turbine locations described by Robert Noun, U.S. Solar Energy Research Institute, speech at Palm Springs Wind Energy Conference.
46. The reliability issue is discussed in "Wind and Utilities," *Wind Power Digest,* Fall 1979, and W. D. Marsh, *Requirements Assessments of Wind Power Plants in Electric Utility Systems* (Palo Alto, Calif.: Electric Power Research Institute, 1979). An energy analyst in Australia has calculated that for small electricity grids wind turbines have approximately the same reliability as a nuclear power plant, as reported in Mark Diesendorf, "Australian Wind Reliability," *Wind Power Digest,* Spring 1981.
47. A plan to integrate wind power and hydropower is described in Clifford Barrett, "Bureau of Reclamation's Wind/Hydroelectric Energy Project," presented to the U.S. Department of Energy's Fourth Biennial

Conference and Workshop on Wind Energy, Washington, D.C., October 29–31, 1979.

48. B. Maribo Pedersen, Technical University of Denmark, private communication, April 6, 1981; Richard Williams, "An Update on Activities at Rocky Flats," presented at the Fourth Biennial Conference and Workshop.
49. U.S. Government funding levels are from Tom Grey, American Wind Energy Association, private communication, September 21, 1982. Some of the commercial wind farm projects in the U.S. are premised on further government support to perfect the "Mod-2" turbine design and develop two more-advanced and less-expensive "Mod-5" turbines designed by Boeing and General Electric. As of late 1982 this program is going ahead on a 50-50 cost sharing basis between the government and the companies according to Tom Grey, but the pace has slowed since 1980. For an overview of U.S. Government support of wind power prior to the 1981 budget cuts, see U.S. Department of Energy, "Federal Wind Energy Program Fact Sheet," Washington, D.C., unpublished, June 25, 1980, and Kent Rissmiller and Larry M. Smukler, "An Analysis of the Legislative Initiatives of the 96th Congress to Accelerate the Development and Deployment of Wind Energy Conversion Systems," Energy Law Institute, Concord, N.H., unpublished, June 17, 1980.
50. California Energy Commission, "Wind Energy Assessment of California;" Nancy Phillips, "Wind Farms Blow U.S. Land Prices Sky High," *Canadian Renewable Energy News,* April 1981.
51. World Meteorological Organization activities are described by Carl Aspliden, U.S. Department of Energy, private communication, June 4, 1981. The World Meteorological Organization map is from Pacific Northwest Laboratory, "Worldwide Wind Energy Resource Distribution Estimates."
52. Information on availability of sufficient wind for wind pumps is from Alan Wyatt, Volunteers in Technical Assistance, and Peter Fraenkel, Intermediate Technology Development Group, private communications, April 28 and June 2, 1981.
53. "Danes Sail Ahead with Wind Power," *New Scientist,* July 2, 1981.
54. Arthur D. Little, Inc., *Near-Term High Potential Counties for SWECS, Final Report* (Springfield, Va.: National Technical Information Service, 1981).
55. The North Sea study was described by David Lindley of Taylor Woodrow Construction Ltd. at the Palm Springs Wind Energy Conference and in Judy Redfearn, "The Prospect for Ocean-Going Windmills," *Nature,* April 24, 1980.

56. Some of the best material on global wind availability is included in Pacific Northwest Laboratory, "World-Wide Wind Energy Resource Distribution Estimates." This is a broad view of wind availability, and more detailed material is available only on a limited, regional basis. The estimates of wind power's potential contribution are the authors', based on regional estimates made in Lockheed California Company, "Wind Energy Mission Analysis," in Matania Ginosar, "A Proposed Large-Scale Wind Energy Program in California," *Energy Sources,* Vol. 5, No. 2, 1980, in Dubey and Coty, *Impact of Large Wind Energy Systems in California,* and in Redfearn, "Prospect for Ocean-Going Windmills." The calculations on wind's energy potential are based on the assumption that wind machines operate on average at 30 percent of rated capacity compared to 45 percent for hydro dams in recent years. In 1979, a hydro capacity of 440.5 gigawatts provided 1.72 million gigawatt hours of electricity. Total world electricity generating capacity in 1979 was 1,914 gigawatts. These numbers are from United Nations, *1979 Yearbook of World Energy Statistics* (New York: 1981). The primary energy figure is derived by assuming that it takes 10.7 exajoules of primary energy (the equivalent of 366 million tons of coal) to generate a million gigawatt hours of electricity.

Chapter 10. Geothermal Energy: The Powering Inferno

1. For a discussion of the origin of the earth's heat, see *Encyclopaedia Britannica,* 15th ed., s.v., "Earth, Heat Flow In."
2. Estimate of current geothermal energy use is derived from Ronald DiPippo, "Geothermal Power Plants: Worldwide Survey of July 1981," presented to the Geothermal Resources Council Fifth Annual Meeting, Houston, Texas, October 25–29, 1981, "Report of the Technical Panel on Geothermal Energy," prepared for the United Nations Conference on New and Renewable Sources of Energy, Nairobi, Kenya, August 10–21, 1981, and Jón Gudmundsson, Department of Petroleum Engineering Stanford University, private communication, November 24, 1981. The estimates on number of houses that could be served assume an average heating requirement of 145 gigajoules per year and an average electricity requirement of 800 kilowatt hours per month. Geothermal power plants are assumed to operate on average at 70 percent of capacity. Total capacity in 1981 was approximately 2500 megawatts.
3. Average thermal gradient is from Donald White, "Characteristics of Geothermal Resources," in Paul Kruger and Carel Otte, eds., *Geothermal Energy* (Stanford, Calif.: Stanford University Press, 1973). Maxi-

mum temperature gradient is from Vasel Roberts, Electric Power Research Institute, private communication, April 26, 1982.

4. The world's geothermal zones are described in Erika Laszlo, "Geothermal Energy: An Old Ally," *Ambio,* Vol. 10, No. 5, 1981.
5. For further discussion of hydrothermal systems, see White, "Characteristics of Geothermal Resources."
6. For further discussion of geopressured systems, see U.S. Congress, General Accounting Office, *Geothermal Energy: Obstacles and Uncertainties Impede Its Widespread Use* (Washington, D.C.: January 18, 1980).
7. Estimating the total amount of useful energy contained in the earth is difficult, but experts agree that the ultimate potential is staggering. Vasel Roberts of the Electric Power Research Institute estimates that the portion that is potentially useable is roughly 25,000 exajoules (EJ) for power generation and 3 million EJ for direct use. About 20 percent of this potential is believed to be exploitable using current technology, or 5,000 EJ for electricity generation and 600,000 EJ for direct use. Together, this is enough energy to provide for our current level of energy use for almost 2000 years. See Vasel Roberts, "Geothermal Energy," in *Advances in Energy Systems and Technology,* Vol. 1, 1978.
8. William W. Eaton, *Geothermal Energy* (Washington, D.C.: U.S. Energy Research and Development Administration, 1975); "Report of the Technical Panel on Geothermal Energy."
9. Description of baths in Japan is from John W. Lund, "Direct Utilization —the International Scene," Geo-Heat Utilization Center, Oregon Institute of Technology, Klamath Falls, Oregon, unpublished, undated. Estimates of energy saved by Japanese baths and geothermal applications in Thailand and Mexico is from Jón Steinar Gudmundsson and Gudmondur Pálmason, "World Survey of Low-Temperature Geothermal Energy Utilization," prepared for the United Nations Conference on New and Renewable Sources of Energy, Nairobi, Kenya, August 10–21, 1981. Geothermal use in Guatemala was noted by Kathleen Courrier, Center for Renewable Resources, private communication, June 12, 1982.
10. Information on Idaho aquaculture facility is from "Energy Researchers Are Getting Into Hot Water," *New York Times,* December 14, 1980. Description of greenhouse applications in Lund, "Direct Utilization," and Gudmundsson and Pálmason, *World Survey.*
11. Industrial uses in Iceland and New Zealand are described in Paul J. Lineau, "Geothermal Resource Utilization," presented to American Association for the Advancement of Science Annual Meeting, Washington, D.C., January 3–8, 1980. Onion dehydration project is described

in Joe Glorioso, "Geothermal Moves Off the Back Burner: Part II," *Energy Management*, October/November, 1980.

12. Paul J. Leinau and John W. Lund, "Utilization and Economics of Geothermal Space Heating in Klamath Falls, Oregon," Geo-heat Utilization Center, Oregon Institute of Technology, Klamath Falls, Oregon, unpublished, undated. The life-cycle cost analysis is from John W. Lund, "Geothermal Energy Utilization for the Homeowner," Geo-Heat Utilization Center, Klamath Falls, Oregon, unpublished, December, 1978.
13. "Basic Statistics of Iceland," Ministry of Foreign Affairs, Iceland, April 1981; Gudmundsson and Pálmason, *World Survey*. The comparison with oil costs is from "Hitaveita Reykjavíkur," a government pamphlet describing Iceland's district heat programs, undated.
14. All existing geothermal district heating systems have proved economical according to Paul J. Lineau, "Space Conditioning with Geothermal Energy," Geo-Heat Utilization Center, Klamath Falls, Oregon, unpublished, undated.
15. Estimate of number of groundwater heat pumps in U.S. is from Robert Hoe, Vanguard Energy systems, private communication, May 12, 1982. Jay Lehr, president of the National Well Water Association, optimistically predicts that by the end of this decade all new free-standing houses in the U.S. built on quarter-acre or larger lots will be heated by groundwater heat pumps in Gene Bylinsky, "Water to Burn," *Fortune*, October 20, 1980. For more information on heat pumps, see John Sumner, *An Introduction to Heat Pumps* (Dorset, England: Prism Press, 1976), Dana M. Armitage et al., "Groundwater Heat Pumps: An Examination of Hydrogeologic, Environmental, Legal, and Economic Factors Affecting Their Use," National Well Water Association, Worthington, Ohio, unpublished, November 12, 1980, and Paul Lineau, "Heat Pumps and Geothermal," *Geo-Heat Utilization Center Quarterly Bulletin*, March 1980.
16. Geothermal power figures are from DiPippo, "Geothermal Power Plants." Although geothermal projects can be as large as 1,000 megawatts, the individual power plants are kept small to avoid the costly and wasteful long distance transportation of hot water and steam.
17. For a thorough discussion of the different electricity-generating technologies and operating experience to date, see Ronald DiPippo, *Geothermal Energy as a Source of Electricity* (Washington, D.C.: U.S. Department of Energy, 1980). Information about the Geysers is from the Pacific Gas & Electric Company, "The Geysers Power Plant Development," unpublished, March 26, 1982. The figures in the table are

conservative since they simply show announced plans for various countries as collected by DiPippo and the UN Technical Panel. The actual total could be much higher as shown by the fact that the Philippines now plans to have 1209 MW by 1989, compared with the earlier estimate of 1225 MW by 2000.

18. The Wairakei plant is described in DiPippo, *Geothermal Energy as a Source of Electricity.*
19. Fred L. Hartley, "The Future of Geothermal Energy as an Alternative Energy Source," presented to the Second ASCOPE Conference and Exhibition, Manila, Philippines, October 10, 1981.
20. The prospects for double flash plants were described by David Anderson, Geothermal Resources Council, private communication, June 18, 1982.
21. The binary plant design is discussed in DiPippo, *Geothermal Energy as a Source of Electricity,* and U.S. Congress, General Accounting Office, *Elimination of Federal Funds for the Heber Project Will Impede Full Development and Use of Hydrothermal Resources* (Washington, D.C.: June 25, 1981).
22. Dual use of geothermal water in Japan is described in Wilson Clark and Jake Page, *Energy, Vulnerability, and War* (New York: W. W. Norton and Co., 1981) and by Jón Gudmundsson, private communication, April 1982.
23. The quality of Reykjavik geothermal water is discussed in "Hitaveita Reykjavíkur." Project closings due to corrosion are described by James Brezee, U.S. Department of Energy, private communication, December 18, 1981. Hydrogen sulfide emissions and control information are described in DiPippo, *Geothermal Energy as a Source of Electricity* and J. Laszlo, "Application of the Stretford Process for H_2S Abatement at the Geysers," Pacific Gas & Electric Co., San Francisco, unpublished, 1976.
24. DiPippo, *Geothermal Energy as a Source of Electricity.*
25. Rate of subsidence at Wairakei is discussed in "Geothermal Energy and Our Environment," U.S. Department of Energy, Washington, D.C., unpublished, undated. For general discussion of the environmental impacts of geothermal development, see M. J. Pasqualetti, "Geothermal Energy and the Environment: The Global Experience," *Energy,* Vol. 5, No. 2, 1980, and A. J. Ellis, "Geothermal Energy Utilization and the Environment," *Mazingira,* Vol. 5, No. 1.
26. "The Hydrothermal Push Cools," *Business Week,* September 14, 1981.
27. Early test results are described in "SRI: Future Dim for Gulf Geothermal Resources," *Oil and Gas Journal,* March 16, 1981. Disappointment in the oil industry was noted by David Anderson, Geothermal Resources

Council, private communication, June 18, 1982. For more information on geopressured resources, see D. G. Bebout and D. R. Gutierrey, "Geopressured Geothermal Resource in Texas and Louisiana, Geological Constraints," presented at the Geothermal Resources Council Fifth Annual Meeting, Houston, Texas, October 25–29, 1981, and O. C. Karkalits, "Economics of Energy from Geopressured Geothermal Reservoirs," also presented at the Geothermal Resources Council Fifth Annual Meeting.

28. The Cornwall project is sponsored by the British Government and the European Economic Commission, while the Fenton Hill Project is being undertaken by the governments of Japan, the United States and West Germany. For more information on the Cornwall project, see Anthony S. Batchelor, "The Status of Hot Dry Rock in the United Kingdom," presented to the Third Annual Los Alamos National Laboratory Hot Dry Rock Geothermal Information Conference, Santa Fe, New Mexico, October 28–29, 1980, and "Energy from Hot Rocks," *Energy Management,* January 1981. For more information on the Fenton Hill project, see G. A. Zyvoloski et al., "Hot Dry Rock Geothermal Energy," *American Scientist,* July/August 1981, and E. L. Kaufman and C. L. B. Siciliano, eds., *Environmental Analysis of the Fenton Hill Hot Dry Rock Geothermal Test Site* (Washington, D.C.: U.S. Department of Energy, May 1979). For further discussion of hot dry rock, see Ronald G. Cummings et al., "Mining Earth's Heat: Hot Dry Rock Geothermal Energy," *Technology Review,* February 1979, and M. C. Smith, "The Future of Hot Dry Rock Geothermal Energy Systems," presented at the Pressure Vessels and Piping Conference, San Francisco, California, June 25–29, 1979.

29. Temperatures of 1000°C to 1200°C have been encountered in volcanic drilling tests in Hawaii according to Jón Gudmundsson, Stanford University, private communication, April 25, 1982. Development plans for the Avachinski Volcano were reported in "5000 MW Geothermal Power Plant?," *Energy in Countries with Planned Economies,* December 14, 1977. Information on use of magma on island of Heimaey is from Gudmundsson, private communication, April 26, 1982. Outlook for magmatic energy is from "Report of the Technical Panel on Geothermal Energy."

30. For an example of a national resource assessment, see L. J. P. Muffler, ed., *Assessment of Geothermal Resources of the United States—1978* (Washington, D.C.: U.S. Geological Survey, 1979). Even in the U.S., which has conducted a relatively thorough resource assessment, an estimated 80 percent of the total resource has not been located according

to the Interagency Geothermal Coordinating Council, *Fifth Annual Report* (Washington, D.C.: July 1981).

31. Legal status of geothermal energy is discussed in "Report of the Technical Panel on Geothermal Energy," and Kenneth A. Wonstolen, "Geothermal Energy: Basic Legal Parameters," presented at the Geothermal Resources Council Fourth Annual Meeting, Salt Lake City, Utah, September 8–11, 1980.
32. The Yellowstone controversy is discussed in Joan Nice, "Energy: Geothermal Lease Plans Threaten Yellowstone's Geysers," *Audubon,* May 1982. Conflicts in Japan are discussed in "What About Geothermal Power," *PRIEE News,* May–June–July, 1982.
33. Union Oil's role in the Philippines program is described in "Pacific's Ring of Fire Spews Geothermal Electric Power," *Christian Science Monitor,* September 18, 1980. For a discussion of bi- and multi-lateral assistance to developing countries, see James B. Koenig, James R. McNitt and Murray C. Garner, "Geothermal Power Developments in the Third World," presented to the Geothermal Resources Council Fifth Annual Meeting, Houston, Texas, October 25–29, 1981.
34. A description of the French program is found in Ministère de l'Industrie du Commerce, et de l''Artisanat, *La Geothermie en France* (Paris: 1978). Iceland's program was described by Jón Gudmundsson, Stanford University, private communication, April 25, 1982. The U.S. program is described in "Federal Cost-Sharing for Exploration of Hydrothermal Reservoirs for Direct Applications," Division of Geothermal Energy, U.S. Department of Energy, Washington, D.C., unpublished, 1980.
35. According to a study by the Electric Power Research Institute, described in Bob Williams, "Action in Geothermal Energy Picking Up Speed in U.S.," *Oil and Gas Journal,* May 3, 1982, loss of federal support could delay indefinitely development of about half of the easily exploitable U.S. geothermal resources. U.S. tax incentives are described in Richard W. Bliss, "Federal Legislation Affecting Geothermal Development," presented to the Conference on Geothermal Energy: The Institutional Maze and Its Changing Structure, Newport Beach, Calif., December 1–2, 1981. The question of the effectiveness and equitability of tax incentives is discussed in Charles V. Higbee, "Pricing Direct-Use Geothermal Energy," presented at the Geothermal Resources Council Fourth Annual Meeting, Salt Lake City, Utah, September 8–11, 1980.
36. Financial and institutional issues surrounding district heat projects are discussed in Diana King, "District Heating: Legal, Institutional, and Public Relations Aspects," presented to the Conference on Geothermal Energy: The Institutional Maze and Its Changing Structure.

37. U.S. utilities' outlook on geothermal projects is described in John T. Nimmons, "Current Public Utility Considerations for Geothermal Power Producers and Direct Heat Distributors," presented at the Geothermal Resources Council Fifth Annual Meeting, Houston, Texas, October 25–29, 1981. Several utilities in the western U.S. have gone to great length to avoid geothermal involvement according to Randy Stephans, U.S. Department of Energy, private communication, December 29, 1981. The point regarding geothermal power's reliability was made by R. P. Wischow, Geysers Project Manager for the Pacific Gas & Electric Company, private communication, August 16, 1982.
38. Rate of growth of geothermal generating capacity since the mid-seventies is derived from Roberts, "Geothermal Energy," and DiPippo, "Geothermal Power Plants." Projections of future use are based on "Report of the Technical Panel on Geothermal Energy," and Roberts, "Geothermal Energy."
39. Current direct use figure is from Gudmundsson and Pálmason, *World Survey,* and Jón Gudmundsson, Stanford University, private communication, November 24, 1981. Direct use projections are from Gudmundsson, private communication, May 24, 1982, and Roberts, "Geothermal Energy." Iceland's residential heating goal is from Gudmundsson and Pálmason, *World Survey.* France's resource potential and utilization goals are from "France and Geothermal Power—A Source with 'Enormous' Potential," *Christian Science Monitor,* October 1, 1980. Future estimates are the authors'.
40. Potentials in Canada, China and the USSR are from Jón Gudmundsson, Stanford University, private communication, November 24, 1981. Number of 'hot spots' in China from "China's Growing Geothermal Use," *Energy in Countries with Planned Economies,* November 2, 1979. Direct use projects in the U.S. are described in Cassel et al., "National Forecast for Geothermal Resource Exploration and Development," prepared for the U.S. Department of Energy, unpublished, March 31, 1982.
41. National plans are from DiPippo, "Geothermal Power Plants." Mexico's plans are described in "Report of the Technical Panel on Geothermal Energy." The El Salvador figure is from DiPippo, *Geothermal Energy as a Source of Electricity.*
42. The Philippines program is discussed in Bob Williams, "Many Countries Tapping Geothermal," *Oil & Gas Journal,* May 10, 1982; see also Philippines Ministry of Energy, *Ten-Year Energy Program: 1980–1989* (Manila: 1980).

Chapter 11. Working Together: Renewable Energy's Potential

1. Authors' estimates are based on data from Organisation for Economic Co-operation and Development (OECD), *Energy Balances of OECD Countries* (Paris: 1978), Joy Dunkerley, ed., *International Comparisons of Energy Consumption* (Washington, D.C.: Resources for the Future, 1978), U.S. Department of Energy, *Annual Report to Congress, 1980* (Washington, D.C.: 1981), and U.S. Congress, Office of Technology Assessment, *Residential Energy Conservation* (Washington, D.C.: 1979).
2. The potential for energy conservation in existing buildings is explored in Robert H. Socolow, ed., *Saving Energy in the Home: Princeton's Experiments at Twin Rivers* (Cambridge, Mass.: Ballinger, 1978). The potential for energy conservation in new houses is explored in William A. Shurcliff, *Superinsulated Houses and Double-Envelope Houses* (Cambridge, Mass.: privately published, 1980). Recent energy trends in buildings are from Eric Hirst and Bruce Hannon, "Effects of Energy Conservation in Residential and Commercial Buildings," *Science*, August 17, 1979, "Energy Conservation," *OECD Observer*, November 1979, Joy Dunkerley, *Trends in Energy Use in Industrial Societies* (Washington, D.C.: Resources for the Future, 1980) and International Energy Agency, *Energy Policies and Programmes of IEA Countries: 1981 Review* (Paris: Organisation for Economic Co-operation and Development, 1982).
3. Passive solar architecture is described in detail in Chapter 3. Throughout this chapter we discuss renewable energy technologies that are explored more fully in earlier chapters and rely on the reader to search out the details. Estimates of future use that are not referenced are the authors' and are based on material in earlier chapters.
4. Davis' programs and achievements are described in David Morris, *Self-Reliant Cities: Energy and the Transformation of Urban America* (San Francisco: Sierra Club Books, 1982).
5. Energy use in Japan's industry is discussed in International Energy Agency, *Energy Policies and Programmes.* Energy use in Soviet industry is discussed in U.S. Congress, Office of Technology Assessment, *Technology and Soviet Energy Availability* (Washington, D.C.: 1981). Energy use in the industries of rapidly industrializing Third World countries is discussed in the World Bank, *Energy in the Developing Countries* (Washington, D.C.: 1980).
6. Spending of energy-intensive U.S. industries is a U.S. Department of

Energy estimate cited in "Energy Conservation: Spawning a Billion-Dollar Business," *Business Week*, April 6, 1981. Energy use trends in the industrial countries are found in International Energy Agency, *Energy Policies and Programmes.*

7. Solar Energy Research Institute, *A New Prosperity: Building A Renewable Energy Future* (Andover, Mass.: Brick House, 1981).
8. Food and Agriculture Organization of the United Nations, *Energy for World Agriculture* (Rome: United Nations, 1979). In developing countries much of the energy used in agriculture is uncounted in these figures. If biomass energy and the work expended by people and draft animals were included, agriculture would claim a substantially larger share of total energy use.
9. The potential uses of renewable energy in U.S. agriculture are discussed in Kevin Finneran, "Solar on the Farm," *Sun Times*, March/April 1982.
10. Walter G. Heid, Jr., U.S. Department of Agriculture, "Using Solar Energy to Dry Grain—An Economic Evaluation," paper presented at the High Plains Energy Forum, Dodge City, Kansas, October 20, 1979; Walter G. Heid, Jr. and Warren K. Trotter, *Progress of Solar Technology and Potential Farm Uses* (Washington, D.C.: U.S. Department of Agriculture, 1982).
11. Robert V. Enochian, *Solar- and Wind-Powered Irrigation Systems* (Washington, D.C.: U.S. Department of Agriculture, 1982).
12. The World Bank, *Renewable Energy Resources in the Developing Countries* (Washington, D.C.: 1980).
13. The results of the Conference are described in United Nations Conference on New and Renewable Sources of Energy, "Programme of Action," Nairobi, Kenya, August 21, 1981. See also Charles Drucker, "UNERG: Unfortunately Unproductive," *Soft Energy Notes*, October/November 1981, and Margaret R. Biswas, "The United Nations Conference on New and Renewable Sources of Energy: A Review," *Mazingira*, Vol. 5, No. 3, 1981.
14. Erik Eckholm, *Planting for the Future: Forestry for Human Needs*, Worldwatch Paper 26 (Washington, D.C.: Worldwatch Institute, February 1979).
15. John Ashworth, "Renewable Energy for the World's Poor," *Technology Review*, November 1979.
16. Although diesel engines are widely used in rural areas, there is a dearth of surveys indicating their numbers and reliability. Anecdotal evidence provided by individuals who have traveled widely in developing countries indicates that the machines are out of order more often than not. This issue deserves study since if initial indications are right, diesel engines

should be introduced more carefully and in many cases be replaced by renewable energy technologies.

17. Potential use of these technologies in the rural Third World is discussed in John H. Ashworth and Jean W. Neuendorffer, *Matching Renewable Energy Systems to Village-Level Energy Needs* (Golden, Colo.: U.S. Solar Energy Research Institute, 1980), Gerald Foley, "The Future of Renewable Energy in Developing Countries," *Ambio*, Vol. 10, No. 5, 1981, and World Bank, "Working Paper on Research and Technological Capacity for the Use of Renewable Energy Resources in Developing Countries," unpublished, October 30, 1980.
18. This figure is based on the world's using about 55 million barrels of oil per day of which around 25 percent goes to ground and air transportation. There are 159 liters in a barrel. For more detailed breakdowns, see U.S. Department of Energy, *1980 International Energy Annual* (Washington, D.C.: 1981).
19. New cars sold in the United States averaged 14.2 miles per gallon in 1974 and 22.4 miles per gallon in 1980 according to Eric Hirst et al., *Energy Use from 1973 to 1980: The Role of Improved Energy Efficiency* (Oak Ridge, Tenn: Oak Ridge National Laboratory, December 1981). The United States uses approximately 100 billion gallons of gasoline each year or 11.25 million gallons each hour at an average price of $1.25 per gallon. These figures can be found in U.S. Department of Energy, *Annual Report to Congress 1980*.
20. Synthetic fuels are discussed in more detail in Chapter 2.
21. The prospects for electricity storage systems and electric cars are examined in U.S. National Research Council, "Energy Storage for Solar Applications," unpublished, 1980, U.S. Congress General Accounting Office, *Electric Vehicles: Limited Range and High Costs Hamper Commercialization* (Washington, D.C.: 1982) and David Bloor, "Plastics that Conduct Electricity," *New Scientist*, March 4, 1982.
22. The prospects for hydrogen fuel are discussed in Peter Hoffman, *The Forever Fuel: The Story of Hydrogen* (Boulder, Colo.: Westview Press, 1981); Rif S. El-Mallakh, "Fuel for Thought: The Hydrogen-Powered Automobile," *Environment*, April 1981. New processes for separating hydrogen using sunlight directly are being developed by several teams of scientists; see for example "LBL Develops Single, Inexpensive Method for Dissociation of Water with Sunlight," *Solar Energy Intelligence Report*," September 27, 1982, and "Texas A & M Hails Cheap Hydrogen," Energy Daily, October 12, 1982.
23. Quote from Christopher Pope, "Slow Down in the Fast Lane: Transportation in the 80's," *New Roots*, Winter 1982.

24. Authors' estimates based on Joy Dunkerley, Trends in *Energy Use in Industrial Societies,* and United Nations, *1979 Yearbook of World Energy Statistics* (New York: 1981).
25. Ellish L. Armstrong, *History of Public Works in the United States, 1776–1976* (Chicago: American Public Works Association, 1976); Howard Brown and Tom Stomolo, *Decentralizing Electrical Production* (New Haven, Conn.: Yale University Press, 1981).
26. A good discussion of the difficulties utilities face and options for alleviating them is found in John Bryson, Chairman, California Public Utilities Commission, "The Future of Electrical Utilities," unpublished, 1981 and James A. Walker, "It Takes Real Money to Run Utilities: Costs of Future Power Options," California Energy Commission, unpublished, September 14, 1982.
27. U.S. utilities' research into renewable energy technologies is described in Elizabeth Baccelli and Karen Gordon, *Electric Utility Solar Energy Activities: 1981 Survey* (Palo Alto, Calif.: Electric Power Research Institute, 1982). California utilities' growing commitment to renewable energy sources is described in Phillip A. Greenberg, "Conservation and Renewable Resources in California's Energy Future," California Governor's Office, unpublished, February 1982 and California Energy Commission, "Exploring New Energy Choices for California: The 1982/1983 Report to the Legislature," unpublished, March 1982.
28. A program for combined development of wind power and hydropower is described in Clifford Barrett, "Bureau of Reclamation's Wind/Hydroelectric Energy Project," presented to the Fourth Biennial Conference and Workshop on Wind Energy, Washington, D.C., October 29–31, 1979.
29. John W. Shupe, "Energy Self-Sufficiency for Hawaii," *Science,* June 11, 1982; Philippines Ministry of Energy, *Ten-Year Energy Program, 1980–1989* (Manila: 1980); Bonneville Power Administration, "A Conservation Manifesto," unpublished, January 26, 1982. Baccelli and Gordon, *Electric Utility Solar Energy Activities.* European wind power development is described in Chapter 9.
30. It is not unusual for electricity costs to vary by a factor of five or ten for individual power plants within the same utility grid. Renewable energy technologies will make their first contribution as substitutes for the most expensive sources and gradually phase out less expensive plants as the new technologies mature.
31. This point is emphasized in World Bank, *Energy in the Developing Countries* (Washington, D.C.: 1980).
32. This estimate and the figures in the table are based on the assessments

of the individual renewable energy sources in earlier chapters. Generally the potential of each energy source is discussed at the end of each chapter.

33. This slow growth in energy use is likely to result from a combination of price-induced conservation and moderate economic growth. In most industrial countries overall energy use will likely grow less than 1 percent each year while in developing countries, 2 to 4 percent per year is the likely growth rate.

Chapter 12. Institutions for the Transition

1. Between 1900 and 1980 the U.S. Government spent $1 billion on all renewable energy sources other than hydropower, $15 billion on hydropower, and $240 billion on fossil and nuclear energy according to B.W. Cone et al., *An Analysis of Federal Incentives Used to Stimulate Energy Production,* Battelle Pacific Northwest Laboratory report to U.S. Department of Energy, February 1980. This lengthy report is summarized in B.W. Cone and R.H. Bezdek, "Federal Incentives for Energy Developments," *Energy,* Vol. 5, No. 5, May 1980. For a description of attitudes toward solar energy in the heyday of atomic energy enthusiasm, see Lamont C. Hempel, "The Politics of Sunshine," Ph.D. Diss., Claremont College, 1982.
2. International Energy Agency, *Energy Research, Development and Demonstration in the IEA Countries, 1981 Review of National Programmes* (Paris: Organisation for Economic Co-operation and Development/ International Energy Agency, 1982).
3. Ibid.
4. For a discussion of budget cutbacks in the U.S., see Jim Harding, "Reagan Cuts Solar," *Soft Energy Notes,* April/May 1981. Budget cuts in the U.K. are described in "Recession Bites into U.K. Solar," *World Solar Markets,* November 1981.
5. For a discussion of the management issues common to all R&D enterprises, see U.S. Congress, Office of Technology Assessment, *Government Involvement in the Innovation Process* (Washington, D.C.: 1979), and J. Herbert Holloman et al., "Government and the Innovation Process," *Technology Review,* May 1979. For a provocative defense of expanded, balanced energy R&D programs, see Don Kash et al., *Our Energy Future: The Role of Research, Development, and Demonstration in Reaching a National Consensus on Energy Supply* (Norman, Okla.: University of Oklahoma Press, 1976).
6. New Energy Development Organization and Japanese External Trade

Organization, *Japanese New Energy Technologies* (Tokyo: 1981). European solar programs are discussed in Sherri Barron, "Almeria," *Canadian Renewable Energy News,* November 1981, and French Solar Energy Authority (Commissariat a l'Energie Solaire), "France and the Development of New and Renewable Energies," prepared for the United Nations Conference on New and Renewable Sources of Energy, Nairobi, Kenya, August 10–21, 1981.

7. For a description of political meddling and vacillation in the U.S. Government's solar energy program, see William Rice, "Cash Flow in Congress," *Solar Age,* October 1979, and U.S. Congress, General Accounting Office, "Loss of Experienced Staff Affects Conservation and Renewable Energy Programs," Washington, D.C., July 1982.
8. Allen L. Hammond and William D. Metz, "Solar Energy Research: Making Solar After the Nuclear Model?," *Science,* July 15, 1977; W. D. Metz, "Solar Thermal Electricity: Power Tower Dominates Research," *Science,* July 22, 1977; Ray Reece, *The Sun Betrayed: A Report on the Corporate Seizure of U.S. Solar Energy Development* (Boston, Mass.: South End Press, 1979).
9. For a discussion of research priorities in biomass, see Roger Revelle, "Energy Dilemmas in Asia: The Needs for Research and Development," *Science,* July 4, 1980, R. A. Yates, *Biomass Production Technology: Current Status and Research Needs* (Booker Agricultural International, Ltd., July 1980, and Committee on Foreign Affairs, U.S. House of Representatives, *Background Papers for Innovative Biological Technologies for the Lesser Developed Countries,* an Office of Technology Assessment Workshop, November 24–25, 1980 (Washington, D.C.: U.S. Government Printing Office, 1981).
10. For a discussion of the small wind turbine test center at Rocky Flats, see Richard Williams, "An Update on Activities at Rocky Flats," presented to the Fourth Biennial Conference and Workshop on Wind Energy, Washington, D.C., October 29–31, 1979. Denmark's small wind turbine test center is described in B. Maribo Pederson, Technical University of Denmark, private communication, April 6, 1981.
11. U.S. Congress, Office of Technology Assessment, *The Role of Demonstrations in Federal R&D Policy* (Washington, D.C.: July 1978); Franklin A. Lindsay, "Financing High Cost, High-Risk Energy Development," *Harvard Business Review,* November/December 1978.
12. R. S. Claassen, "Materials for Advanced Energy Technologies," *Science,* February 20, 1976; "GAO Finds Supply of Nine Minerals 'Critical' to Development of Alternative Energy," *Energy & Minerals Resources,*

September 18, 1981; W. B. Hilliq, "New Materials and Composites," *Science,* February 20, 1976; Harry Z. Tabor "Materials Technology In The Harnessing of Solar Energy," *Sunworld*, April 1982.

13. For a discussion of Third World research priorities, see World Bank, *Mobilizing Renewable Energy Technology in Developing Countries: Strengthening Local Capabilities and Research* (Washington, D.C.: July 1981).
14. For a discussion of resource assessment, see National Academy of Sciences, *Proceedings: International Workshop on Energy Survey Methodologies for Developing Countries,* January 21–25, 1980 (Washington, D.C.: National Academy Press, 1980).
15. Ivan Illich, "Vernacular Values," *CoEvolution Quarterly,* Summer 1980; Laura Nades et al., "Belief, Behavior and Technologies as Driving Forces in Transitional Stages—The People Problem in Dispersed Energy Futures," in *Centralized vs Decentralized Energy Systems: Diverging or Parallel Roads?,* a report prepared for the Subcommittee on Energy and Power of the Committee on Interstate and Foreign Commerce, U.S. House of Representatives (Washington, D.C.: Congressional Research Service, May 1979).
16. "Field Test Finds Solar Panels Fail Too Fast," *The Energy Daily,* August 12, 1981; "High Winds Topple, Smash Mirrors at France's Prized Solar Station," *The Energy Daily,* January 21, 1982.
17. U.S. Congress, General Accounting Office, *Federal Demonstrations of Solar Heating and Cooling on Commercial Buildings Have Not Been Very Effective* (Washington, D.C.: April 15, 1980).
18. For discussion of the Sanman Gorge Dam and solar collectors in the Sahel, see Chapters 8 and 4.
19. John Ashworth, "Technology Diffusion Through Foreign Assistance: Making Renewable Energy Sources Available to the World's Poor, *Policy Sciences,* Summer 1980; P. Greenwood and C.W. Perrett, "Generating the Links Between Engineers and Rural Villages, Papua New Guinea," *Appropriate Technology,* March 1982.
20. For a discussion of the role of "do-it-yourselfers" and tinkerers, see Harry C. Boyte, *The Backyard Revolution: Understanding the New Citizens' Movement* (Philadelphia: Temple University Press, 1980), David Brand, "Some Small Innovators Heat Homes by Sun, Light Them by Wind," *Wall Street Journal,* March 18, 1975, and Stephen Salter, "The Perils of Being Simple," *New Scientist,* February 25, 1982.
21. For more details on the China/India biogas comparison, see Chapter 7.
22. Caroline E. Mayer, "CPSC: Wood-Stove Fires Double," *Washington*

Post, December 18, 1981; Christopher Pope, "Specialization, Standards and Speed, Hallmarks of Maturing Solar Installers Industry," *Renewable Energy News,* May 1982.

23. William Ramsay and Elizabeth Shue, "Infrastructure Problems for Rural New and Renewable Energy Systems," *Journal of Energy and Development,* Spring 1981.
24. The difficulties in regulating dispersed sources of pollution are described in Steve Plotkin, "Biomass Energy and the Environment," *Environment,* June 1980. For a discussion of the problems with operating catalytic combusters, see Elissa Krzeminski, "The Catalytic Combuster," *New Roots,* January 1982.
25. U.S. Congress, General Accounting Office, "New England Can Reduce Its Oil Dependence Through Conservation and Renewable Resource Development" (Washington, D.C.: June 1981), summarized in Rushworth M. Kidder, "New England Shows U.S. How to Cut Back on Use of Energy," *Christian Science Monitor,* June 24, 1981.
26. For an overview of North American oil pricing issues, see John Kraft, "Crude Oil Price Controls: Their Purpose and Impact," *Denver Journal of International Law and Policy,* Winter 1979, U.S. Congressional Budget Office, *The Decontrol of Domestic Oil Prices: An Overview* (Washington, D.C.: Congressional Budget Office, May 1979) and Robert Stobaugh and Daniel Yergin, "Decontrol the Price of Oil—Now," *New York Times,* January 18, 1981.
27. U.S. Congress, Office of Technology Assessment, *Technology and Soviet Energy Availability* (Washington, D.C.: 1981); Marshall I. Goldman, *The Enigma of Soviet Petroleum: Half-Full or Half-Empty?* (Boston: George Allen & Unwin, 1980); Marshall I. Goldman, "Energy Policy in the Soviet Union and China," in Hans H. Landsberg, ed., *Selected Studies on Energy, Background Papers for Energy: The Next Twenty Years* (Cambridge, Mass.: Ballinger, 1980).
28. Dennis W. Bakke, "Energy in Non-OPEC Developing Countries," in Hans H. Landsberg, ed., *Selected Studies on Energy: The Next Twenty Years* (Cambridge, Mass.: Ballinger, 1980); Douglas Martin, "The Very Mixed Blessings of Pure Liquid Gold," *New York Times,* May 23, 1982; Douglas Martin, "Saudi Craze for Car is Both a Blessing and a Blight," *New York Times,* January 22, 1982; Jonathan Spivak, "Nigeria, Hit by Plunge in Petroleum Exports, Faces Painful Changes," *Wall Street Journal,* August 2, 1982; Dial Torgerson, "Mexico's Gas-Guzzling Cars Ambushed by Doubled Price at the Pump," *Los Angeles Times,* January 3, 1982.
29. Amulya Kumar N. Reddy, "Alternative Energy Policies for Developing

Countries: A Case Study of India," in Robert A. Bohm et al., eds., *World Energy Production and Productivity, Proceedings of the International Energy Symposium I,* (Cambridge, Mass.: Ballinger, 1981).

30. Hans H. Lansberg and Joseph M. Dukert, *High Energy Costs: Uneven, Unfair, Unavoidable?* (Baltimore, Md.: The Johns Hopkins University Press, 1981); Elizabeth Cecelski et al., *Household Energy and the Poor in the Third World* (Washington, D.C.: Resources for the Future, 1979); Douglas R. Sease, "High Gasoline Prices are Hurting the Rural Poor, Who Often Face Long Drives to Jobs, Medical Care," *Wall Street Journal,* August 17, 1979; James O'Toole, *Energy & Social Change* (Cambridge, Mass.: MIT Press, 1978).
31. For a discussion of the possible uses of oil revenues for energy development in the U.S., see The White House, *The Energy Security Act: A National Imperative* (Washington, D.C.: June 1979). Sri Lanka's program is discussed in detail in Chapter 6.
32. "Venezuela Sets Five-Year Plan for New Energy Development," *Latin American Energy Report,* August 13, 1981; Bob Shallit, "Alaska Hydroelectric Plans Aren't Simply Water Over a Dam," *Wall Street Journal,* July 8, 1981; "Saudi Arabia and Solar Energy," *ARAMCO World,* September/October 1981; David B. Ottaway, "Saudis Plot Oil Profits Into Agriculture," *Washington Post,* December 7, 1981.
33. For an overview of the lost momentum in the West in the early eighties, see Daniel Yergin and Martin Hillenbrand, eds., *Global Insecurity: A Strategy for Energy and Economic Renewal* (Boston: Houghton Mifflin Co., 1982), "Energy-Sufficient Britain faces a Crisis of Identity," *Financial Times Energy Economist,* April 1982, and "Holland Studying Its Natural Gas Policy," *Journal of Commerce,* August 31, 1982.
34. Mahbub ul Haq, "Financing the Energy Transition," presented to the North-South Roundtable Seminar at the United Nations Conference on New and Renewable Sources of Energy, Nairobi, Kenya, August 10–21, 1981; World Bank, *Energy in the Developing Countries* (Washington, D.C.: August 1989); World Bank, *Renewable Energy Resources in the Developing Countries* (Washington, D.C.: November 1980); Maurice Strong and Mahbub ul Haq, *The Castel Gandolfo Report on Renewable Energy: Policies and Options,* presented to the North-South Roundtable Seminar at the United Nations Conference on New and Renewable Sources of Energy, Nairobi, Kenya, August 10–21, 1981.
35. Colin Norman, "U.S. Derails Energy Plan for Third World," *Science,* April 3, 1981; Roger S. Leeds, *Co-Financing for Development: Why Not More?* (Washington, D.C.: Overseas Development Council, 1982).
36. Thomas Hoffman and Brian Johnson, *The World Energy Triangle: A*

Strategy for Cooperation (Cambridge, Mass.: Ballinger, 1981); Andrew MacKillop, "Energy for the Developing World: A Critique of the New Wisdom," *Energy Policy,* December 1980; Andrew MacKillop, "Global Economic Change and New Energy," *Energy Policy,* December 1981.

37. For a description of China's mobilization of labor, see Robert P. Taylor, *Rural Energy Development in China* (Washington, D.C.: Resources for the Future, 1982). The Chinese have been less successful in harnessing labor for large-scale projects; see, for example, James Sterba, "Huge Dam Tames Yangtze and Gives Chinese a Lift," *New York Times,* May 20, 1981.
38. "Tax Facts," *Soft Energy Notes,* October/November 1981; D. Chapman, *Taxation and Solar Energy* (Sacramento, Calif.: California Energy Commission, April 1978).
39. Leonard Rodberg and Meg Schachter, *State Conservation and Solar Energy Tax Programs: Incentives or Windfalls?* (Washington, D.C.: The Council of State Planning Agencies, 1980). Brazil's subsidies for alcohol fuel are discussed in "What is the Real Price of 'green gasoline'?," *Financial Times Energy Economist,* April 1982. Steven Ferrey, "Solar Banking: Constructing New Solutions to the Urban Energy Crisis," *Harvard Journal of Legislation,* Summer 1981.
40. Jonathan Lloyd-Owen, "Japan," *Canadian Renewable Energy News,* August 1981; New Energy Development Organization, *Japanese New Energy Technologies* (Tokyo: Japan External Trade Organization, 1981).
41. *Canadian Home Insulation Program,* prepared for the use of the Committee on Interstate and Foreign Commerce and the Committee on Banking Finance and Urban Affairs, U.S. House of Representatives (Washington, D.C.: U.S. Government Printing Office, November 2, 1979). The Forestry Industry Renewable Energy program is discussed in Chapter 6, note 20.
42. Vince Taylor, "Electric Utilities: A Time of Transition," *Environment,* May 1981; Alvin L. Alm and Daniel A. Dreyfus, *Utilities in Crisis: A Problem in Governance* (New York: Aspen Institute for Humanistic Studies, 1982); Anthony J. Parisi, "Utilities Have Cause To Thank Their Critics," *New York Times,* September 7, 1980.
43. Richard Corrigan, "Utilities Paying Price for Counting on Demand Growth that Never Came," *National Journal,* October 17, 1981; Jim Harding, "Electric Forecasting: Round to the Nearest Trillion," *Soft Energy News,* October/November 1980; Raphael Papadopoulos, "Communications on Energy: Growth and Overcapacity in the UK Electricity Industry," *Energy Policy,* June 1981; John E. Bryson, "Electric Utilities:

The Next Ten Years," presented to the California Public Utilities Commission Symposium on Energy Utilities, Stanford, Calif., March 27, 1981.

44. Albin J. Dahl, "California's Lifeline Policy," *Public Utilities Fortnightly*, August 31, 1978.
45. Demand management successes are discussed in J.G. Asbury et al., "Electric Heat: The Right Price at the Right Time," *Technology Review*, December 1979/January 1980.
46. For an overview of Third World utility policies, see Mohan Munasinghe and Jeremy J. Warford, *Electricity Pricing: Theory and Case Studies* (Baltimore, Md.: The Johns Hopkins University Press, 1982). The need for demand management and improved systems management is discussed in Eric Jeffe, "Energy Profile of Brazil," *Energy International*, September 1976. Central Intelligence Agency, *Electric Power for China's Modernization: The Hydroelectric Option* (Washington, D.C.: 1980).
47. Barry Satlow, "The Energy Security Act and Public Utilities: A Yellow Light for Utility Solar Financing and Marketing," *Solar Law Reporter*, January/February 1981; Jim Harding, "Selling Savings," *Soft Energy Notes*, August/September 1980.
48. David Talbot and Richard E. Morgan, *Power & Light: Political Strategies for the Solar Transition (New York: The Pilgrim Press, 1981)*; George Sterzinger, "Why Utilities Can't be Conservationists," *Working Papers*, September/October 1981 and reply by David Morris, "Utilities and Conservation," *Working Papers*, November/December 1981.
49. Walter J. Primeux, Jr., "The Decline in Electric Utility Competition," *Land Economics*, May 1975; Jeffrey L. Harrison, "Yardstick Competition: A Prematurely Discarded Form of Regulatory Relief," *Tulane Law Review*, February 1979.
50. Christopher Pope, "PURPA," *Canadian Renewable Energy News*, June 1981; Waring Patridge, "A Road Map to Title I of the Public Utility Regulatory Policies Act of 1978," *Public Utilities Fortnightly*, January 18, 1979; *Promoting Small Power Production: Implementing Section 210 of PURPA*, (Washington, D.C.: Solar Lobby/Center for Renewable Resources, 1981).
51. Roger D. Colton, "Mandatory Utility Financing of Conservation and Solar Measures," *Solar Law Reporter*, January/February 1982; Elizabeth Bacelli and Karen Gordon, *Electric Utility Solar Energy Activities: 1981 Survey* (Palo Alto, Calif.: Electric Power Research Institute, 1982).
52. For history of scale in utility industry, see M. Messing et al., *Centralized Power: The Politics of Scale in Electricity Generation* (Cambridge,

Mass.: Oelgeschlager, Funn & Hain, 1979), and N. B. Guyol, *The World Electric Power Industries* (Berkeley, Calif.: University of California Press, 1969).

53. "The Utilities Are Building Small," *Business Week*, March 17, 1980.
54. Jack Doyle, *Line Across the Land, Rural Electric Cooperatives: The Changing Politics of Energy in Rural America* (Washington, D.C.: Rural Land and Energy Project, Environmental Policy Institute, 1979); "Small Utilities Put Wind, Other Solar Among Top Priorities for Long Term," *Solar Energy Intelligence Report*, May 3, 1982; Daniel Deudney, "Public Power Lost: The Metamorphosis of Rural Electrification 1935–1980," *Working Papers*, September/October 1982.
55. For a discussion of the link between Brazil's alcohol fuels program and the politically powerful sugar producers, see Hal Berton, William Kovarik, and Scott Sklar, *The Forbidden Fuel: Power Alcohol in the Twentieth Century* (New York: Boyd Griffin, Inc., 1982). For an overview of the politics of energy, see David Howard Davis, *Energy Politics*, 2nd ed. (New York: St. Martin's Press, 1978), and Leon N. Lendberg, *The Energy Syndrome: Comparing National Responses to the Energy Crisis* (Lexington, Mass.: Lexington Books, 1977).
56. Kathleen Courrier, "Americans Want Solar Energy," *Christian Science Monitor*, June 30, 1981; "Americans Prefer Conservation to Nuclear by Ratio of 7:2, New Survey Indicates," *Solar Energy Intelligence Report*, January 11, 1982.
57. *Domestic Policy Review of Solar Energy*, A Memorandum to the President of the United States (Washington, D.C.: Government Printing Office, February 1979); "Policy Review Boosts Solar as a Near-Term Energy Option," *Science*, January 19, 1979; U.S. Congress, General Accounting Office, "20 Percent Solar Energy Goal—Is There a Plan to Attain It?," Washington, D.C., March 31, 1980.
58. Center for Renewable Resources, *Shining Examples: Model Projects Using Renewable Resources* (Washington, D.C.: 1980); National Governors' Association Energy and Natural Resources Program, *Ensuring Our Energy Future: State Initiatives for the 80's* (Washington, D.C.: National Governors Association, August 1980).
59. D. Pomerantz et al., *Franklin County Energy Study: A Renewable Energy Scenario for the Future* (Greenfield, Mass.: Franklin County Energy Project, 1979), summarized in G. Coates, ed., *Resettling America* (Andover, Mass.: Brick House, 1981).
60. M. N. Corbett and T. Hayden, "Local Action for a Solar Future," *Solar Law Reporter*, January/February 1981.
61. Ronald D. Brunner, "Decentralized Energy Policies," *Public Policy*,

Winter 1980; M. Hunt and D. Bainbridge, "The Davis Experience," *Solar Age,* May 1978; E. Vine, *Solarizing America: The Davis Experience,* (Washington, D.C.: Conference on Alternative State and Local Policies, 1981); Peter Calthorpe with Susan Benson, "Beyond Solar: Design for Sustainable Communities" in Gary J. Coates, ed., *Resettling America: Energy Ecology and Community* (Andover, Mass.: Brick House, 1981).

62. For a discussion of energy resources in the hands of local governments, see David Morris, *Self-Reliant Cities: Energy and the Transformation of Urban America* (San Francisco: Sierra Club Books, 1982).
63. Colin Norman, "Renewable Power Sparks Financial Interest," *Science,* June 26, 1981.
64. Lisa Atchin, "Can Solar Entrepreneurs Survive Government Help?," *Successful Business,* Spring 1979.

Chapter 13. Shapes of a Renewable Society

1. For a discussion of social and civil liberties "fall out" from nuclear power, see Alvin Weinberg, "Social Institutions and Nuclear Energy," *Science,* July 7, 1972, Robert Jungk, *The New Tyranny* (New York: Grossett & Dunlop, 1980), Russell W. Ayres, "Policing Plutonium: The Civil Liberties Fallout," *Harvard Civil Rights—Civil Liberties Law Review,* Vol. 10, 1975, John Shattuck, "Nuclear Power and the Constitution," *The Nation,* November 3, 1979, and Gerald Garvey, *Nuclear Power and Social Planning* (Lexington, Mass.: Lexington Books, 1978).
2. Catherine Caufield, "Energy Threat to Valuable Land," *New Scientist,* March 11, 1982; Roderick Nash, "Problems in Paradise," *Environment,* July/August 1979.
3. Steve Myers, "Debunking the Myth of Solar Sprawl," *Soft Energy Notes,* June/July 1980; R.H. Twiss et al., "Land Use and Environmental Impacts of Decentralized Solar Energy Use," Energy and Environment Division, Lawrence Berkeley Laboratory, Berkeley, Calif., 1980.
4. "Popular Planting Plans Take Root in Maryland," *Journal of Commerce,* March 18, 1982; D.R. DeWalle and G.M. Heisler, "Landscaping to Reduce Year-Round Energy Bills," in U.S. Department of Agriculture, *Cutting Energy Costs* (Washington, D.C.: U.S. Government Printing Office, 1980). Researchers at the U.S. Department of Agriculture estimate that winter heating bills may be reduced as much as 15 percent, while summer cooling energy needs may be cut 50 percent or more.
5. An excellent discussion of humanity's ability to work with rather than against the landscape is Rene Dubos, "Humanizing the Earth," *Science,*

February 23, 1979. For an early view of renewable energy's ability to blend into the landscape, see Lamont C. Hempel, "The Original Blueprint for a Solar America," *Environment*, March 1982.

6. Aesthetic dimensions of renewable energy development are discussed in Nash, "Problems in Paradise," and Leo Marx, *The Machine in the Garden: Technology and the Pastoral Idea in America* (New York: Oxford University Press, 1964).
7. For an overview of world employment trends, see Kathleen Newland, *Global Employment and Economic Justice: The Policy Challenge*, Worldwatch Paper 28 (Washington, D.C.: Worldwatch Institute, April 1979).
8. Bruce Hannon, "Energy, Labor, and the Conserver Society," *Technology Review*, March/April 1977; Leonard Rodberg, "Employment Impact of the Solar Transition," prepared for the Subcommittee on Energy of the Joint Economic Committee, U.S. Congress, April 6, 1979.
9. Steven Buchsbaum et al., *Jobs and Energy: The Employment and Economic Impacts of Nuclear Power, Conservation, and Other Energy Options* (New York: Council on Economic Priorities, 1979); B. Mason, G. Ferris and B. Burns, *Solar Energy Commercialization and the Labor Market* (Golden, Colo.: Solar Energy Research Institute, 1978); Leonard Rodberg, "More Jobs Under the Sun: Solar Power and Employment," *Social Policy*, May/June 1980; U.S. Congress, Office of Technology Assessment, *Energy from Biological Processes* (Washington, D.C.: September 1980); Richard Grossman and Gail Daneker, *Jobs and Energy* (Washington, D.C.: Environmentalists for Full Employment, 1977); Mary Schifflett and John V. Zuckerman, "Who Will be Working in Solar Energy Jobs?," *Solar Engineering*, May 1979; California Public Policy Center, *Jobs from the Sun: Employment Development in the California Solar Energy Industry* (Los Angeles: 1978).
10. The employment impact of methanol development in Canada is discussed in "Major Study Finds Enormous Potential in Canadian Biomass," *Soft Energy Notes*, October 1978.
11. To create jobs in rural areas of the Philippines, the International Labor Organization is promoting the replacement of chain saws with more labor-intensive traditional tools as described in "Trees and Jobs in Philippines—ILO," *Development Forum*, January/February 1982.
12. World trends in urbanization are discussed in Kathleen Newland, *City Limits: Emerging Constraints on Urban Growth*, Worldwatch Paper 38, (Washington, D.C.: Worldwatch Institute, August 1980). For a discussion of the need for new rural development strategies, see M.P. Todaro

and J. Stilkind, "City Bias and Rural Neglect: The Dilemma of Urban Development," The Population Council, New York, unpublished, 1981.

13. John S. Steinhart et al., *A Low Energy Scenario for the United States: 1975–2050* (Madison, Wisc.: Institute for Environmental Studies, 1977).
14. Paolo Soleri, *Arcology: The City in the Image of Man* (Cambridge, Mass.: MIT Press, 1979).
15. New York City Energy Office, *Energy Consumption in New York City: Patterns and Opportunities* (New York: 1981); U.S. House of Representatives, Subcommittee on the City, Committee on Banking, Finance, and Urban Affairs, *Compact Cities: Energy Saving Strategies for the Eighties*, Committee print, July 1980; Jon Van Til, "A New Type of City for an Energy-Short World," *The Futurist*, June 1980; U.S. Congress, Office of Technology Assessment, *Energy Efficiency of Buildings in Cities* (Washington, D.C., March 1982).
16. United Nations (UN), Department of International Economic and Social Affairs, Statistical Office, *1979 Yearbook of World Energy Statistics* (New York: 1976); U.N. Department of Economic and Social Affairs, Statistical Office, *World Energy Supplies, 1950–1974* (New York: 1976).
17. David Morris, *Self-Reliant Cities: Energy and the Transformation of Urban America* (San Francisco: Sierra Club Books, 1982).
18. Wilson Clark and Jake Page, *Energy, Vulnerability and War: Alternatives for America* (New York: W.W. Norton & Co., 1981); Amory B. Lovins and L. Hunter Lovins, *Brittle Power* (Andover, Mass.: Brick House, 1982).
19. For a discussion of the role of hydropower in relations between Argentina and Brazil, see Winthrop P. Carty, "A Farewell to Arms?," *Americas Magazine*, August 1981.
20. Ericsson quoted in Ethan B. Kapstein, "The Transition to Solar Energy: An Historical Approach," in Lewis Perelman et al., eds., *Energy Transitions: Long-Term Perspectives* (Boulder, Colo.: Westview Press, 1981).
21. For a discussion of the role of hydropower in settlement patterns, see Joseph Ermenc, "Small Hydraulic Prime Movers for Rural Areas of Developing Countries: A Look at the Past," in N.L. Brown, ed., *Renewable Energy Resources and Rural Applications in the Developing World* (Boulder, Colo.: Westview Press, 1978).
22. For details of the project and its impact on Quebec, see "La Grande Complex," Societé d'énergie de la Baie James, Montreal, unpublished, 1978, Clifford D. O. May, "The Power and the Glory in Quebec,"

GEO, December 1979, E. J. Dionne, Jr., "Quebec's Profit May be New York's Gain," *New York Times*, August 15, 1982, and Clayton Jones, "Quebec Turns Water Into Gold," *Christian Science Monitor*, July 30, 1980.

23. For aluminum's relation to hydropower and its role in industrial societies, see Thomas Canby, "Aluminum, the Magic Metal," *National Geographic*, August 1978; Aluminum Association, "Energy and the Aluminum Industry," Washington, D.C., April 1980; Dan Morgan, "Aluminum Industry's Factor in Northwest Power Dispute," *Washington Post*, October 29, 1979; Kai Lee and Donna Lee Klemka, *Electric Power and the Future of the Pacific Northwest* (Pullman, Wash.: State of Washington Water Resource Center, March 1980).
24. For a discussion of hydropower's impact on mineral and aluminum projects, see Chapter 8, footnote 35, and Harafumi Mochizuki, "Plight of Basic Materials Industries," *Journal of Japanese Trade & Industry*, Vol. 2, No. 1, 1982.
25. Amal Nag, "Rising Nationalism in Host Countries Threatens U.S. Control of Aluminum," *Wall Street Journal*, January 20, 1981.
26. For a penetrating overview of the equity implications of energy trends and policies, see Ivan Illich, *Energy and Equity* (New York: Harper & Row, 1974).
27. Robert Dunsmore, San Luis Valley Solar Energy Association, private communication, July 19, 1982; see also Jeffrey Ruth, "Harvesting the Sun in an Agrarian Desert: Colorado's San Luis Valley," *Sun Times*, March/April 1982.
28. Third World renewable energy leadership is discussed in Jose Goldemberg, "Brazil: Energy Options and Current Outlook," *Science*, April 14, 1978, and Brian Murphy, "Third World Looks to Brazil, Philippines as Biomass Leaders," *Renewable Energy News*, April 1982.

Note on Energy Units

The reader deserves a brief explanation of the energy units used here. This is an area of considerable confusion since different organizations and countries continue to measure energy in different units. The United Nations uses *million metric tons of coal equivalent;* the United States uses *quadrillion British Thermal Units* (quads); and many oil companies use *barrels of oil equivalent.*

We have adopted *exajoules* because it is a standard metric unit of energy not tied to any particular energy source. An exajoule is a large amount of energy—equivalent to 34 million metric tons of coal or 163 million barrels of oil. (For the convenience of Americans, there happen to be 1.06 exajoules in a quad.) An exajoule of energy is sufficient to heat and cool approximately 7 million modern single-family residences for a year, and the world uses approximately 350 exajoules of energy annually.

Unless specified otherwise, all energy totals used here indicate primary energy—that is the amount of energy contained in a particular fuel before it is burned (perhaps inefficiently) in an engine or furnace. For a purely electricity-generating technology such as a nuclear plant or a wind turbine, the primary energy figure in exajoules indicates the amount of fuel that would have been burned in a typical coal-fired power plant to generate that much electricity.

Electricity is commonly measured throughout the world in kilowatts or megawatts (1000 kilowatts). These figures specify only the capacity to produce electricity or the output at any given moment. By knowing the proportion of the time that a power plant is operating, one can calculate the amount of electricity generated in kilowatt-hours or megawatt-hours. Different types of power plants operate at different average levels of capacity—known as capacity factors (ranging from 20 to 90 percent). As a result, it is impossible to know automatically how much electricity a megawatt of generating capacity represents. The world now has approximately 2 million megawatts of generating capacity, and, assuming a capacity factor of 50 percent, that yields approximately 9 billion megawatt-hours of electricity each year.

Selected References

Chapter 2. Energy at the Crossroads

1. Bupp, Irvin C., and Derian, Jean-Claude. *The Failed Promise of Nuclear Power: The Story of Light Water.* New York: Basic Books, 1978.
2. Committee on Nuclear and Alternative Energy Systems. *Energy in Transition: 1985–2010.* San Francisco: W.H. Freeman, 1979.
3. Council on Environmental Quality. *Global Energy Futures and the Carbon Dioxide Problem.* Washington, D.C.: U.S. Government Printing Office, January 1981.
4. Daly, Herman E., and Umana, Alvaro F. *Energy, Economics, and the Environment.* Boulder, Colo.: Westview Press, 1981.
5. Gibbons, John H., and Chandler, William U. *Energy: The Conservation Revolution.* New York: Plenum Press, 1981.

6. Howard, Ross, and Perley, Michael. *Acid Rain.* New York: McGraw-Hill, 1982.
7. International Energy Agency. *Natural Gas: Prospects to 2000.* Paris: 1982.
8. International Energy Agency. *World Energy Outlook.* Paris: 1982.
9. International Institute for Applied Systems Analysis. *Energy in a Finite World.* Laxenburg, Austria: 1981.
10. Komanoff, Charles. *Power Plant Cost Escalation: Nuclear and Coal Capital Costs, Regulation, and Economics.* New York: Komanoff Energy Associates, 1981.
11. Lovins, Amory B. *Soft Energy Paths: Toward a Durable Peace.* Cambridge, Mass.: Ballinger, 1977.
12. Nuclear Energy Policy Study Group. *Nuclear Power: Issues and Choices.* Cambridge, Mass.: Ballinger, 1977.
13. Resources for the Future. *Energy: The Next Twenty Years.* Cambridge, Mass.: Ballinger, 1979.
14. Ross, Marc H., and Williams, Robert H. *Our Energy: Regaining Control.* New York: McGraw-Hill, 1981.
15. Smil, Vaclav. *China's Energy.* New York: Praeger Publishers, 1976.
16. Solar Energy Research Institute. *A New Prosperity: Building a Sustainable Future.* Andover, Mass.: Brick House, 1981.
18. Stobaugh, Robert, and Yergin, Daniel, eds. *Energy Future: Report of the Energy Project of the Harvard Business School.* New York: Random House, 1979.
19. U.N. Department of International Economic and Social Affairs. *World Energy Supplies, 1973–1978.* New York: 1979.
20. U.S. Congress, Office of Technology Assessment. *Technology and Soviet Energy Availability.* Washington, D.C.: 1981.
21. U.S. Congress, Office of Technology Assessment. *World Petroleum Availability 1980–2000.* Washington, D.C.: 1980.
22. U.S. Department of Energy. *1980 International Energy Annual.* Washington, D.C.: 1981.
23. World Bank. *Energy in the Developing Countries.* Washington, D.C.: 1980.
24. World Coal Study. *Coal—Bridge to the Future.* Cambridge, Mass.: Ballinger, 1980.

Chapter 3. Building with the Sun

1. American Section of the International Solar Energy Society. *Proceedings of the National Passive Solar Conference.* Philadelphia, Pa.: annual.

2. Anderson, Bruce. *The Solar Home Book: Heating, Cooling, and Designing with the Sun.* Andover, Mass.: Brick House, 1976.
3. Butti, Ken, and Perlin, John. *A Golden Thread: 2000 Years of Solar Architecture and Technology.* New York: Van Nostrand Reinhold, 1980.
4. Farallones Institute. *The Integral Urban House.* San Francisco, Calif.: Sierra Club Books, 1979.
5. Flavin, Christopher. *Energy and Architecture: The Solar and Conservation Potential,* Worldwatch Paper 40. Washington, D.C.: Worldwatch Institute, November 1980.
6. Hayes, Gail Boyer. *Solar Access Law.* Cambridge, Mass.: Ballinger, 1979.
7. Mazria, Edward. *The Passive Solar Energy Book.* Emmaus, Pa.: Rodale Press, 1979.
8. Shurcliff, William A. *Superinsulated and Double-Envelope Houses.* Cambridge, Mass.: privately published, 1980.
9. Stein, Richard G. *Architecture and Energy.* Garden City, N.Y.: Anchor Press/Doubleday, 1978.
10. Thompson, Grant P. *Building to Save Energy: Legal and Regulatory Approaches.* Cambridge, Mass.: Ballinger, 1980.
11. U.S. Congress, Office of Technology Assessment. *Energy Efficiency of Buildings in Cities.* Washington, D.C.: 1982.
12. U.S. Congress, Office of Technology Assessment. *Residential Energy Conservation.* Washington, D.C.: 1979.

Chapter 4. Solar Collection

1. Austrialian Academy of Science. *Liquid Fuels: What Can Australia Do?* Canberra: 1981.
2. Charters, William W.S., and Pryor, Trevor L. *Solar Energy: An Introduction to the Principles and Applications.* West Heidelburg, Austrialia: Beatrice Publishing, 1981.
3. Heid, Walter G., and Trotter, Warren K. *Progress of Solar Technology and Potential Farm Uses.* Washington, D.C.: U.S. Dept. of Agriculture, 1982.
4. International Solar Energy Society. *Sun: Mankind's Future Source of Energy.* New York: Pergamon Press, 1978.
5. Jet Propulsion Laboratory, California Institute of Technology. *Regional Applicability and Potential of Salt-Gradient Solar Ponds in the United States.* Pasadena, Calif.: 1981.
6. Metz, William D., and Hammond, Allen L. *Solar Energy in America.*

New York: American Association for the Advancement of Science, 1978.

7. Solar Work Institute. *Status Report on California's Solar Collector Industry.* Sacramento, Calif.: Office of Appropriate Technology, 1982.
8. U.S. Congress, Office of Technology Assessment. *Ocean Thermal Energy Conversion.* Washington, D.C.: 1979.
9. U.S. Department of Energy. *A Response Memorandum to the President, Domestic Policy Review of Solar Energy.* Washington, D.C., February 1979.
10. Wich, Gerald L., and Schmitt, Walter R., eds. *Harvesting Ocean Energy.* Paris: The UNESCO Press, 1981.
11. Williams, Robert H., ed. *Toward a Solar Civilization.* Cambridge, Mass.: MIT Press, 1978.

Chapter 5. Sunlight to Electricity: The New Alchemy

1. Flavin, Christopher. *Electricity from Sunlight: The Future of Photovoltaics,* Worldwatch Paper 52. Washington, D.C.: Worldwatch Institute, December 1982.
2. Institute of Electrical and Electronics Engineers. *Proceedings of the IEEE Photovoltaic Specialists Conference.* United States: annual.
3. Maycock, Paul D., and Stirewalt, Edward N. *Photovoltaics: Sunlight to Electricity in One Step.* Andover, Mass.: Brick House, 1981.
4. Monegon, Ltd. *The Future of Photovoltaic Solar Electricity.* Gaithersburg, Md.: 1982.
5. National Science Foundation. *Electric Power from Orbit: A Critique of a Satellite Solar Power System.* Washington, D.C.: 1981.
6. Pacific Northwest Laboratory. *Photovoltaic Product Directory and Buyers Guide.* Washington, D.C.: U.S. Department of Energy, 1981.
7. Reece, Ray. *The Sun Betrayed: A Report on the Corporate Seizure of U.S. Solar Energy Development.* Boston: South End Press, 1979.
8. Solar Energy Research Institute, *Basic Photovoltaic Principles and Methods.* Golden, Colo.: 1982.
9. Science Applications, Inc. *Characterization and Assessment of Potential European and Japanese Competition in Photovoltaics.* Springfield, Va.: National Technical Information Service, 1979.
10. Stambler, Barrett, and Stambler, Lyndon. *Competition in the Photovoltaics Industry: A Question of Balance.* Washington, D.C.: Center for Renewable Resources, 1982.
11. U.S. Department of Energy. *Photovoltaic Energy Systems Program Summary.* Washington, D.C.: 1982.

Chapter 6. Wood Crisis, Wood Renaissance

1. Agarwal, Bina. *The Wood Fuel Problem and the Diffusion of Rural Innovations.* Sussex: University of Sussex, Science Policy Research Unit, October 1980.
2. Council on Environmental Quality and U.S. Department of State. *The Global 2000 Report to the President.* Washington, D.C.: U.S. Government Printing Office, 1980.
3. Eckholm, Erik P. *Losing Ground: Environmental Stress and World Food Prospects.* New York: W.W. Norton, 1976.
4. Eckholm, Erik P. *Planting for the Future: Forestry for Human Needs,* Worldwatch Paper 26. Washington, D.C.: Worldwatch Institute, February 1979.
5. Hewett, Charles E., and High, Colin J. *Construction and Operation of Small, Dispersed, Wood-fired Power Plants.* Hanover, N.H.: Thayer School of Engineering, Resource Policy Center, September 1978.
6. Knowland, B., and Ulinski, C. *Traditional Fuels: Present Data, Past Experience and Possible Strategies.* Washington, D.C.: Agency for International Development, 1979.
7. Love, Peter, and Overend, Ralph. *Tree Power: An Assessment of the Energy Potential of Forest Biomass in Canada.* Ottawa: Ministry of Energy, Mines and Resources, 1978.
8. National Academy of Sciences. *Firewood Crops: Shrub and Tree Species for Energy Production.* Washington, D.C.: National Academy Press, 1980.
9. National Academy of Sciences. *Leucaena: Promising Forage and Tree Crops for the Tropics.* Washington, D.C.: National Academy Press, 1977.
10. Smith, Nigel. *Wood: An Ancient Fuel With a New Future,* Worldwatch Paper 42. Washington, D.C.: Worldwatch Institute, January 1981.
11. Tillman, David A. *Wood as an Energy Resource.* New York: Academic Press, 1978.
12. United Nations Conference on New and Renewable Sources of Energy. "Report of the Technical Panel on Fuelwood and Charcoal." Nairobi, Kenya: 1981.
13. World Bank. *Forestry Sector Policy Paper.* Washington, D.C.: 1978.

Chapter 7. Growing Fuels: Energy from Crops and Waste

1. Barnett, Andrew; Pyle, Leo; and Subramanian, S.K. *Biogas Technology in the Third World: A Multidisciplinary Review.* Ottawa: International Development Research Centre, 1978.

2. Bernton, Hal; Kovarik, William; and Sklar, Scott. *The Forbidden Fuel: Power Alcohol in the Twentieth Century.* New York: Boyd Griffin, 1982.
3. Brown, Lester R. *Food or Fuel: New Competition for the World's Cropland,* Worldwatch Paper 35. Washington, D.C.: Worldwatch Institute, March 1980.
4. Hall, D.O.; Barnard, G.W.; and Moss, P.A. *Biomass for Energy in the Developing Countries.* Oxford: Pergamon Press, 1982.
5. Pimentel, David, and Pimentel, Marcia. *Food, Energy and Society.* New York: John Wiley & Sons, 1979.
6. Smith, Russel J. *Tree Crops—A Permanent Agriculture.* New York: Harper & Row, 1978.
7. U.S. Congress, Office of Technology Assessment. *Gasohol, A Technical Memorandum.* Washington, D.C.: September 1979.
8. U.S. Congress, Office of Technology Assessment. *Energy from Biological Processes.* Washington, D.C.: 1980.
9. U.S. Congress, Office of Technology Assessment. *Materials and Energy from Municipal Waste.* Washington, D.C.: 1979.
10. U.S. National Alcohol Fuels Commission. *Fuel Alcohol: An Energy Alternative for the 1980's.* Washington, D.C.: 1981.
11. Vogler, John. *Work from Waste: Recycling Wastes to Create Employment.* London: Intermediate Technology Publications and Oxfam, 1981.

Chapter 8. Rivers of Energy

1. Ackermann, William, et al., eds. *Man-Made Lakes: Their Problems and Environmental Effects.* Washington, D.C.: American Geophysical Union, 1973.
2. Cavanaugh, Ralph, et al. *Choosing an Electrical Energy Future for the Pacific Northwest: An Alternative Scenario.* San Francisco: Natural Resources Defense Council, August 1980.
3. Central Intelligence Agency. *Electric Power for China's Modernization: The Hydroelectric Option.* Washington, D.C.: May 1980.
4. Committee on Environmental Effects of the U.S. Committee on Large Dams. *Environmental Effects of Large Dams.* New York: American Society of Civil Engineers, 1978.
5. Deudney, Daniel. *Rivers of Energy: The Hydropower Potential,* Worldwatch Paper 44. Washington, D.C.: Worldwatch Institute, June 1981.
6. Kirkpatrick, J.B. *Hydro-Electric Development and Wilderness in Tasmania.* Hobart, Tasmania: Department of the Environment, 1979.
7. Klotz, Louis H., ed. *Energy Sources: The Promises and Problems.* Dur-

ham, N.H.: Center for Industrial and Institutional Development, Univ. of New Hampshire, 1980.
8. McPhee, John. *Encounters with the Archdruid.* New York: Doubleday, 1975.
9. Smith, Nigel J.H. *Man, Fishes and the Amazon.* New York: Columbia Univ. Press, 1981.
10. Smith, Norman A.F. *Man and Water: A History of Hydro-Technology.* London: Peter Davies, 1975.
11. Waterbury, John. *Hydropolitics of the Nile Valley.* Syracuse, N.Y.: Syracuse Univ. Press, 1980.

Chapter 9. Wind Power: A Turning Point

1. Bergeson, Lloyd. *Wind Propulsion for Ships of the American Merchant Marine.* Washington, D.C.: U.S. Maritime Agency, 1981.
2. Bollmeier, W.S., et al. *Small Wind Systems Technology Assessment: State of the Art and Near Term Goals.* Springfield, Va.: National Technical Information Service, 1980.
3. Dubey, Michael, and Coty, Ugo. *Impact of Large Wind Energy Systems in California.* Sacramento, Calif.: California Energy Commission, 1981.
4. Flavin, Christopher. *Wind Power: A Turning Point,* Worldwatch Paper 45. Washington, D.C.: Worldwatch Institute, July 1981.
5. Hunt, Daniel V. *Windpower: A Handbook on Wind Energy Conversion Systems.* New York: Van Nostrand Reinhold, 1981.
6. Lockheed California Company. *Wind Energy Mission Analysis.* Burbank, Calif.: October 1976.
7. Naar, Jon. *The New Wind Power.* New York: Penguin Books, 1981.
8. NASA Lewis Research Center. *Wind Energy Developments in the Twentieth Century.* Cleveland, Ohio: National Aeronautics and Space Administration, 1979.
9. Pacific Northwest Laboratory. "World-Wide Wind Energy Resource Distribution Estimates." A map prepared for the World Meteorological Organization, 1981.
10. Park, Jack. *The Wind Power Book.* Palo Alto, Calif.: Cheshire Books, 1981.
11. United Nations Conference on New and Renewable Sources of Energy. "Report of the Technical Panel on Wind Energy." Nairobi, Kenya: 1981.

Chapter 10. Geothermal Energy: The Powering Inferno

1. DiPippo, Ronald. *Geothermal Energy as a Source of Electricity.* Washington, D.C.: U.S. Department of Energy, 1980.
2. Eaton, William W. *Geothermal Energy.* Washington, D.C.: U.S. Energy Research and Development Administration, 1975.
3. Kruger, Paul, and Otte, Carel, eds. *Geothermal Energy.* Stanford, Calif.: Stanford Univ. Press, 1973.
4. Ministere de l'Industrie du Commerce, et de l'Artisanat. *La Geothermie en France.* Paris: 1978.
5. Muffler, L.J.P. *Assessment of Geothermal Resources of the United States—1978.* Washington, D.C.: U.S. Geological Survey, 1979.
6. Geothermal Resources Council Annual Meeting. Proceedings. United States: annual.
7. United Nations Conference on New and Renewable Sources of Energy. "Report of the Technical Panel on Geothermal Energy." Nairobi, Kenya: 1981.
8. U.S. Congress, General Accounting Office. *Geothermal Energy: Obstacles and Uncertainties Impede Its Widespread Use.* Washington, D.C.: January 18, 1980.

Chapter 11. Working Together: Renewable Energy's Potential

1. Ashworth, John H., and Neuendorffer, Jean W. *Matching Renewable Energy Systems to Village-Level Energy Needs.* Golden, Colo.: U.S. Solar Energy Research Institute, 1980.
2. Baccelli, Elizabeth, and Gordon, Karen. *Electric Utility Solar Energy Activities: 1981 Survey.* Palo Alto, Calif.: Electric Power Research Institute, 1982.
3. Brown, Norman L., ed. *Renewable Energy Resources and Rural Applications in the Developing World.* Boulder, Colo.: Westview Press, 1978.
4. Dunkerley, Joy, et al. *Energy Strategies for Developing Nations.* Baltimore, Md.: Johns Hopkins Univ. Press, 1981.
4. Food and Agriculture Organization of the United Nations. *Energy for World Agriculture.* Rome: United Nations, 1979.
5. Hirst, Eric, et al. *Energy Use from 1973 to 1980: The Role of Improved Energy Efficiency.* Oak Ridge, Tenn.: Oak Ridge National Laboratory, December 1981.
6. Hoffman, Peter. *The Forever Fuel: The Story of Hydrogen.* Boulder, Colo.: Westview Press, 1981.

7. Johansson, Thomas B., and Steen, Peter. *Solar Sweden: An Outline to a Renewable Energy System.* Stockholm: Secretariat for Future Studies, 1979.
7. National Research Council. *Energy for Rural Development.* Washington, D.C.: National Academy Press, 1981.
9. North-South Roundtable. *Energy for Development: An International Challenge.* New York: Praeger Publishers, 1981.
10. Philippines Ministry of Energy. *Ten-Year Energy Program, 1980–1989.* Manila: 1980.
11. Solar Energy Research Institute. *A New Prosperity: Building a Renewable Energy Future.* Andover, Mass.: Brick House, 1981.
12. United Nations Conference on New and Renewable Sources of Energy, "Programme of Action." Nairobi, Kenya: August 21, 1981.
13. World Bank. *Renewable Energy Resources in the Developing Countries.* Washington, D.C.: 1980.

Chapter 12. Institutions for the Transition

1. Alm, Alvin L., and Dreyfus, Daniel A. *Utilities in Crisis: A Problem of Governance.* New York: Aspen Institute for Humanistic Studies, 1982.
2. Center for Renewable Resources. *Shining Examples: Model Projects Using Renewable Resources.* Washington, D.C.: 1980.
3. Center for Renewable Resources. *The Solar Agenda: Progress and Prospects.* Washington, D.C.: 1982.
4. Committee for Economic Development and the Conservation Foundation. *Energy Prices and Public Policy.* Washington, D.C.: 1982.
5. Cone, B.W., et al. *An Analysis of Federal Incentives Used to Stimulate Energy Production.* Washington, D.C.: U.S. Department of Energy, 1980.
6. Davis, David Howard. *Energy Politics,* 2nd ed. New York: St. Martin's Press, 1978.
7. Doyle, Jack. *Line Across the Land, Rural Electric Cooperatives: The Changing Politics of Energy in Rural America.* Washington, D.C.: Rural Land and Energy Project, Environmental Policy Institute, 1979.
8. Hempel, Lamont C. *The Politics of Sunshine.* Doctoral dissertation of the Public Policy Program, Claremont Graduate School. Claremont, Calif.: 1982.
9. Henderson, Hazel. *The Politics of the Solar Age: Alternatives to Economics.* Garden City, N.Y.: Anchor Press/Doubleday, 1981.
10. Hoffman, Thomas, and Johnson, Brian. *The World Energy Triangle: A Strategy for Cooperation.* Cambridge, Mass.: Ballinger, 1981.

11. International Energy Agency. *Energy Research, Development and Demonstration in the IEA Countries.* Paris: Organization for Economic Cooperation and Development/International Energy Agency, 1982.
12. International Energy Agency. *Energy Policies and Programs of IEA Countries, 1981 Review.* Paris: Organization for Economic Co-operation and Development/International Energy Agency, 1982.
13. Messing, M., et al. *Centralized Power: The Politics of Scale in Electricity Generation.* Cambridge, Mass.: Oelgeschlager, Funn & Hain, 1979.
14. Pomerantz, D., et al. *Franklin County Energy Study: A Renewable Energy Scenario for the Future.* Greenfield, Mass.: Franklin County Energy Project, 1979.
15. Rodberg, Leonard, and Schachter, Meg. *State Conservation and Solar Energy Tax Programs: Incentives or Windfalls?* Washington, D.C.: The Council of State Planning Agencies, 1980.
16. Strong, Maurice, and Haq, Mahbub ul. *The Castel Gandolfo Report on Renewable Energy: Policies and Options,* presented to the North South Roundtable Seminar at the United Nations Conference on New and Renewable Sources of Energy. Nairobi, Kenya: 1981.
17. Talbot, David, and Morgan, Richard E. *Power & Light: Political Strategies for the Solar Transition.* New York: The Pilgrim Press, 1981.
18. Taylor, Robert P. *Rural Energy Development in China.* Washington, D.C.: Resources for the Future, 1982.
19. World Bank. *Mobilizing Renewable Energy Technology in Developing Countries: Strengthening Local Capabilities and Research.* Washington, D.C.: July 1981.
20. Yergin, Daniel, and Hillenbrand, Martin, eds. *Global Insecurity: A Strategy for Energy and Economic Renewal.* Boston: Houghton Mifflin, 1982.

Chapter 13. Shapes of a Renewable Society

1. Buchsbaum, Steven, et al. *Jobs and Energy: The Employment and Economic Impacts of Nuclear Power, Conservation, and Other Energy Options.* New York: Council on Economic Priorities, 1979.
2. Clark, Wilson, and Page, Jake. *Energy, Vulnerability and War: Alternatives for America.* New York: W.W. Norton, 1981.
3. Clark, Wilson. *Energy for Survival.* Garden City, N.Y.: Anchor Press-/Doubleday, 1974.
4. Coates, Gary J., ed. *Resettling America: Energy, Ecology and Community.* Andover, Mass.: Brick House, 1981.

5. Deese, David A., and Nye, Joseph S., eds. *Energy and Security.* Cambridge, Mass.: Ballinger, 1981.
6. Grossman, Richard, and Daneker, Gail. *Jobs and Energy.* Washington, D.C.: Environmentalists for Full Employment, 1977.
7. Hempel, Lamont C. "The Original Blueprint for a Solar America." *Environment,* March 1982.
8. Illich, Ivan. *Energy and Equity.* New York: Harper & Row, 1974.
9. Jungk, Robert. *The New Tyranny.* New York: Grossett & Dunlap, 1980.
10. Lovins, Amory B., and Lovins, L. Hunter. *Brittle Power: Energy Strategy for National Security.* Andover, Mass.: Brick House, 1982.
11. Marx, Leo. *The Machine in the Garden: Technology and the Pastoral Idea in America.* New York: Oxford Univ. Press, 1964.
12. Morris, David. *Self-Reliant Cities: Energy and the Transformation of Urban America.* New York: W.W. Norton, 1981.
13. Newland, Kathleen. *City Limits: Emerging Constraints on Urban Growth,* Worldwatch Paper 38. Washington, D.C.: Worldwatch Institute, August 1980.
14. New York City Energy Office, *Energy Consumption in New York City: Patterns and Opportunities.* New York: 1981.
15. Perelman, Lewis, et al., eds. *Energy Transitions: Long-Term Perspectives.* Boulder, Colo.: Westview Press, 1981.
16. Ridgeway, James. *Energy-Efficient Community Planning.* Emmaus, Pa.: The JG Press, 1979.
17. Soleri, Paolo. *Arcology: The City in the Image of Man.* Cambridge, Mass.: MIT Press, 1979.
18. Steinhart, John S., et al. *A Low Energy Scenario for the United States: 1975–2050.* Madison, Wis.: Institute for Environmental Studies, 1977.
19. Van Til, Jon. *Living with Energy Shortfall: A Future for American Towns and Cities.* Boulder, Colo.: Westview Press, 1982.

Periodicals

Alternative Sources of Energy, bimonthly
    107 S. Central Ave., Milaca, MN 56353
Appropriate Technology, quarterly
    IT Publications Ltd., 9 King St., London WC2E 8HN, U.K.
Biofuels Report, weekly
    Pasha Publications, 1828 L St., N.W., Washington, D.C. 20036
California Energy Commission News & Comment, monthly
    California Energy Commission, 1516 Ninth St., Sacramento, CA 95814

Critical Mass Energy Journal, monthly
Critical Mass Energy Project, 215 Pennsylvania Ave., S.E., Washington, D.C. 20003
Earth Shelter Digest & Energy Report, bimonthly
1701 E. Cope, St. Paul, MN 55109
Energy Conservation Digest, biweekly
Dulles International, P.O. Box 17346, Washington, D.C. 20041
European Energy Report, biweekly
Financial Times Business Information Ltd., Bracken House, 10 Cannon St., London EC4P 4BY, U.K.
The Geyser: International Geothermal Energy Newsletter, monthly
P.O. Box 1738, Santa Monica, CA 90406
In Review: SERI Research Update, bimonthly
Solar Energy Research Institute, 1617 Cole Blvd., Golden, CO 80401
Latin American Energy Report, biweekly
Business Publishers, Inc., 951 Pershing Dr., Silver Spring, MD 20910
Photovoltaic Insider's Report, monthly
1011 W. Colorado Blvd., Dallas, TX 75208
Photovoltaics, bimonthly
Fore Publishers, Inc., P.O. Box 3269, Scottsdale, AZ 85257
Renewable Energy News, monthly
P.O. Box 4869 Stn. E., Ottawa, Canada, K1S 5J1
Soft Energy Notes, bimonthly
Friends of the Earth Foundation, 124 Spear St., San Francisco, CA 94105
Solaire 1, bimonthly
57, rue Escudier, 92100 Boulogne, France
Solar Age, monthly
SolarVision, Inc., Harrisville, NH 03450
Solar Energy Intelligence Report, weekly
Business Publishers, Inc., 951 Pershing Dr., Silver Spring, MD 20910
Solar Engineering & Contracting, monthly
Business News Publishing Co., P.O. Box 3600, 755 W. Big Beaver Rd., Troy, MI 48099
Solar Law Reporter, bimonthly
Solar Energy Research Institute, 1617 Cole Blvd., Golden, CO 80401
Solar Magazine, bimonthly
P.O. Box A, Del Mar, CA 92014
Sun Times, bimonthly
Solar Lobby, 1001 Connecticut Ave., N.W., Washington, D.C. 20036

Sunworld, bimonthly
International Solar Energy Society, P.O. Box 26, Highett, Victoria 3190, Australia
Unasylva: International Journal of Forestry, quarterly
United Nations Food and Agriculture Organization, Via delle Terme di Caracalla, 00100 Rome, Italy
VITA News, quarterly
Volunteers in Technical Assistance, 1815 North Lynn St., Arlington, VA 22209
Wind Energy Report: International Newsletter, monthly
189 Sunrise Highway, Rockville Centre, NY 11570
Wind Power Digest, quarterly
398 E. Tiffin St., Bascom, Ohio 44809
Wood 'n Energy: Professional Solid Fuel Journal, monthly
13 Depot St., P.O. Box 2008, Concord, NH 03301
World Solar Markets, monthly
Financial Times Business Information Ltd., Bracken House, 10 Cannon St., London EC4P 4BY U.K.

Index